Lecture Notes in Physics

Lecture Notes
in Physics

Edited by H. Araki, Kyoto, J. Ehlers, München, K. Hepp, Zürich
R. Kippenhahn, München, H. A. Weidenmüller, Heidelberg
and J. Zittartz, Köln
Managing Editor: W. Beiglböck

233

High Resolution
in Solar Physics

Proceedings of a Specialized Session of the
Eighth IAU European Regional Astronomy Meeting
Toulouse, September 17–21, 1984

Edited by R. Muller

Springer-Verlag
Berlin Heidelberg GmbH

Editor

Richard Muller
Observatoires du Pic du Midi et de Toulouse
F-65200 Bagnères-di-Bigorre

ISBN 978-3-540-15678-9   ISBN 978-3-540-39608-6 (eBook)
DOI 10.1007/978-3-540-39608-6

Originally published by Springer-Verlag Berlin Heidelberg New York Tokyo in 1985

2153/3140-543210

PREFACE

The 8th European Regional Assembly of the International Astronomical Union held in Toulouse September 17-21, 1984, was an excellent opportunity to organize a specialized session on "High Resolution in Solar Physics". The Pic du Midi Observatory is well known for its fine seeing allowing for very high resolution observations, especially of the solar surface. The Solar Session was dedicated to Professor J. J. RÖSCH, who definitely ascribed to the Pic du Midi Observatory high angular resolution of the Sun as well as of night sky objects as its most specific task. In particular, he designed the so strangely shaped but so efficient "Coupole-Tourelle" associated with the 50 cm solar refractor. Professor RÖSCH is newly retired, but he remains very active at the Observatory, designing and setting up the Heliometer which will be used for the study of the fluctuations of the extreme limb profile as well as of the shape of the Sun and of its short and long period variations. The contributions of Professor RÖSCH and of the Pic du Midi Observatory to high resolution studies of the solar photosphere was presented by R.B. DUNN in an introductory talk to the meeting.

Apart from a few contributions, the meeting was restricted to photospheric features. A number of physical processes taking place in the solar photosphere are highly structured with features of size smaller than 1"; that is the case, for example, for the granular convection, the small-scale end of the spectrum of sizes of emerging magnetic tubes, the concentration of magnetic fields in the photospheric network and faculæ, for sunspots' umbræ and penumbræ. The chromosphere, the transition zone and even the corona are highly structured, too, with features of size of the order of 1" or smaller. It is clear that observations of resolution of 1" and even, very often, much better, are needed to improve our knowledge of the Sun (and stars).

New instrumentation, consisting of ground-based as well as space telescopes, and advanced technology like active and adaptative optics, are in full development. They will allow us to reach a resolution of 0.1" in the near future. The major projects are reviewed here in Session 1. The posters displayed in the Poster Session, and published in these proceedings, were devoted to analysis procedures of high resolution observations. The fine structure of the solar photosphere as well as the various techniques of magnetic field strength determination are described in Session II; new results are presented in the contributions. Session III deals with the fine structure of the photosphere from a theoretical point of view, emphasizing still unresolved questions. Surface small-scale features can be used as powerful tools for the study of phenomena originating in deeper layers; that is the topic of Session IV.

Only those participants who formally registered for the Solar Session are listed; but the session was also attended by others participating in the general meeting.

I wish to thank Dr. J.L. LEROY, Mr. J.C. AUGSITROU and Ms. M. MAURUC, for their help in the organization of the Solar Session, which was only a part of their tasks among the other specialized sessions and the general meeting, and also Ms. H. CLOS who provided invaluable assistance in preparing these proceedings for publication.

Toulouse, June 1985   R. Muller

TABLE OF CONTENTS

INTRODUCTION

PROFESSOR RÖSCH, PIC DU MIDI AND HIGH RESOLUTION

Richard B. Dunn

National Solar Observatory

National Optical Astronomy Observatories

To speak on behalf of Rösch's work one must speak about the Pic du Midi! I have always been close in spirit with Pic du Midi. Lyot invented the birefringent filter and used it at the Pic in conjunction with the coronagraph, which he also invented. I started in astronomy by building birefringent filters and used them to study the corona with the 10-cm and 40-cm coronagraphs at Sac Peak. I remember how Dolfus, who observed the corona at the Pic, and I used to compare our coronal pictures!

Nowadays the corona is studied primarily from space by means of such satellites as the Solar Maximum Mission, although we still continue to observe coronal transients from the ground. But studies of the corona represent only a small part of the current research in solar physics. We are also interested in solar activity, global oscillations, and the solar-stellar connections. Many of these studies would benefit from higher resolution observations, which is the subject of this symposium. We all now take the importance of high resolution in solar physics for granted because solar astronomers now believe that the fundamental physical processes on the Sun occur on a small scale in the presence of magnetic fields and convection. This is the justification for high-resolution solar telescopes in space.

High resolution is a relatively new field in solar physics. Until 15 years ago Zirin and I, and perhaps Rösch, were called solar "dermatologists"! Theory had not advanced to the point where the details were important! But now they are! High resolution has a wide following and the theorists want many more details and accurate measurements. I feel that the United States' interest in the field was established when Gene Parker supported the funding of Sac Peak by the National Science Foundation (NSF) after the U. S. Air Force decided to reduce its support. He said that we must continue the work on high resolution.

High resolution has also been one of the primary interests of Prof. Rosch with the Pic du Midi solar telescopes for the last 30 years. The site is good and the telescopes are also very good. At the IAU meeting in Brighton in 1970 I remember Prof. Rösch comparing granulation pictures from Pic du Midi to those taken from a balloon telescope flown in Russia by Krat and Karpinsky. Those from the Pic were easily as good as those from the balloon.

Most telescopes need improvement if they are to meet their diffraction limit. There are always problems in the lens or mirror "figure," heating of windows, internal seeing, seeing in the dome, vibration, servos, etc. When the telescope is improved, the quality of the observations also improves. At the present time the SPO telescope suffers from vibration and from thermal problems in the window. Our granulation pictures are not as good as those from the Pic. I am convinced that the problem is not with the Earth's atmosphere, but with the telescope. I must improve it.

There is a long history of improvements to the Pic du Midi telescopes. They started with a 38-cm aperture and now have a 50. They rebuilt the dome to eliminate internal seeing. They even made some tests with a 60-cm refracto-reflector. Mehltretter, who was on leave from Freiburg, and whom we all remember as a charming man, was familiar with the thermal problems of the window at Sac Peak, and he analyzed the 50-cm and was able to improve it by placing a light shield in front of the telescope. I believe that there are major problems with all current solar telescopes. Perhaps even the 50-cm at the Pic could be improved, but it does appear to be perfect at times. I am not convinced that the compound telescopes have ever been aligned well enough and figured well enough to give the best results. Have you compared your granulation pictures to those from the Pic? Perhaps your seeing is better than you think, and the lack of detail is caused by problems in your telecope or with lack of aperture. If the picture is uniform but soft, then the troubles are close to the telescope or within the telescope itself.

The turbulence in the Earth's atmosphere sometimes seems insuperable, but do not forget that solar astronomers do have some tests to see if their pictures are accurate.

- One could compare the results from ground-based observations to those from balloon flights or, later on, from space.

- Alternatively, compare your pictures to those from Pic du Midi. I always have! This works best for granulation pictures; and now there are some excellent H-alpha pictures. But you must enlarge the pictures to the same scale and use the same contrast, because small-scale pictures always look good!

- Take bursts of pictures and select the best frame. Do two pictures separated in time show the same features? Take a burst movie and print the best frames as a movie. The eye can reject the film grain and flaws. It can follow the evolution of a granule or bright point.

Soon there may be some new tools! At this meeting you will hear about progress in adaptive optics which may someday correct the aberrations caused by the Earth's atmosphere. You will also see an example of the automatic selection and registration of pictures by computer. Perhaps this technique will noticeably improve movies taken under less-than-perfect conditions.

Concentrate your resources! Rösch showed great wisdom in concentrating on granulation studies in the early years. This works especially well if you have limited resources. Improve your telescope, look for the best film, improve the shutter, study the seeing in the dome and telescope, and voila!--you have the best granulation pictures. This approach has led to important research papers from the Pic du Midi, including:

- The best granulation pictures
- Studies of the explosion of granules
- Evolution of granules - granules become brighter and enlarge; then a hole develops and the granule breaks up...
- Studies of elongated granules near the limb
- Changes in the center-to-center distance of granules with the spot cycle (with Macris)
- Bright points and their relation to granulation and chromospheric structures (Muller)
- More recently Rösch has started a program for solar oblateness measurement, which branched off towards solar oscillations (Yerle).

But do not concentrate too much! For example,

- We need more movies of these phenomena. Only one movie is available that shows the relationship between granules, bright points, filigree, crinkles, spicules and micropores, and that movie is over 10 years old and leaves much to be desired! It is unfortunate that the Pic did not have better movie equipment and copying cameras and a larger film budget. Perhaps if it did, by now we would have a "reference" movie of all these phenomena, and the dermatologists and the theorists would be satisfied!
- Pic du Midi should have its own birefringent filter with the proper beamsplitters that permit simultaneous observa-

tions at several wavelengths -- like, for example, the one used by Lyot in 1944 to photograph the corona. Then the observations of the granulation and bright points could be extended to the filigree and magnetic fields. I have seen some recent H-alpha pictures from the Pic that are promising, but I understand that the filter was borrowed. The Zeiss Universal Birefringent Filter (UBF) operated at the Pic by the Italian group under the direction of Righini will help in this regard, as will the double-pass spectrograph designed by Pierre Mein.

- The observations must be extended to polarization measurements. We must know the state of the magnetic field. This is the reason for THEMIS, but perhaps its focal-plane instrumentation could be used initially at the Pic.

Because of the problems with the Earth's atmosphere and the need to be patient and wait for the best moments of seeing, a telescope dedicated to high resolution is needed.

Finally, we need people dedicated to high-resolution observations. Perhaps if Prof. Rösch continues to work for another 30 years, he will obtain the additional high-resolution observations that we need to understand the Sun.

ACKNOWLEDGEMENTS AND COMMENTS

Jean Rösch

Observatoires du Pic du Midi et de Toulouse

I would like to say, first, how grateful I am to the colleagues who
decided to have this Colloquium devoted to high-angular-resolution
in solar physics, and particularly to my old friend Dick Dunn who
accepted to travel eight thousand kilometers and back to summarize
here what I have done in that field, where he is a prominent <u>connois-
seur</u>.

Needless to say, I deeply appreciated his recommendations for testing
solar pictures : "You could compare your results to those from space
when available; alternatively, compare your pictures to those from
Pic-du-Midi"; and I am happy, too, to learn that until 15 years ago,
Zirin and himself were called "dermatologists"; it did not happen to
me, contrary to what he suspects, but that was for the simple reason
that people, in France, did use, instead, the more trivial words
"post-card maker". And I remember that almost thirty years ago, more
than one famous theorist positively denied the existence of some of
the sharp features recently detected at the Pic, because they contra-
dicted the theory !

So, may I risk a question : if things have changed, is it because the
theorists refined their theories and needed sharper observations to
support them, or because the bulk of high resolution pictures in the
market forced them to refine their theories ?

I shall come back to such matters soon. At the moment, I would simply
express my strong support to Dick Dunn when he says that so much is
still to be expected from high resolution observations - except that
he may be slightly too optimistic when relying upon my activity for the
thirty years to come ...

After Dick Dunn has so nicely commented about myself, it is my own
duty, now,to pay a tribute of gratitude to all those without whom I
could not have carried out the achievements which he spoke about.
Numerous as they have been along the years, I shall name only a few of
them. First of all, as you may guess, my dear master Bernard Lyot, who
welcomed me in his Meudon laboratory exactly forty nine years ago, as
his first disciple in time. How could I forget, from his example, that

the discovery of the Universe as it is primarily based upon well-thought instruments and careful observations ?

Then, I would like to cite four men who played a major role around myself in the developpement of high resolution solar programs at the Pic - not including Richard Muller, whom I shall call "the leader of the second generation" -. These four are : Marcel Gentili, such a dedicated observer as a friend of Lyot since their youth could be; Marcel Hugon, with his dependable experience in classical optics and photography; Jacques Pageault, who entirely designed,after my general scheme, the "Turret-Dome", directed its construction, and cared for its health for years; and François Chauveau who has been so efficient in the definition of observing procedures and data processing. Gentili and Hugon, alas ,passed away during the last decade. But after thirty years, I have been fortunate enough to have still with me Pageault for the mechanics design and Chauveau for the optics of the "scanning heliometer" which some of you may discover as a new-born next saturday at the Pic. While Jacques Pageault is untimely retired, François Chauveau is present in this room to-day, and I would be very pleased with you applause for himself and for the others I have named.

And now, let me express some simple ideas - which it may not be useless to remind, however - about the philosophy of high angular resolution.

I happened to read, some months ago, in a "scientific magazine" - not a very reliable one, I must say - that astronomy is no more interested, nowadays with the morphology, but definitely with the physics, of celestial objects. An astonishing non-sense, indeed. Since Aristoteles, physics means nature of the object; and how can you discover and understand its nature if you deliberately ignore its shape and space structure ? Let us go further into the analysis.

Distant as they are, celestial objects are primarily observed as two-dimensional features. Until Galileo came, their optical image was directly projected by the optics of the eye onto the surface of the retina. We believe we have invented the bi-dimensional detectors, but the first one did exist already some one or two million years ago : the detectors of the central part of the retina of superior animals (the cones) are individually connected (through some relays) to the cortex of the hind part of the brain by nerve fibers, just as the diodes of an array are individually connected to the computer. But something else does exist too, which we have not yet been able to

re-invent : through such fibers, there is a point-to-point correspondance between the retinal area and the visual cortex area, and therefore a true bi-dimensional projection of the image onto the brain, in
contact with all the other cerebral centers, so that the brain is
instantaneously able to compare with previously memorized images, to
analyze, to think, to interpretate, and to find the meaning of what
the eye is presently seeing. The most convincing demonstration of the
fundamental necessity of such a marvelous device is the fact that in the
spectral domains where neither the eye nor equivalent bi-dimensional
detectors like the photographic film can be used, constant efforts
have been made to reconstruct indirect images, be it by scanning
procedures or by interferometry, so as, finally and unavoidably, to
display a bi-dimensional image for the eye to look at. Now, what does
a bi-dimensional image contain ? It gives, for every point defined by
two geometrical coordinates, $\underline{u}$ and $\underline{v}$, the energy E measured by the
detector at that point. In addition, one has to specify in which wave-
length, or, rather, specify E for each wave-length. Therefore, E is a
three-variable function, $E(\underline{u},\underline{v},\lambda)$ which cannot be obtained at once on
a bi-dimensional detector, and which,anyhow, we need for the knowledge
of the physics of the object. Thus, we have to choose a compromise
between two extremes : a bi-dimensional image in a given spectral range,
or a spreading side by side over the bi-dimensional detector, according
to their wave-length, or monochromatic images of a narrow strip cut
across the full image by a slit - in other words, make a spectrum -.
What to begin with ? Obviously a wide band full image, since it provides a first knowledge of the morphology of the object, whereas a
spectrum through a slit placed at random somewhere on an unknown object
has a large probability being useless. Please note that,strictly
speaking, the same holds even for a field of point sources, since it
is nothing else than an extended object where $E(\underline{u},\underline{v},\lambda)$ is zero
everywhere, except for a number of pairs of values of u and v which
you must know before any spectral analysis or discrimination; think
about HR diagrams of globular clusters, or about identification of
optical counterparts of radiosources.
Now, the amount of information provided by a bi-dimensional image is
a direct function of the inverse of the area $(\Delta u, \Delta v)$ within which,
centered at any point of the image, no variation of E can be detected.
$\Delta \underline{u}$ and $\Delta \underline{v}$ just express the angular resolution of the image along two
directions; the smaller they are, the larger the amount of information.
But that is only mathematics. The effective information provided by a
better resolved image contains considerably more than what is merely

measured by a number of pixels, because it opens at once a new insight
into the objects and the phenomena. There are so many striking examples that I shall only pick up an almost trivial one. For millenia of
naked-eye astronomy, Saturn has been taken as a dimensionless bright
"erratic" star. In january 1610, Galileo, using a one-inch telescope,
described it as a strange "triple star". Fourty seven years later,
Huygens'eye, through a somewhat larger telescope, saw a more complex
picture, and Huygens' brain immediately understood that it was a
spherical globe surrounded by a flat circular ring. Could you tell me
which computer, in the future generations, on line with a CCD recei-
ving the image of this unexpected object, will immediately display on
its console the words : "Sphere circled by flat ring "?
Finally, what about high angular resolution in solar physics ? There
are several reasons which make the solar case quite specific. Given
the consensus that the best knowledge of our sun is fundamental for the
general knowledge of stars, because it can be studied in so much more
details, the importance of high resolution is evident.
Indeed, it shows a tremendous variety of structures of all sizes, with
intricated connexions and time evolution, which one would like to
explore with more and more sharpness. Moreover, and fortunately, no
other celestial body pours so much energy onto our telescopes, and this
is a priceless advantage when looking for high resolution from ground-
based optical telescopes.
Therefore, it is not surprising that within the considerable effort
developed during the last decades in solar research, a fairly large
part dealt with high resolution in the optical range, the one where it
reaches its optimum value. But the very fortunate fact is the unusual
amount of non-predicted structures or phenomena which were thus revealed. A number of them have been listed by Gene Parker (again !) in a
very interesting "Guest Comment" published in the September 1979 issue
of "Physics Today", where he has beautifully explained the deep reason
for this situation :

   "The sun, he writes, is an obstinate reminder that while we possess
   all the basic partial differential equations of classical and
   quantum physics, the rich variety of solutions of those equations
   extends far beyond present knowledge and imagination".

And next :

   "The problem is to guess from the observations what the underlying
   physical effects may be, and then to establish the ideas firmly
   with the appropriate theoretical studies, and further observations".

A kind of answer to the question I risked a moment ago.

Which stronger encouragement for instrumentists and observers than such a definite statement by as bright and renowned a theorist as Parker is ? And, by the way, which stronger recommendation for the sponsors to finance not only gigantic computers, but also powerful telescopes - not to forget the men to serve them ?

As for myself, I must again express my heartiest thanks to those who chose to hold this Colloquium as a recognition of my contribution in the subject, since not any reward for my efforts could be both more friendly and more constructive. And I do hope the papers and discussions to come these days will help in awaking a general interest in the policy I mentioned a minute ago after Parker's remarks.

1. INSTRUMENTATION : DEVELOPMENTS

THE LARGE EUROPEAN SOLAR TELESCOPE

O. Engvold

Institute of Theoretical Astrophysics
University of Oslo
P.O.Box 1029 Blindern
N-0315 Oslo 3, Norway

1   INTRODUCTION

The quest for better spatial resolution and for high precision
polarimetric measurements have led to refurbishment of existing telescopes and to the development of new telescopes and observatories. A
number of solar telescopes in use, under construction, and in the
planning are given in Table 1. The internal telescope seeing is eliminated effectively in telescopes aiming for high spatial resolution
by evacuating the light path. The aperture size of vacuum telescopes
is limited essentially by the tolerable thickness of the glass (Dunn
1984). The largest vacuum telescope in use has aperture of 76 cm
(Table 1). A substantial improvement in spatial resolution in the
visible and UV will be achieved through the realization of the 125 cm
Solar Optical Telescope (SOT) of NASA for the 1990's. Instrumental
polarization caused by thermal stresses in entrance windows and by
reflections from variably inclined optical surfaces are difficult to
compensate properly and will therefore degrade the measurements. When
it comes to reducing instrumental effects in polarimetric measurements designs that are based on direct pointing Gregorian or
Cassegrain systems seem to offer the most promising solution. Existing and known telescope designs do not offer the possibility of
high spatial resolution and low instrumental polarization at the same
time. The need to combine the two requirements has been decisive in
the development of the design of the Large European Solar Telescope
(LEST).

The objectives of the LEST project are justified in publications by

Table 1

SOME SOLAR TELESCOPES IN USE

Telescope	Feed system	Aperture diameter (cm)	f/D of final beam	Number of optical components Mirrors	Windows (Lenses)	Tower height (m)	Altitude of site (m)
McMath Solar Telescope, National Solar Obs., Kitt Peak, USA	Heliostat	152	54	3	1	31	2060
Crimea Solar Telescope USSR	Coelostat Cassegrain	90	50, 70	5	-	25	350
Vacuum Tower Telescope, National Solar Obs., Sacramento Peak, USA	Alt/az - turret	76	72	4	2	41.5	2810
Vacuum Tower Telescope, National Solar Obs. Kitt Peak USA	Coelostat	70	60	4	2	23	2060
Aerospace Solar Obs., Calif., USA	Cassegrain-Gregorian-Coude	61	20	7	2	14	-
Domeless Solar Tower Telescope of Hida Obs. Japan	Gregorian	60	53.7	4	2	23	1300
Solar Tower Telescope, Meudon, France	Coelostat	60	75	5	2	35	-
Turret Dome of Pic-du-Midi, France	Refractor	50	12.5	(2)	1	-	2860
Solar Telescope of Pulkov Astronomical Obs., Pamir, USSR	Cassegrain (Open)	50	120				4300
Big Bear Solar Obs. California, USA	Several	41-23	35-14	(4)	(2)	9	2000
Solar Telescope of Yunnan Observatory, China	Coelostat Horizontal tel.	40	40	(4)	-	-	2000
German-Spanish Solar Tel., Izaña, Spain	Newton reflector	40	(7)	3	2	12.5	2387
German Domeless Refr., Anacapri, Italy	Coudé	35	12.1	2	1	10	-

Table 1 (cont.)

SOME SOLAR TELESCOPES IN USE

Telescope	Feed system	Aperture diameter (cm)	f/D of final beam	Number of optical components Mirrors	Windows (Lenses)	Tower height (m)	Altitude of site (m)
Oslo Solar Observatory, Harestua, Norway	Coelostat	35	86	4	-	12	580
The Solar Telescope of Nanjing Univ., China	Coelostat	33	65	5	-	21	36
CSIRO Solar Obs., Australia	Refractor	30	10.2	-	3	15	-
Mt.Wilson Solar Tower, California, USA	Coelostat	30	150	2	1	46	1740
Ottawa River Solar Observatory, Canada	Refractor	25	17	1	1	5	58
Swedish Tower Telescope, La Palma, Spain	Heliostat	25	75	2	2	16	2360
SOLAR TELESCOPES UNDER CONSTRUCTION							
Solar Telescope of Göttingen, FRG, at Izaña, Spain	Gregory - Coudé	45	60(Coudé)	3	1	18.7	2400
Vacuum Tower Telescope, of Freiburg, FRG, at Izaña, Spain	Coelostat	60	64	4	2	33.5	2400
Swedish Tower Telescope, La Palma, Spain	Alt/az - turret	50	45	3	2	16	2360
THEMIS of Paris Observatory, France (Izaña)	Ritchey-Chretien	90	(16.7)	2	(2)	20	2400
Solar telescope-magnetograph of Beijing Observatory, Huairou Reservoir, China	Refractor	35	8	-	4	23	(900)
Open Solar Telescope of Utrecht, The Netherlands (La Palma)	Gregorian (modified)	45		3	3	(15)	-
Big Solar Vacuum Telescope, SibIZMIR, USSR	Heliostat	76	53	3	3	31	-

JOSO (JOSO Annual Report 1970, Zwaan et al. 1982) and by the LEST Foundation (Wyller 1983).

The Canary Islands of La Palma and Tenerife have been the subjects of extensive solar and stellar site testing campaigns. National solar telescopes are presently being errected at the sites and still others are planned for the near future (see Table 1). The 25 cm aperture heliostat tower telescope of the Swedish station have produced very nearly diffraction limited images of the Sun at several occasions. A

Figure 1   Photograph of a model for LEST.

reconstructed and enlarged version of that telescope will be operational in summer 1985. The high quality solar observations of the

Canary Islands sites will presumably be manifested as the new and larger telescopes are put into operation. The final selection of the site for LEST will be based largely on the results obtained from the new, medium large national instruments.

The legal body for the LEST project, the LEST Foundation, was established in 1983 in Stockholm. By the end of 1984 the following 7 countries have joined the project:

FRG	ISRAEL	ITALY
NORWAY	SWEDEN	SWITZERLAND
	AUSTRALIA	

The design study of LEST has recently been completed (Andersen et al. 1984). The overall and detailed design of the telescope and tower is estimated to take 1½-2 years. We shall here review briefly the main features of the proposed design.

The optical system of LEST is planned for the use of <u>active</u> and eventually <u>adaptive</u> optics (cf. von der Lühe, 1983) which may permit 0.1 arcsec resolution of details in the solar atmosphere. The telescope will have low instrumental polarization that makes it suitable for high precision measurements of solar magnetic fields. The diameter of the main mirror will be 2.4 m, which will give the telescope a very high photon collecting power. Given LEST the astronomers will be able to perform simultaneous observations with high spectral, spatial and temporal resolution.

Figure 1 shows a photograph of a model of the outer tower and "domeless" structure of the proposed LEST. The telescope is placed on top of a tower so that the influence of ground turbulence is reduced. The telescope has an alt/az mount with the azimuth axis located within one of the forks. With this telescope mounting only five mirrors are needed to bring the telescope light to the instrument ports at the base of the tower. The telescope is sealed with an entrance window so that its entire optical path may be filled with helium to reduce the internal seeing effects (Engvold et al. 1983).

2 OPTICS

The proposed optical configuration for LEST is displayed in Figure 2. The telescope is a Gregorian system with an additional concave mirror behind the primary. Most of the energy is absorbed by a heat shield at the prime focus of the f/2.3 parabolic objective mirror. The third mirror re-images the Gregorian focus to the instruments at the base of the tower that supports the telescope. Two flat mirrors reflect the light out along the elevation and azimuth axes of the telescope mounting so that the final image is fixed (except in rotation about its center) with respect to its auxiliary instrumentation. The vertical beam of light coming from the telescope focuses on a rotating table ("rotator") at ground level.

The image at the f/2.3 parabolic primary mirror focus is strongly aberrated by coma and offers virtually no usable field. The coma at the Gregorian focus exceeds the diffraction limit at about 20 arcsec off the central axis. The third and ellipsoidal mirror corrects largely the aberrations of the secondary focus. It's diffraction limited field of view of about ±12 arcmin exceeds by far the angular field of interest for LEST.

The prime focus field stop and heat rejection system shall remove an excess solar flux of about 5.5 kW and ensures an acceptably low heat load on the optical surfaces of the telescope. It is proposed to use a liquid-cooled tank with a slightly curved entrance window and a central hole, mounted on the spider arm unit in the prime focus.

The quality requirements of the optical surfaces for LEST leads to a peak to peak error amplitude of $\lambda/25$, or $\lambda/70$ rms, at $\lambda=5000\text{Å}$. It is assumed that the primary mirror for LEST will be made from a conventional heavy blank.

There are basically four types of misalignments in the Gregorian system:

- inclination of the axis of the primary with respect to the system axis
- inclination of the axis of the secondary
- pure lateral displacement of the two axes
- axial shift of the secondary

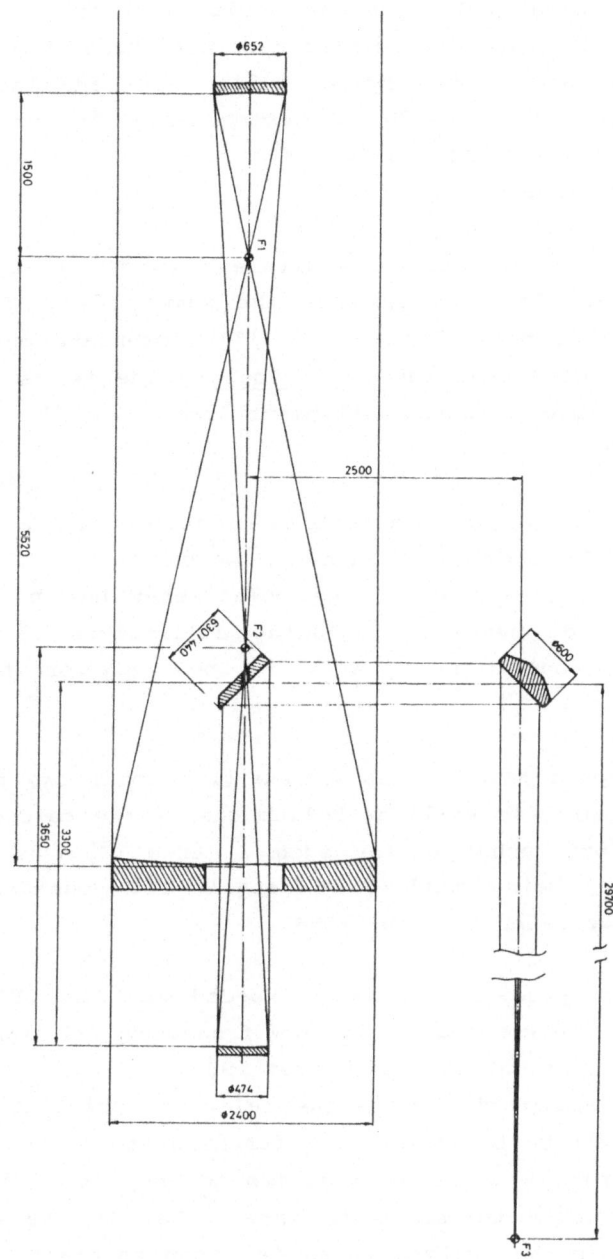

Figure 2. The optical system of LEST.

The alignment tolerances in positioning errors are very small. The tilt of the secondary mirror must not exceed 9 arcsec relative to the central axis of the system. It's axial and lateral shifts must be less than 13 μm and 70 μm, respectively. The positional accuracy for the tertiary mirror is less strict and its adjustments may be solved by passive means.

A small inclination is equivalent to a corresponding lateral displacement. Both errors have the same effect; they produce a field independent coma. The errors will compensate mutually and the image will be good when the conic foci of the two successive mirrors coincide. The proposed alignment system shall make the conic foci coincide.

A compactly built polarization modulator may be inserted in the beam before the hole of mirror 4, which is prior to any oblique reflection. A version of the polarimeter has been suggested that is only 20 cm overall, including collimators (Stenflo 1984). Piezoelastic modulation appears to be superior to other modulation methods.

LEST should have a thin window (1-2 cm thick) preferably made from fuzed quartz to seal the helium gas of the telescope light path. The production technique for such a large (⌀2.4 m) and thin window is presently being studied in detail. For considerations on the LEST window we refer to Dunn (1984).

The high precision tracking needed for the LEST will be achieved through the combination of low frequency and high frequency response systems (see Engvold and Hefter 1982):

- The slow drifts in the telescope pointing (frequencies < 1Hz) will be taken out by referencing to the positions of the solar limb. A solar limb guider in the focal plane of an auxiliary tracker telescope is interlocked to the declination and the azimuthal drives which then can maintain a steady and correct pointing.
- The high frequency image motions (>1Hz), due to vibrations of the instrument and by the seeing, will be corrected by reference to the actual small scale structure being observed. The error signal from the image displacement thus detected is used

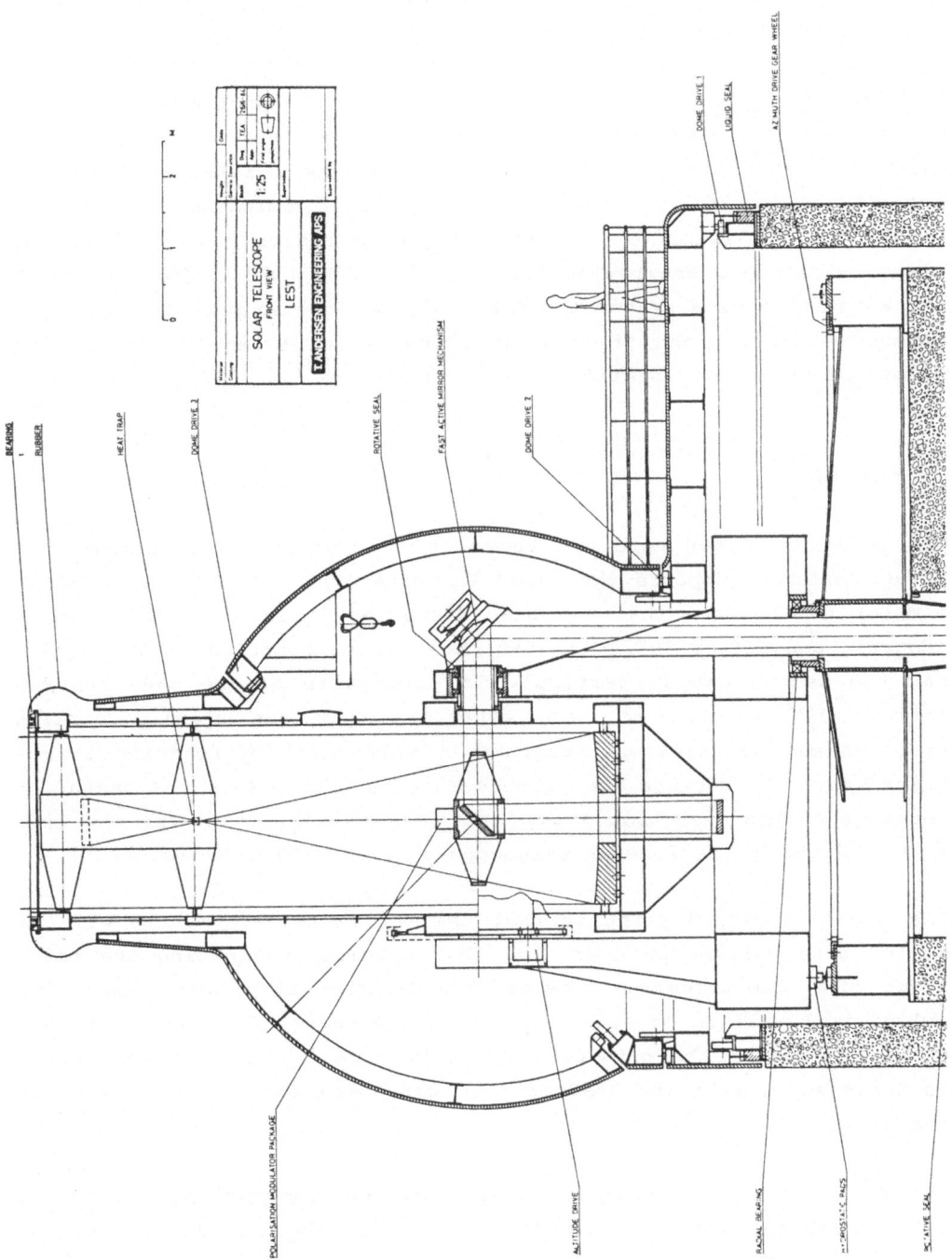

Figure 3. The telescope tube, mounting and dome structure.

to bring the image back to its correct position by a small tilt
of the agile, <u>active</u> mirror 5.

An <u>adaptive optics</u> system will be developed to compensate distortion
of the observed light phase front in real-time, i.e. compensation of
image blurring. Existing systems for use in solar telescopes are yet
far from being fully developed (Hardy 1981, Smithson and Tarbell
1984). The possibilities of using adaptive optics system in ground
based telescopes, are reviewed by von der Lühe (1983). The design of
the adaptive optics system for LEST will be done in parallel with the
telescope design, and at that time the actual scheme for real-time
wavefront error compensation will be chosen.

3 MECHANICS

The solution chosen here and shown in the Figure 3 has a symmetric
structure that supports the tube. The dome that has been selected is
similar to the dome that has been discussed in earlier LEST reports
(Engvold and Hefter 1982). It has one axis of rotation at 45 degrees
and the second one is vertical. The two parts of the dome are rotating with respect to each other so that the opening remains
centered on the telescope aperture. A third axis of rotation is required for LEST because the azimuth axis is offset from the center of
rotation of the elevation axis. This dome is similar to the one used
with success at the 2 meter telescope at Pic-du-Midi (Rösch 1981).

The other important point is that the entire telescope tube will be
filled with helium, an operation that requires evacuating the telescope tube. The telescope tube is thus designed as a vacuum tank. The
cylindrical wall is of steel and has a thickness of 12 mm. There are
multiple ribs on the outside to prevent buckling. An aperture cover
is designed to take the load of the atmosphere when the telescope is
evacuated.

A ring on top of the inner tower provides the horizontal bearing
surface for the hydrostatic pads. It contains channels for collecting
the oil from the hydrostatic pads which rotates with the azimuth base
triangle.

Three radial struts that are displaced 120 degrees from each other, support the radial azimuth bearing at the center of the ring.

The secondary, tertiary and mirror 4 are small and the design of the support systems is not critical. Mirror 5 is "active" i.e. it can move quickly to compensate for image motion. To obtain a high servo bandwidth, it is important to avoid coupling between the mirror servo and the telescope structure. This is accomplished by using a counter-moving counter-weight in a fashion similar to the systems used for chopping secondaries in infrared telescopes.

All five mirrors can be adjusted in tilt by means of small adjust-mentdrives. For the primary, this is most easily obtained by moving the three defining points. Pneumatic supports will automatically adjust themselves to the desired mirror position.

4 BUILDINGS

All facilities such as offices, laboratories, workshops, kitchen, etc., that are not absolutely required near the telescope, are situ-ated in a separate maintenance building at some distance from the LEST tower.

The telescope aperture must be above the ground turbulence. Indi-cations from various sources show that a height of 20-30 m is ac-ceptable. A height of approximately 25 m has been chosen for the LEST tower, the exact value being dictated by optical requirements.

The shape of the dome is a feature that is important to telescope seeing. Experience from various observatories shows that there is an advantage in letting the telescope tube stick out through the dome, preferably protected by an exterior dome tube structure.

The dome which is made from three parts, each with their own axis of rotation, has a special advantage from the point of view of space (see Figure 3). It can be designed as a self-supporting shell structure, for instance of a sandwich material, or alternatively a large steel structure covered by spherical panels, for instance of Kevlar reinforced polyester.

The entrance window must be protected whenever the telescope is not

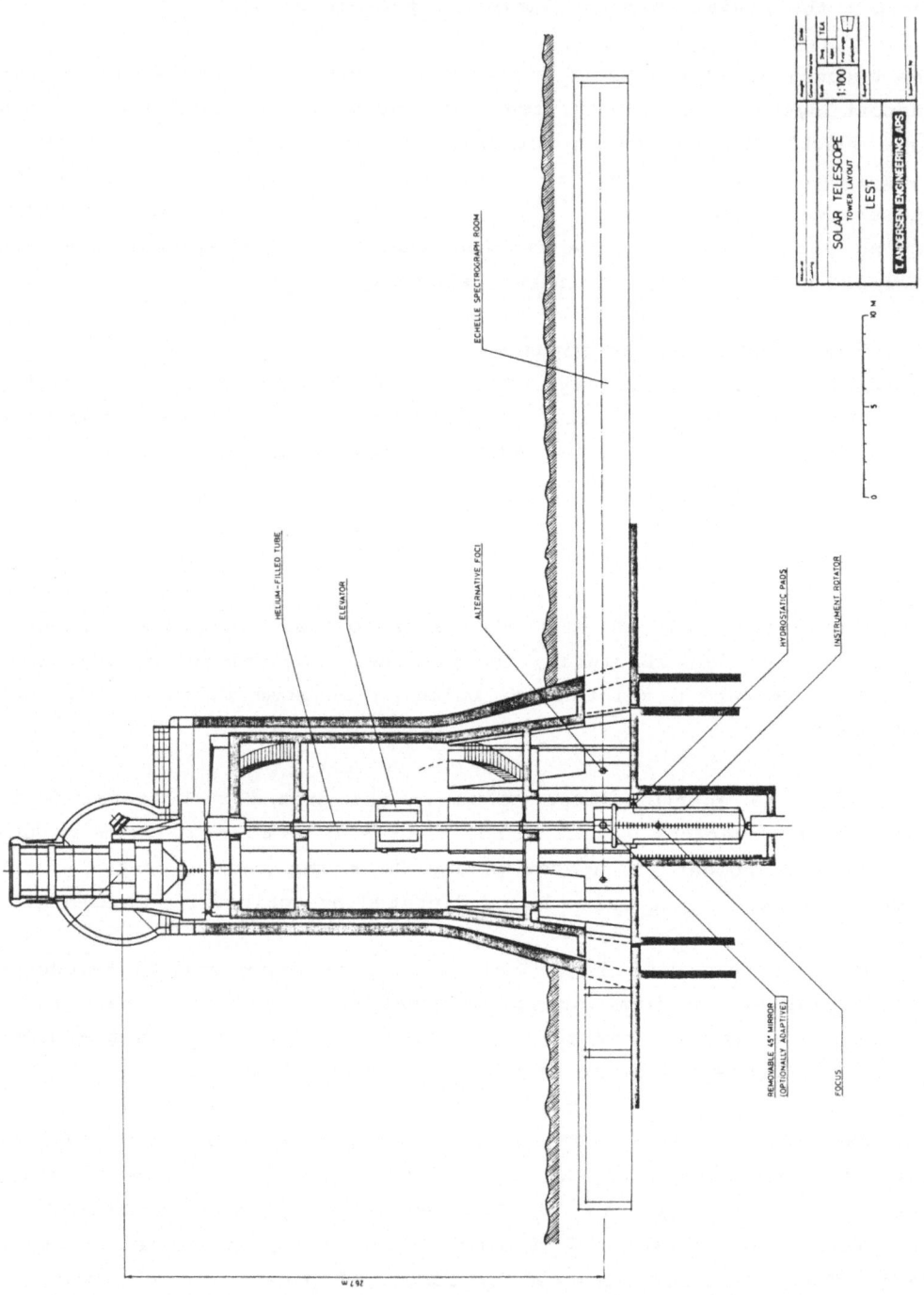

Figure 4.   Cross-section of the double tower and the telescope.

in use and an automatic cover mechanism is required. As mentioned earlier the cover must take the load of the atmosphere when the telescope is evacuated. The same cover performs both of these tasks in the design chosen for LEST.

The instrument stations and other auxiliary telescope facilities are placed underground. This solution minimizes the ground induced wind-turbulence and establishes a thermally stable environment for the instrument stations, which are placed in the center of the tower at 6 locations distributed radially around the central axis. Five of the stations may have permanently installed equipment (cf. Wöhl et al. 1984).

5 TELESCOPE OPERATION

The LEST will be a complicated and expensive facility. It is therefore essential that the operator be experienced. Operation should be performed exclusively by specially trained personnel according to the instructions of the observing astronomer assisted by a resident astronomer of the LEST observatory.

Three observing schemes are anticipated:

- Observing astronomer present at LEST site
- Observing astromer present in one of several
  control stations in the member countries
- Observations are performed by local staff according
  to written instructions given by the astronomer.

It is proposed to have various remote control stations in the LEST member states. For this purpose, communication links between the LEST installation on the observatory site and the respective control rooms at the home bases are required.

REFERENCES

Andersen, T.E.: 1984, LEST Foundation Technical Report No. 1.

Andersen, T.E., Dunn, R.B., and Engvold, O.: 1984, LEST Foundation Technical Report No. 7.

Dunn, R.B.: 1984, LEST Foundation Technical Report No. 3

Engvold, O. and Hefter, M: 1982, "Phase A Feasibility Study on Principal Aspects on Design of a High Resolution Solar Telescope", JOSO Study Report, Institute of Theoretical Astrophysics, University of Oslo, and ARNE JOHNSON Consulting Engineering, Stockholm.

Engvold, O., Dunn, R.B., Livingston, W., and Smartt, R.N.: 1983, Applied Optics, 22, 10.

Fried, D.L.: 1978, J. Opt. Soc. A.. 68, 1651.

Hardy, J.W.: 1981, "Solar Instrumentation: What's Next?", SPO Conference, Oct. 14-17, Proceedings, Ed. R.B.Dunn, p. 421.

von der Lühe, O.: 1983, LEST Foundation Technical Report No. 2.

Rösch, J.: 1981, "Instrumentation for Astronomy with Large Telescopes", IAU Coll. No.67, Zelenchukskaya, USSR, 8-10 Sept., Proceedings Ed. C.M.Humpries, p.79.

Smithson, R.V. and Tarbell, T.D.: 1984, (Private communications).

Stenflo, J.O.: 1984, LEST Foundation Technical Report No. 4.

Wöhl, H., Huber, M.C.E., Mein, P., and Smaldone, L.: 1984, LEST Foundation Technical Report No. 5.

Wyller, A.A.: 1983, "LEST - Large European Solar Telescope", Executive Summary The Royal Swedish Academy of Sciences.

Zwaan, C., Deubner, F.L. and Mein, P.: 1982, "Notes on Scientific Priorities for LEST", JOSO Annual Report p.20.

DISCUSSION

J. Rösch: Did you compute the frequencies of the vibrations of the tower and building of LEST?

O. Engvold: Yes. The eigenfrequency of the outer tower structure is 7-8 Hz, and for the inner structure it is about 15% less.

L.M.B.C.   Concerning vibrations, you mentioned that the structure
 Campos: would be fairly rigid, to reduce oscillation amplitudes. Did you take any special measures concerning vibration damping? Also, which wind speed do you assume for aerodynamic loads?

O. Engvold: Our calculations were made for wind speed of 30 m/s. The amplitudes appear to be comfortably small, only about 0.07 arcsec for the outer tower and 0.002 arcsec for the inner tower structure.

G. Elste: If you observe the limb does the light returned from the secondary fall on the primary?

O. Engvold: No. The heat trap in the prime focus permits observation to about 1 R_o beyond the solar limb before that happens.

G. Elste: If you wish to have the slit of a large spectrograph parallel to the limb at the solar rotational axis you would need an image rotator?

O.Engvold: Yes, unless your instrument could be placed on the vertical rotating table of the LEST focus.

H.U. Schmidt: Could you give me an indication of what is meant by a reactionless mirror?

R.B. Dunn: A reactionless mirror has a second mass mounted in such a way that it cancels the inertias of the main mirror.

Thus, there are no moments transmitted to the supporting structure. These are in use as "chopping secondarys" on large telescopes. They hav bandwidths approaching 200 Hz.

A. Righini: How large is the amount of straylight you estimate due to the reflections in the optical path?

O. Engvold: We have not yet calculated the point spread function for the entire system.

(A. Righini: What is the influence of the Caldera on the La Palma site?)

U. Kusoffsky: What we believe can be a possible Caldera-effect is only seen when there are weak southern winds, or no wind at all. I would like to mention the ongoing campaign for the Nordic Optical Telescope with trail telescopes pointing both to the north and to the south on 3 different sites along the Caldera. Preliminary results show no difference in seeing in the north direction. One should remember from earlier campaigns that good night-time seeing is always followed by good day-time seeing.

A. Dollfus: (1) On the LEST instrument, why a Gregorian rather than a Cassegrain design? (2) What about the need for a field rotator?

O. Engvold: (1) The Gregorian system is chosen for several reasons: It removes the heat load on the secondary mirror and on the polarization modulator unit, which is located near the secondary focus. The Gregorian system offers a good possibility for control of its alignment (see LEST Foundation Technical Report No. 7), and it appears to be advantageous with regards to reduction of straylight. (2) LEST will observe small angular fields presumably less than 1-2 arcmin, most likely only 10-20 arcsec in a given exposure. Hence, there will be no smearing of the images due to field rotaion for exposure times up to several minutes. To obtain a par ticular orientation with respect to the Sun one will either mount the focal plane instrument in the rotating table, or insert a field rotator into the beam.

Question: Will there be thermal problems with the entrance window?

O. Engvold: The window will have a thickness of only 1-2 cm and therefore absorb less heat than a conventional and thicker vacuum window. The cooling effect of the helium in the telescope tube will reduce internal thermal gradients in the window.

R.B. Dunn: Mehltretter showed that the same center-to-edge difference in temperature a thin window would have a smaller effect on the image than a thick window. This comes from two effects: First, the optical path is proportional to the thickness and, second, the temperature distribution is more uniform and with the thin window the front and and back surface more effectively radiate the heat conducted in from the edge.

THE EUROPEAN OBSERVATORY AT THE CANARY ISLANDS

J. RAYROLE
Observatoire de Paris-Meudon
92195 Meudon Principal Cedex

Near the Western African coast, at 28 degrees of northern latitude a new astronomical
observatory is growing. In the next few years, it will be one of the most important
in the world by the size of its telescopes as well as by the diversity of its equipments both for Stellar and Solar works.
It is the result of more than 10 years of considerable work from European stellar and
solar astronomers. Considerable work during the site testing campaigns as well as to
discuss international agreements which are the basis of its reality.
The Canary Archipelago is composed of seven islands. From West to East we have : La
Palma, Hiero, Gomera, Tenerife, Gran Canaria, Fuerteventura, Lanzarote.

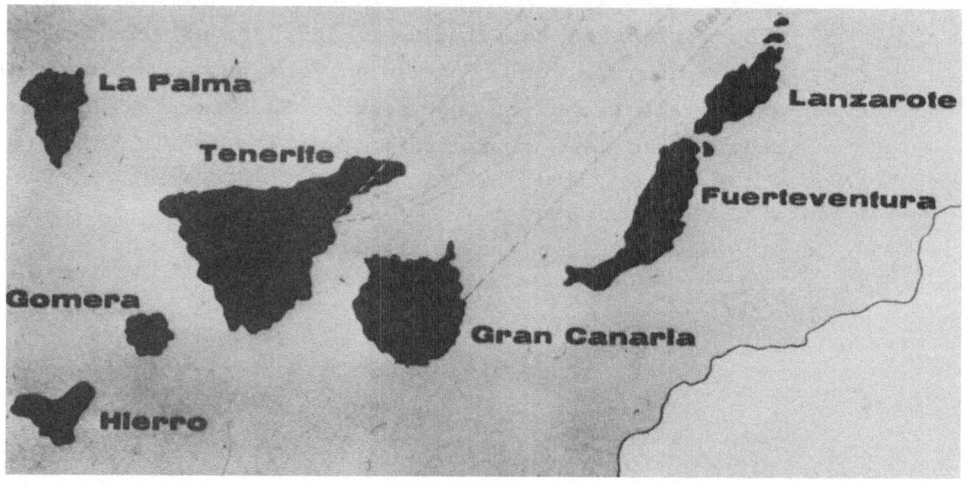

The general meteorological situation at the Canary Islands is characterized by the
Azores high pressure system and the frequent occurence of a stable inversion layers
at a height between 1 and 2 thousands meters above which a stable subsiding airmass
is encountered. Stratocumulus clouds take form there and the higher summits emerge in
a clear sky (Fig. 1, 2). For about 20% of the sunny time the sky is coronal and very
pure for 55%. We may rely on 3500 hours of sunshine and even in winter bad weather
does not occur more than four consecutive days. Furthermore reaching the Canary islands is easy from whatever European country in less than half a day.

Figure 1 : Isla de La Palma. The Caldera de Taburiente full of
clouds.

Figure 2 : The South part of La Palma as seen from Roque de Los
Muchachos at sunset. Note the "cloud falls" often
visible.

The Canary Island Observatory is a Spanish institute which depends of the "Instituto
de Astrofisica de Canarias".

1975   :   CREATION OF THE IAC
            Instituto de Astrofisica de Canarias

1979   May 26   : DENMARK, GREAT BRITAIN, SPAIN, SWEDEN
        signed
        . An Agreement on Co-operation in Astrophysics
          GOVERNMENTAL LEVEL (LEVEL 1)

        . A Protocol on Co-operation in Astrophysical research
          in Spain
          NATIONAL AGENCY LEVEL (LEVEL 2)

        These agreements opened
        El Observatorio del Roque de los Muchachos (La Palma)
        to telescope installations from foreign countries

1983   April 8
        THE FEDERAL REPUBLIC OF GERMANY
        signed
        . Level 1 and 2 agreements
        DENMARK, FEDERAL REPUBLIC OF GERMANY, GREAT BRITAIN, SPAIN,
        AND SWEDEN
        signed
        . An Addendum to the Protocol (Level 2)
        This addendum opened
        El Observatorio del Teide (Tzana : Tenerife)
        to telescope installations from foreign countries

For solar observations point of view, the two Canarian sites on Tenerife and La Palma
have been compared in 1979 with two medium size telescope built by the Kienpenheuer
Institut (Freiburg) and the University of Göttingen.
This one year campaign confirmed clearly the superb atmospheric qualities of airmasses
above Canary Islands and showed up orographic effects due to the Caldeira de Taburiente on La Palma site.
The choice of a location at La Palma is very difficult. The topography of La Palma is
dominated by an enormous crater "Caldera de Taburiente". The site is located at the
northern crater rim. The mountain drops sharply making La Palma the steepest major
island in the world (Fig. 3a).
The western portion of the site between the Roque de Los Muchachos and the western
end of the observatory is certainly poor because the line of sight for the early
hours is above the Caldeira (Fig. 3b).

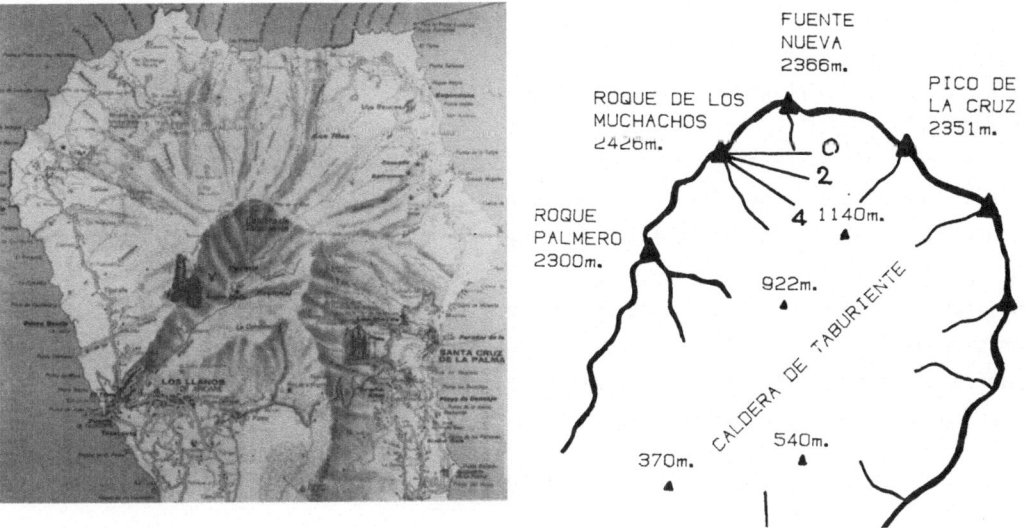

Figure 3 :   a) Map of La Palma (Northern part).

b) The "Caldera de Taburiente" and the line of sight for the early hours (0, 2, 4 hours after sunset).

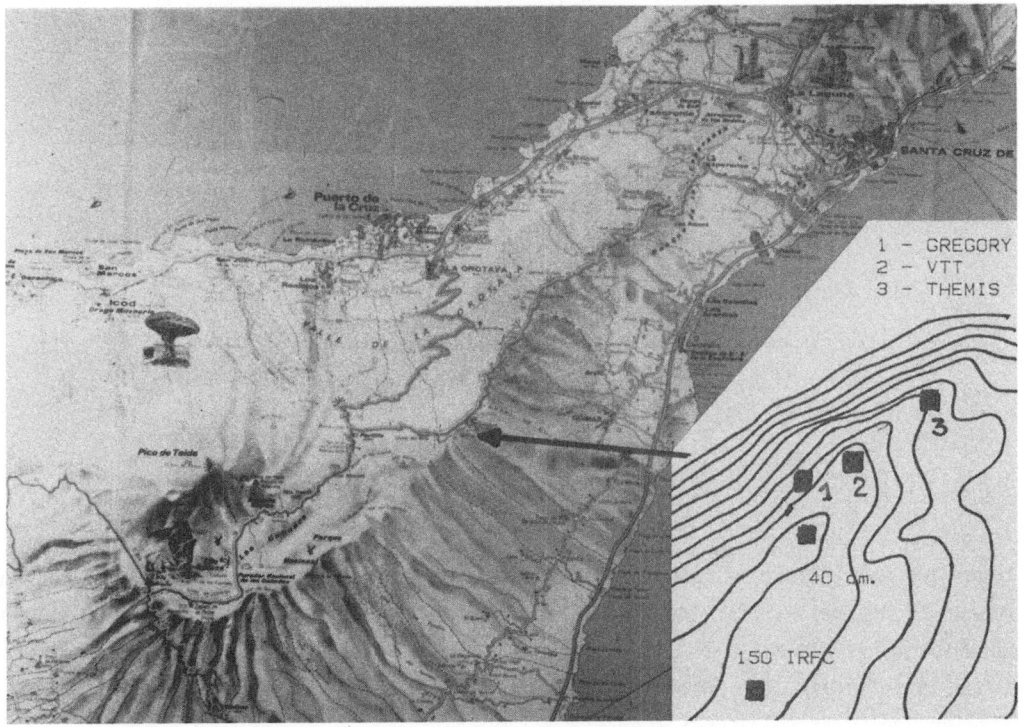

Figure 4 :   a) Map of Tenerife (Central part).

b) Orographic effects at Izana are the same, all points considered.

The eastern portion, Fuente Nueva, is without doubt the best location.

At Izana, on the other hand, the site is smaller and the relief is more or less the same, all points considered. But the useful portion does not leave room for many telescope (Fig. 4b).

The observatory is situated on a volcanic ridge near the centre of the island. Tenerife is one of the larger islands in the Canary archipelago (Fig. 4a).

Before the international agreements, some telescope was running in a regular way in Canary Island.

The excellent quality of the Teide observatory for solar observations has been known for several decades owing to the observations of the chromosphere conducted by our Spanish colleagues with a Razdow telescope and more recently with the 40 cm Newton telescope by our German colleagues.

    I /   OBSERVATORIO DEL TEIDE :   Izana (Tenerife)

       . 150 cm IRFC Telescope
        Imperial College of Science and Technology (London)
        Since 1980 (IAC)

       . 50 cm Telescope
        Mons University (Belgium)

       - Hα Patrol Telescope (Razdow) 1960
        Instituto de Astrofisica de Canarias (IAC)

       - 40 cm Newton-Vacuum Telescope 1972
        Federal Republic of Germany (Freiburg)
        Spain (IAC)
        Used in 1979 for Site testing

    II/ OBSERVATORIO DEL ROQUE DE LOS MUCHACHOS (La Palma)

       - 45 cm Gregory-Newton-Vacuum Telescope (1979)
        Federal Republic of Germany (Göttingen)
        Used for Site testing

Table 2 : Telescopes in use before 1980

Since 1980 many new telescopes are in use or under construction. At "Roque de Los Muchachos" observatory, for the most part, equipments are stellar. We can see that the observatory is already a reality with many various telescopes in use and many others under construction. The Swedish Academy is the only one institution to operate both stellar and solar telescopes.

NEW TELESCOPES IN USE OR UNDER CONSTRUCTION

1981 - * 60 cm Cassegrain Telescope
          Swedish Academy of Science
          Moved from Capri

1981 - ⊖ 25 cm Vacuum Solar Tower
          Swedish Academy of Science
          Planed to increase to 50 cm (1985)
          Moved from Capri

1984 - * 100 cm Cassegrain Telescope
          Great Britain - Netherlands

1984 - * 250 cm Cassegrain-Coude Telescope
          Great Britain-Netherlands
          Isaac-Newton Telescope moved from Herstmonceux

1984 - Transit-Circle
          Denmark-Great Britain
          Moved from Brofelde

1986 - * 420 cm Alt-Azimutal Telescope
          Great Britain-Netherlands

      - ⊖ 45 cm Open Telescope
          Netherlands (Utrecht)

1990 - * 250 cm Alt-Azimutal Telescope
          Nordic Optical Telescope
          Denmark . Finland . Norway . Sweden

          Millimeter RADIO TELESCOPE
          Great Britain-Netherlands

Table 3 : Observatorio del Roque de Los Muchachos

For the most part, new solar telescopes will be located at the Teide observatory.
This is due to the fact that orographic effects locally perturb in daytime the stability of the air masses which arrive over the Roque de Los Muchachos.

NEW TELESCOPE IN USE OR UNDER CONSTRUCTION

1985 - ⊖ 45 cm Gregory Vacuum Telescope
          Federal Republic of Germany (Göttingen)
          Moved from Locarno

1985 - ⊖ 60 cm Vacuum Tower Telescope VTT
          Federal Republic of Germany (Freiburg)

1986 - * 80 cm Cassegrain Telescope
          Spain (IAC)

1989 - ⊖ 90 cm Vacuum Telescope
          Magnetograph THEMIS
          France (INAG)

Table 4 : Observatorio del Teide (Tenerife)

SOLAR INSTRUMENTATION IN CANARY ISLANDS

1. The Swedish Solar Telescope

The Swedish Telescope installation is part of the Astrophysic Research station of the Royal Swedish Academy of Science which, from 1961 to 1978, maintained a Solar Research station on the island of Capri (Italy). After which time the station relocated to Fuente Nueva (La Palma) in 1981.

The original solar telescope consists of a classical coelostat mounted on top of a 16 m high solar tower. The vertical solar parallel beam is directed, by a 80 cm diameter flat mirror, down into a 12 m long vacuum tube which accomodate a 44 cm Cassegrain telescope.

However, because of scattered light problems the Cassegrain optics has been replaced by 25 cm diameter sigle lens which forms a 20 cm solar image that can be directed into two observing rooms (Fig. 5).

The first results obtained with this instrument confirm the good quality of the Fuente Nueva site.

Reduced to zenith r_0 values, averaged over 15 seconds, rise 7.1 cm with peaks up to 30 cm, giving long exposure-time available for high resolution spectroscopy.

White light photographs of solar granulation and Hα pictures taken with the existing single lens show diffraction limited resolution of .5 arc sec.

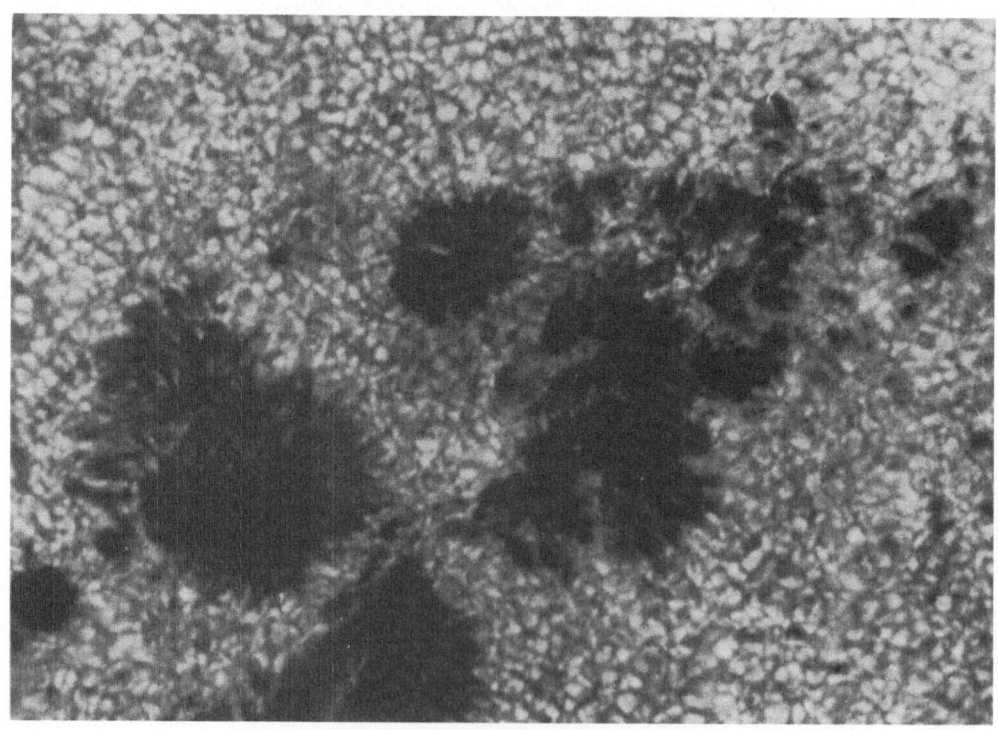

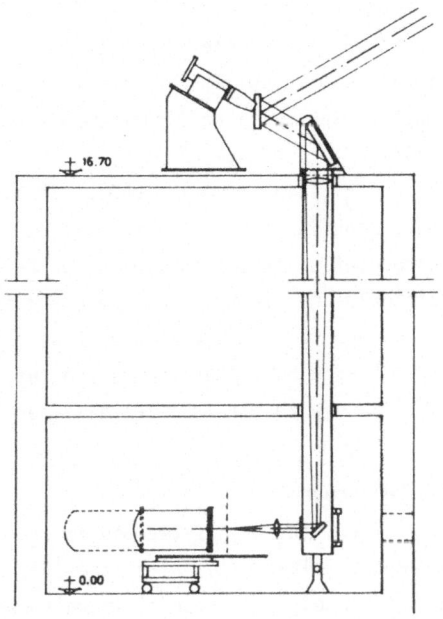

Figure 5a : Optical Scheme of the Swedish Solar Tower.

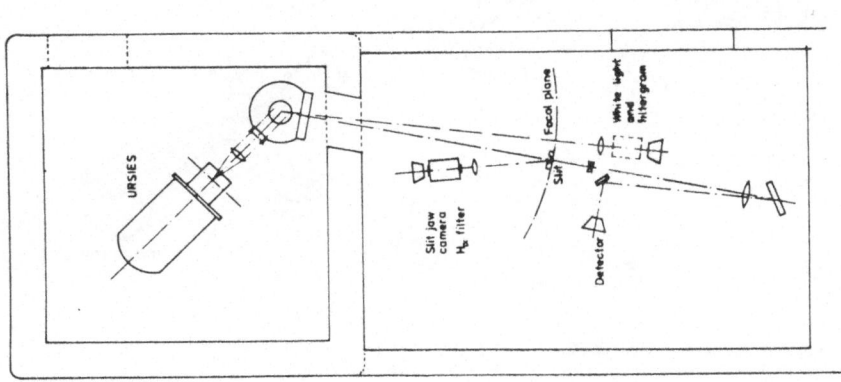

Figure 5b : The observing rooms.

To take advantage of these good seeing conditions the single lens will be changed to an achromatic lens doublet with 50 cm diameter in 1985.
In the same way the classical coelostat will be replaced by an Alt azimuthal system of two flat mirrors. The doublet lens will replace the entrance window of the present vacuum tank which will be filled with helium (Fig. 5).
The auxiliary instrumentation facilities comprise URSIES in its special single purpose room. URSIES is a pressure-scanning echelle + Fabry Perot interferometer spectrograph with two channels photocounting system on line to a PDP 1134 mini-computer.
With two photoelastic modulator placed in front of the entrance pinhole, URSIES works as a Stokesmeter. In the adjoining second spectrograph room a littrow spectrograph (dispersion 2 mm/A) and its slitjaw Hα filter camera are set. White light and filtergram eaquipment are also available with either a Zeiss-filter (.25 A bandpass) or a Halle-filter (.5 A bandpass).

2. The 45 cm Gregory Vacuum Telescope

Some time ago the University of Göttingen has decided to transfer the facilities of the Locarno station to the Canary Island site. After the 1979 site testing campaign decision was taken to move the Gregory Telescope to Izana (Observatorio del Teide). Many Solar Astronomers are familiar with this instrument and its powerful auxiliary facilities (Fig. 6).

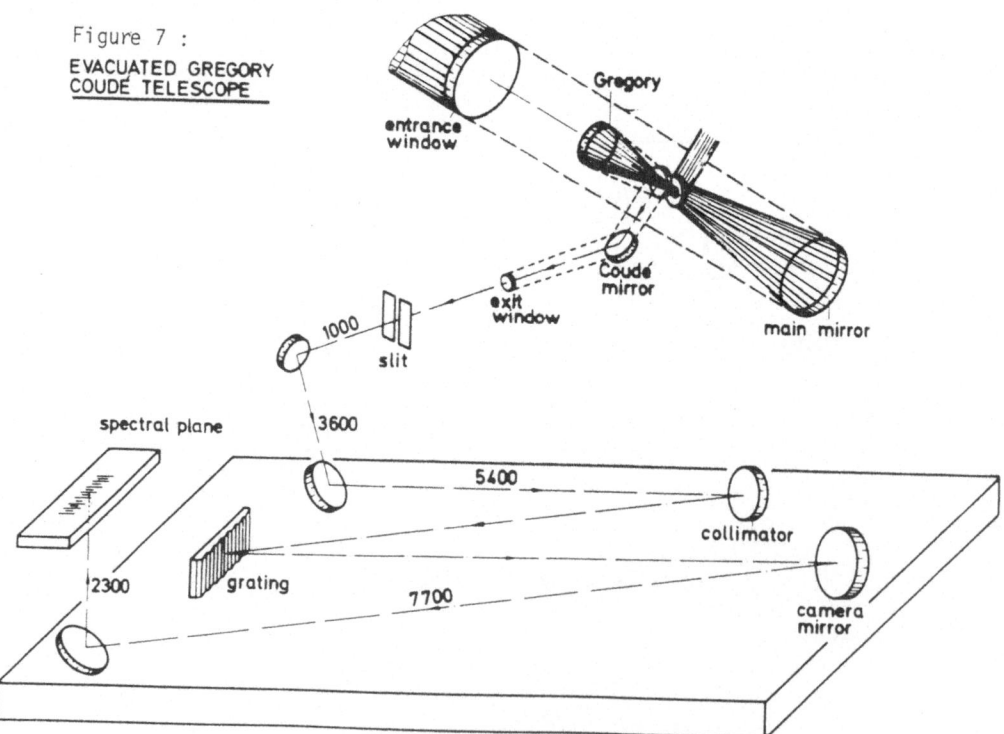

Figure 7 :
EVACUATED GREGORY
COUDÉ TELESCOPE

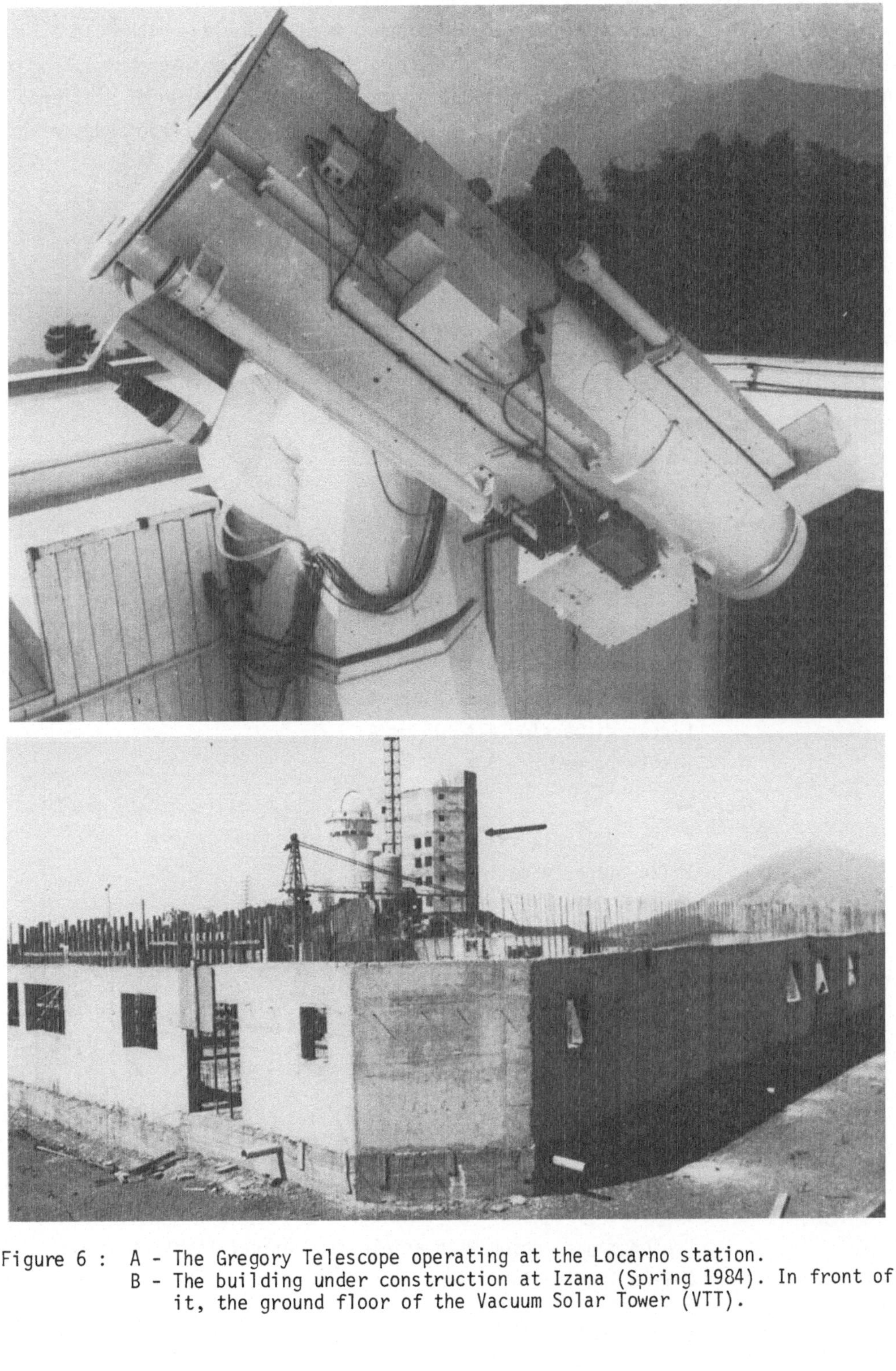

Figure 6 : A - The Gregory Telescope operating at the Locarno station.
            B - The building under construction at Izana (Spring 1984). In front of
                it, the ground floor of the Vacuum Solar Tower (VTT).

It is an all mirror evacuated Gregory Coude Telescope with 45 cm prime aperture and 24 m effective focal length. A water-cooled diaphragm selects via a pinhole a 3.5 arc min field of view and the no used light is refelected out of the tube. Only 1% of the incident beam illuminates the secondary optic, resulting in a minimum of scattered light. Automatic guiding is performed by mean of an auxiliary telescope mounted along-side the main tube (Fig. 7).

All kind of scans can be performed on the sun with an accuracy of 0.2 arc. sec.

In the observing room we have :

- the spectrographs,
- the slight jaw camera device,
- the computer unit with all the necessary electronics for several kind of receivers and the magnetograph unit.

The slit jaw imaging device provides simultaneous pictures in Ca^+ K line, $H\alpha$ line and continuum window either

- on 35 mm film cameras
- or on TV screen
- or on a projection screen

AUXILIARY FACILITIES

Echelle Grating Spectrograph

. Focal length 10 m
. Grating 300 groves/mm. Blaze 61 degrees
. Linear dispersion 4 to 15 mm/A
. Resolving power $> 5 \times 10^5$

Receivers

. 6 Photomultipliers
. 2 Two-Stage image intensifiers
. Several infrared receivers
. 1 Linear Reticon 128 diodes array
. 1 Two-dimensional 100 x 100 diodes array

Computer Unit : Honeywell H316

. 16 Analog input-output channels
. 32 Digital input channels
. 16 Digital output channels
. Magnetic tape mass storage

Magnetograph Unit

. Babcock-Kiepenheuer longitudinal magnetograph
. Vector magnetograph with an electro-optical light modulator
. 3-line longitudinal magnetograph for simultaneous measurements in 3 pre-selected lines
. Photographic Stokes meter . Photoelectric Stokes polarimeter

The spectrum can be recorded with computer controlled film cameras or scanned photo-electrically and recorded either on analog device or digitalized on magnetic tape.
For the purpose of solar magnetic field measurement, method to compensate for instrumental polarization has been developped.
Magnetograph unit can work in different ways :

       . Longitudinal magnetograph.

       . Vector magnetograph.

       . Stokesmeter.

3. The German 60 cm Vacuum Solar Telescope

Just now, the Kiepenheuer Institut stars the final phase of its Vacuum Tower Telescope (VTT). At the end of 1984 the telescope will be put in the building. In 1985 the testing and alignment will take place and from 1986 the telescope and its auxiliary facilities will be running in a regular way.
This telescope, including the spectrograph, has been designed by the late Peter Melhtretter. Project managing and the manufacturing of many components are supported by manpower of the Kiepenheuer Institut.
A Classical 80 cm coelostat, mounted on top of a 38 meter high tower directs down the solar beam into a Herschel vacuum telescope with a focal length of 46 meters (Fig. 8). The image is focussed either on the entrance slit of the vertical main spectrograph or is directed by an additional flat mirror in one of three optical laboratories. Guiding and scanning systems are done by using the second coelostat mirror as an error correcting tilting mirror. The error signal is detected in the focus of an auxiliary telescope using a small fraction of the telescope aperture. Beneath the telescope the vertical spectrograph tank is turnalble to allow for arbitrary orientation of the entrance slit on the sun.

AUXILIARY FACILITIES

Spectrograph

  . Slit length 0.125 m (600 arc Sec)

  . Predisperser F = 4 m

  . Main spectrograph F = 15 m

  . Grating 79 grooves/mm. Blaze 64.5 degrees

  . Typical linear dispersion 12.5 mm/Å

  . MSDP (Meudon)

Computer Unit

  . 4 Megabyte RAM

  . 100 Megabyte Disk-drive

Receivers

  . Several 70 mm film cameras

  . Cooled photomultipliers

  . Photo-counting devices

  . 1 linear cooled Reticon 1024 diodes array

  . 1 CCD Camera (RCA 53612 x 0 : 320 x 512 pixels)

  . Doppler-compensator device

  . Servo-controlled Universal Zeiss filter (ARCETRI)

Table 6

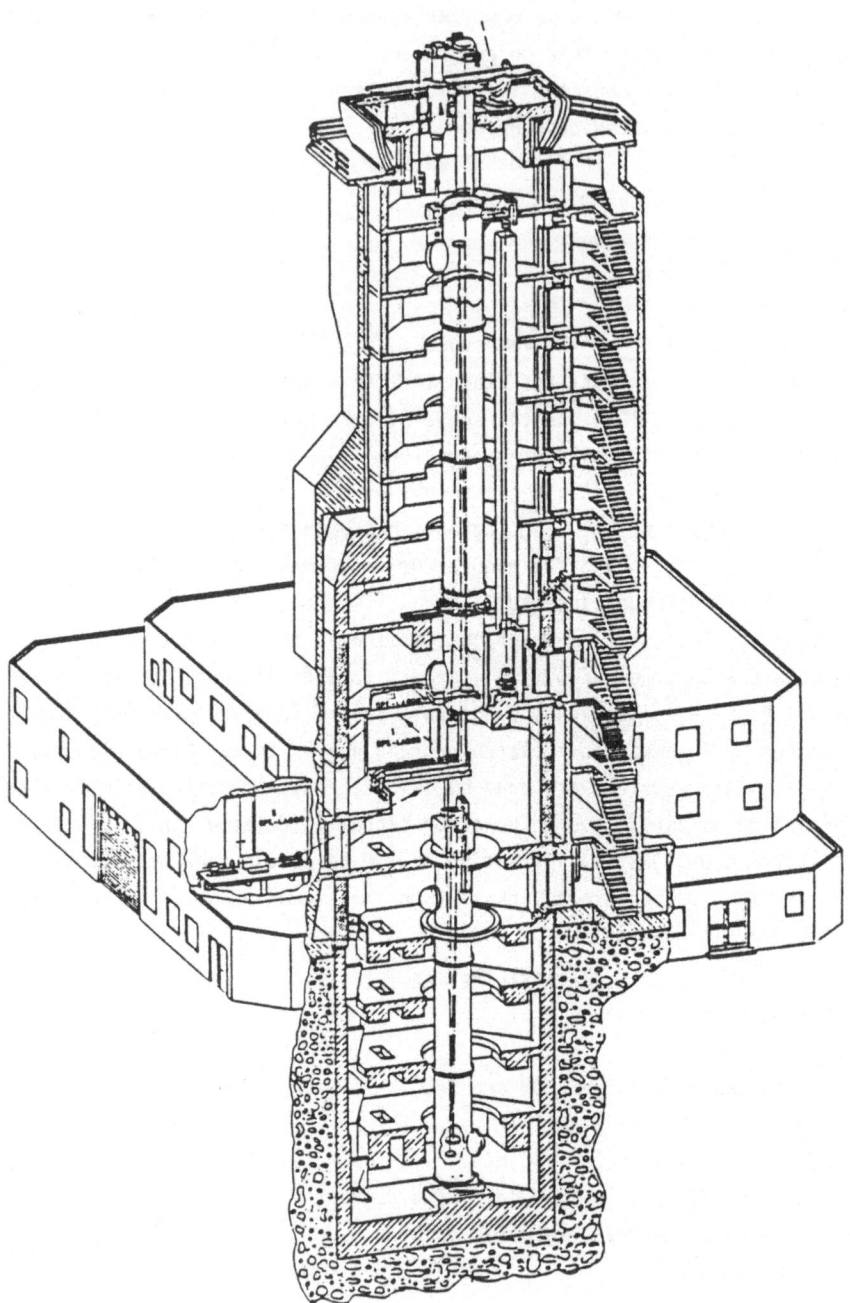

Figure 8 : The VTT Telescope : optical scheme and building.

The spectrograph uses a high order echelle grating and a predisperser to select the desired wavelength region. In case the observer prefers the use of low order grating the tank can be turned. Then a second entrance slit, with interference filters right above the collimator of the main spectrograph, is active. High precision spectrum scanning unit moves the detector head with an accuracy of 2 microns over a distance of 150 mm corresponding to 15°A. Observers can control observations by mean of a slit jaw camera device which provides pictures in Ca$^+$ K line, Hα line and continuum window. Two post focus devices will be installed by guest groups.

- The Arcetri Group will bring their bidimensional Spectroscopic unit in one of the three available optical laboratories. This device will provide narrowband filtergrams in arbitrary wavelengths.

- The Meudon Group is ready to instal its Multichannel Subtractive Double Pass device (MSDP) to the spectrograph. This device will enable to produce up to nine filtergrams simultaneously in adjacent wavelength covering a broad line profile.

4. The French Magnetograph THEMIS

THEMIS is a national project supported by all the solar groups in France. For a long time a new concept of solar magnetograph has been proposed to the French sponsors without success. After hard discussions the project has been accepted and planned to be running in 1989 on the Izana site.
THEMIS has been designed for high accuracy measurements of magnetic field inside fine structures of the solar atmosphere.
In modern solar magnetometry we have to solve two difficult problems :

- High spatial and spectral resolution measurements of the state of polarization inside line profiles.
- Interpretation of these observed values in term of vector magnetic field components in all the layers which can be observed.

To solve the first problem, we need large aperture telescope, free of instrumental polarization as far as the polarimeter package, and located in a good site.
For the second, we need large spectral range in order to enable the simultaneous observation of several spectral lines forming at different levels. Large spectral range is also necessary to observe at each level all the lines necessary to separate the magnetic field measurements from temperature and density variations.
Large spectral range imposes a limit to the telescope aperture by the fact that atmospheric disturbances depend on wavelength (Fig. 9).
90 cm aperture telescope is a good compromise if we take into account the entrance flux, the diffraction limited resolution and the available spectral range for the best conditions we can hope on the site.

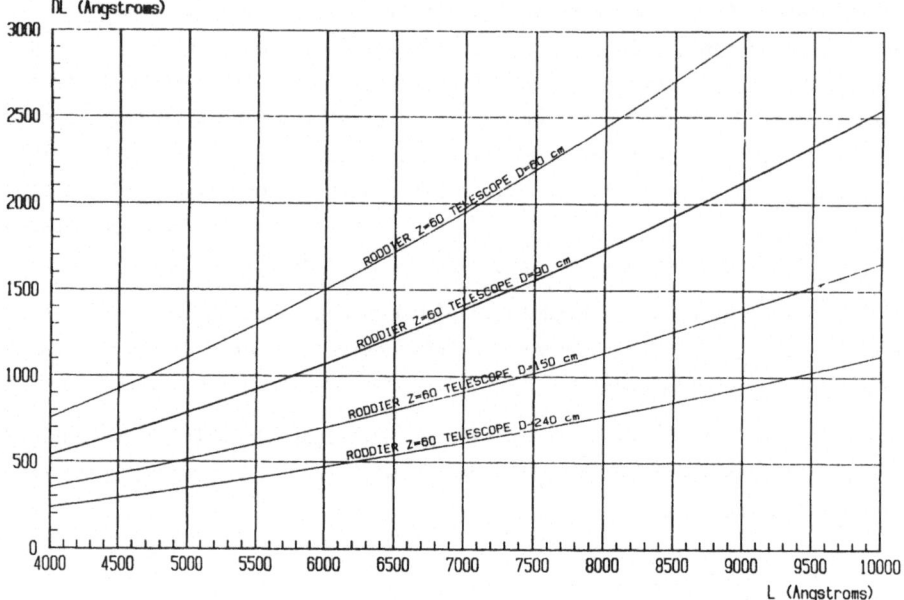

Figure 9 : Coherent spectral range available (DL) as function of the wavelength (L) for different telescope aperture size (60 cm, 150 cm,
240 cm). These curves have been computed for 60 degrees Zenital
distance and a Zenith R_o value of 45 cm for L = 6000 Å.

It is not possible to list here all the problems involved in magnetic field measurements. However it is usefull to specify the choice for the polarization analyzer device. The lifetime of atmospheric disturbances rises to an upper limit of 200 m sec.
so we must observe all the necessary parameters in such a time interval.
As long as large two-dimensional array detectors cannot accomodate high frequency
read out and as long as electro-optical polarizing modulators cannot be used simultaneously with different wavelengths, such a type of polarimetric device cannot be useful for high resolution observations with several spectral lines.
The THEMIS analyzer of polarization will be built with three achromatic retardation
plates which can be quickly inserted on the beam and coupled with a polarizing beam
splitter.
THEMIS is specifically designed for low instrumental polarization. This is achieved
with rotationally symetric optics before the package analyzing the polarization (Fig.
11).
A Ritchey-Chrétien telescope directed at the sun focuses a solar image on the spectrograph slit, behind which is placed the polarimeter package (Achromatic Birefringent plates and Polarizing beam splitter). In this way nothing can alter the line

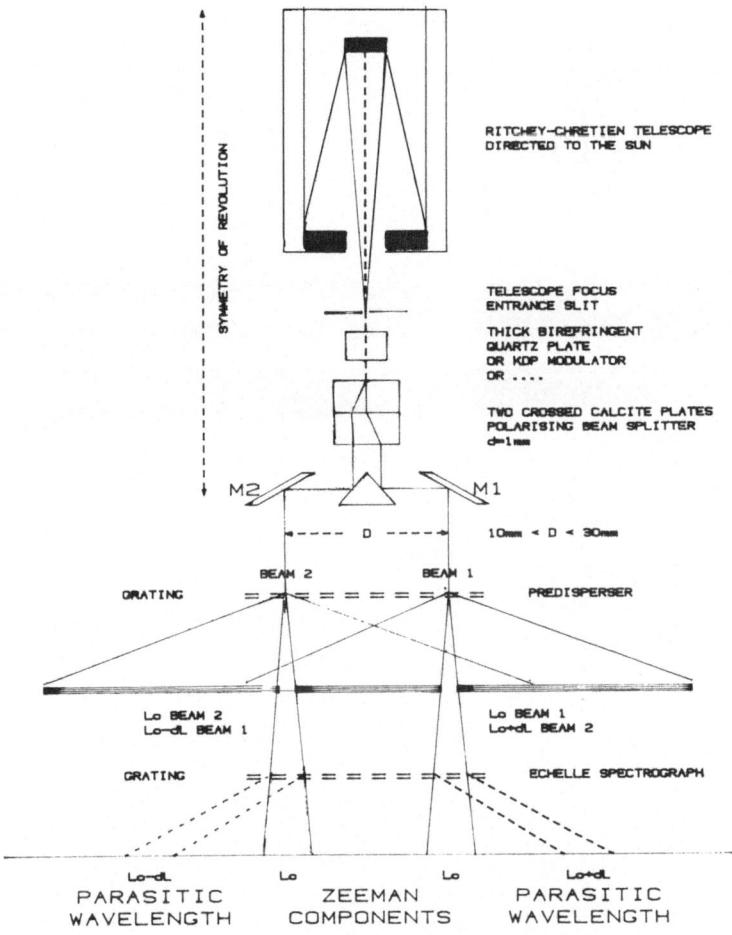

Figure 10 : Principle of THEMIS Analyzer of polarization.

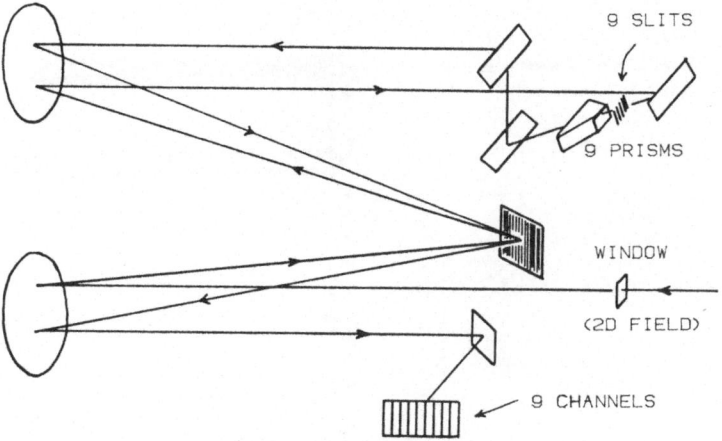

Figure 12 : MSDP Optical scheme.

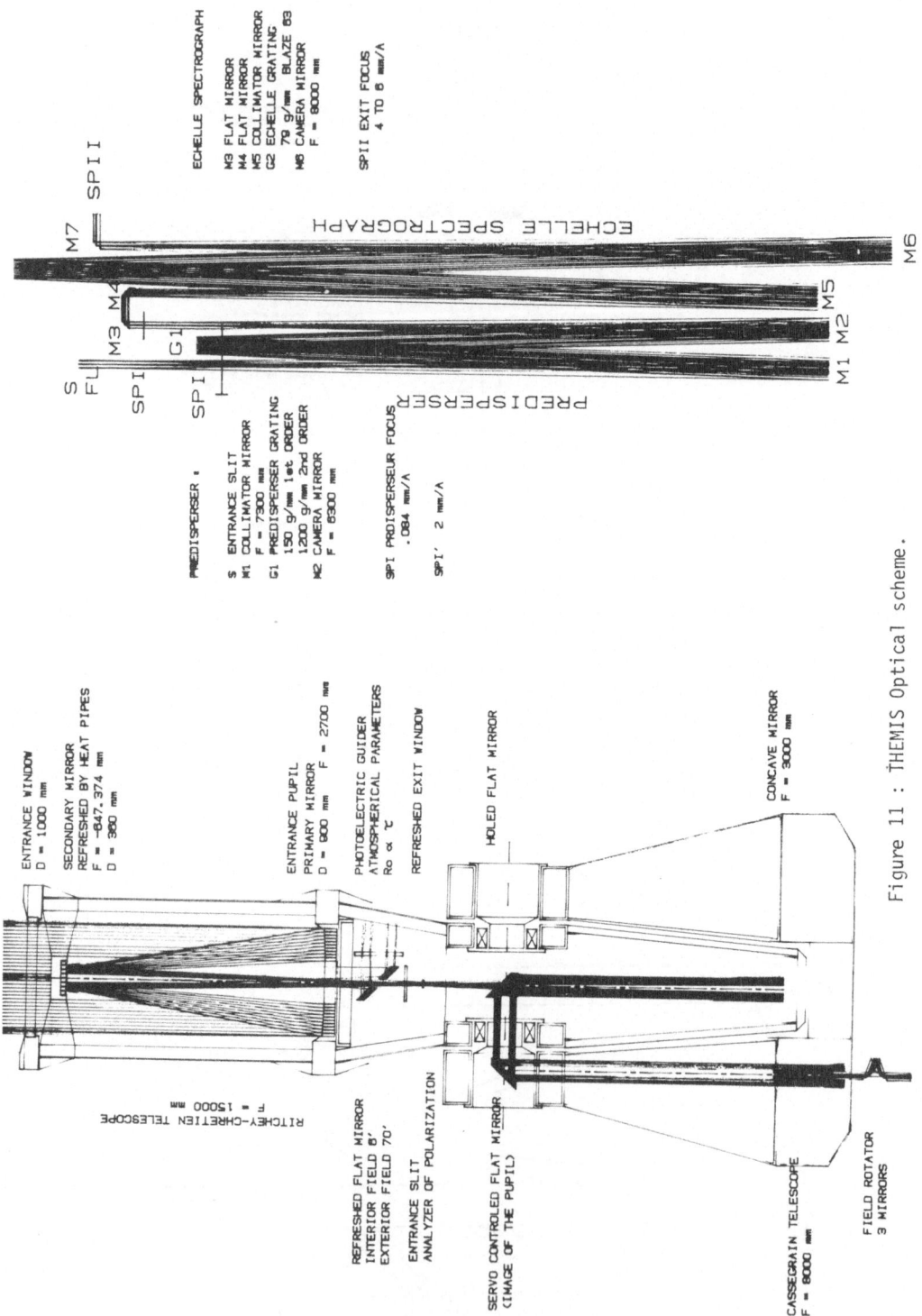

Figure 11 : THEMIS Optical scheme.

10 LINES WITH 2 STATES OF POLARIZATION

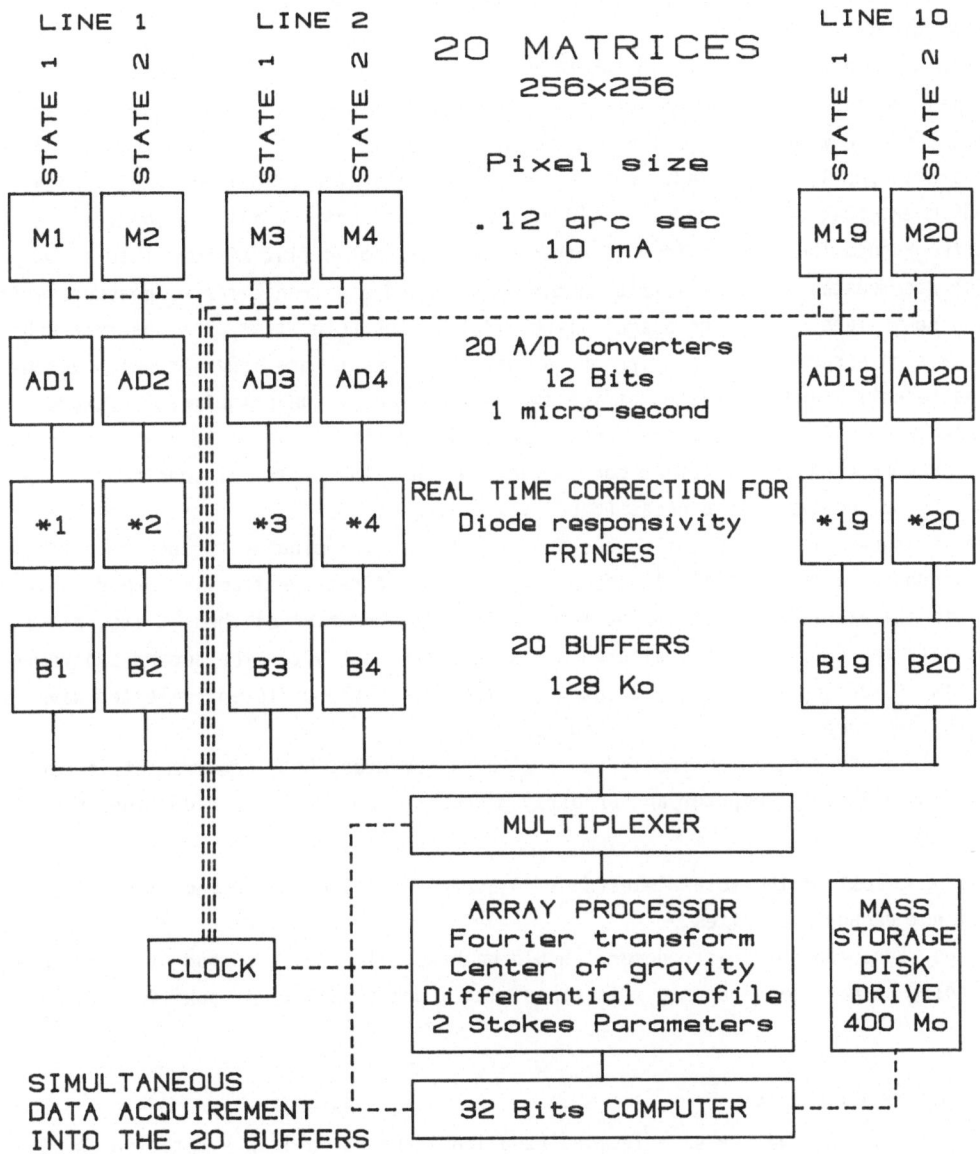

5120 LINE PROFILES EVERY 3 SECONDS FOR 256 SOLAR POINTS

Figure 13 : Example of data acquirement and real time
reduction process with THEMIS.

profiles, if the two beams produced by the polarizing beam splitter can be completely separated. The predisperser characteristics and the echelle spectrograph allow
this separation without any pollution for any line and simultaneously for a great
number of lines (Fig. 10).

Such kind of analyzer is running at Meudon since 1982.

The main telescope with an effective focal length of 15 m is mounted as an alt-azi-
muth. Entrance window, primary and secondary mirror as well as the polarimeter package are thermally controlled. Transfer optics increase the effective focal length up
to 60 m and focuses a 6 arc min field of view on the entrance slit of the vertical
spectrographs. The unused light is directed, via a cooled flat mirror, into a box
which accomodates a photoelectric guiding system and electronic for r_o measuerements.
Scanning is made by turning a small light flat mirror (Fig. 11a). The spectrograph
uses a high order echelle grating and a long focal length predisperser which allows
lines selection and separation of the two beams given by the polarimeter package.
3 selected gratings for the predisperser associated with the echelle gratings of the
main spectrograph give a lot of combinations which allow classical spectroscopy or
Multichannel Double pass process (MSDP) (Fig. 11b).

When it is used with MSDP optics the spectrograph is illuminated through a rectangular window as entrance slit. In the plane where the first spectrum is formed, several intermediate slits select different wavelengths. In front of these slits, prisms
translate the different pre-selected beams in order that after the second subtractive dispersion the images of the entrance windows for the different selected wavelengths are completely separated (Fig. 12).

The whole instrument is controlled by a 16 bits computer. Data acquirement is performed by a 32 bits computer and an array processor for real time deduction (Fig. 13)

Up to now, most of the solar telescopes in Europe were located in bad sites or at
least not enough good sites.

Just now European solar astronomers should be happy. They have a good site and they
are going to have many running telescopes with a great variety of auxiliary facilities.

QUESTION FROM : RIGHINI   TO : RAYROLE

Question : During the '79 site testing campaign in La Palma it was detected an outflow of hot and humid air from the Caldera. Do you know if the Swedish colleagues
has the same experience ?

ANSWER FROM RAYROLE

I did not have informations about that. I do not know if Dr. Kusoffsky has any data
but it is certain that "Caldera effect" is less important at Fuente Nueva.

HIGH RESOLUTION SOLAR OBSERVATIONS

Alan Title
Solar and Optical Physics Department
Lockheed Palo Alto Research Laboratory
Palo Alto, CA 94304, USA.

Abstract.- Traditionally the way to get high quality images has been
to find a good seeing site and establish a high quality observatory. It
now appears that this procedure can be improved by installing active
mirrors which correct real time wavefront tilt, and adaptive mirrors
which correct, at least for a limited field of view, wavefront distortions introduced by the atmosphere. Space telescopes offer the further
advantage of completely eliminating the blurring and distortions introduced by the atmosphere. Further space operations offer the possibility
of uninterrupted observing sequences of many days or weeks.

Introduction.- Significant developments in optical technology, high
speed electronics, and lifting capability of space payloads have occurred in the last decade. Currently there is a world wide effort to deve-
lop optical technology required for large diffraction limited telescopes that must operate with high optical fluxes. These developments can
be used to significantly improve high resolution solar telescopes both
on the ground and in space. When looking at the problem of high resolution observations it is essential to keep in mind that a diffraction
limited telescope is an interferometer. Even a 30 cm aperture telescope,
which is small for high resolution observations, is a big interferometer. Meter class and above diffraction limited telescopes can be expected to be very unforgiving of inattention to details.

Unfortunately, even when an earth based telescope has perfect optics
there are still problems with the quality of its optical path. The
optical path includes not only the interior of the telescope, but also
the immediate interface between the telescope and the atmosphere, and
finally the atmosphere itself. The control of the spatial positioning
of the telescope becomes increasingly critical as the resolution increases because the stability during exposure should be at least a

tenth the resolution. For magnetic and velocity measurements which
require differencing pairs of images, the pointing stability requirements can be even more severe. Pointing and positioning of the telescope is complicated by shake induced by the motions of the telescope
drives, motions coupled into the telescope via a dome if there is one,
vibrations introduced by equipment on the telescope and in its building,
and finally motions introduced by the operators. It should be noted that
computers with high speed disk and tape drives has increased the problem of mechanical coupling from equipment.

All the above mentioned problems have to be solved in order to achieve
high resolution solar observations. To date the best solution has been
to build a very good optical telescope, to show exquisite attention to
details of solar heating of the optics and the immediate interface to
the local atmosphere, to locate the telescope at a exceptional seeing
site, and to show extreme attention to the details of the entire observatory, so that it is virtually always able to achieve its best performance. Probably most important is to exhibit patience. Nowhere in the
world has the above formula been followed better than at the Observatoire du Pic-du-Midi. Furthermore, none of the new techniques that I
will now describe have yet done better than has been demonstrated at
Pic-du-Midi.

The fact that solar images with a quality of 0.2 arc seconds have been
obtained is a clear bench mark that any new telescope will have to sur-
pass. Clearly, it is necessary to remove degradations caused by the
atmosphere, atmosphere-telescope interface, and telescope shake in
order to do significantly better than has been accomplished. This can
be partially done by active optics which removes the mean displacement
of the image; by adaptive optics which removes, at least for some
angular size, the distortion of the wavefront, and by altitude displacement of the telescope. In general a combination of these are required.
At present altitude displacement means putting the telescope into an
orbit in space.

Our laboratory has been working on the problems of image stabilization
via active mirrors, wavefront correction via adaptive mirrors, and
space solar experiments via SOUP (Solar Optical Universal Polarimeter)
and SOT (Solar Optical Telescope) to obtain high resolution solar
images. I will to review the status of these projects below.

Definitions.-   Before proceeding further it is necessary to define two
terms normally used to describe optical properties of the atmosphere.
The first is size scale over which the wavefront is locally plane,

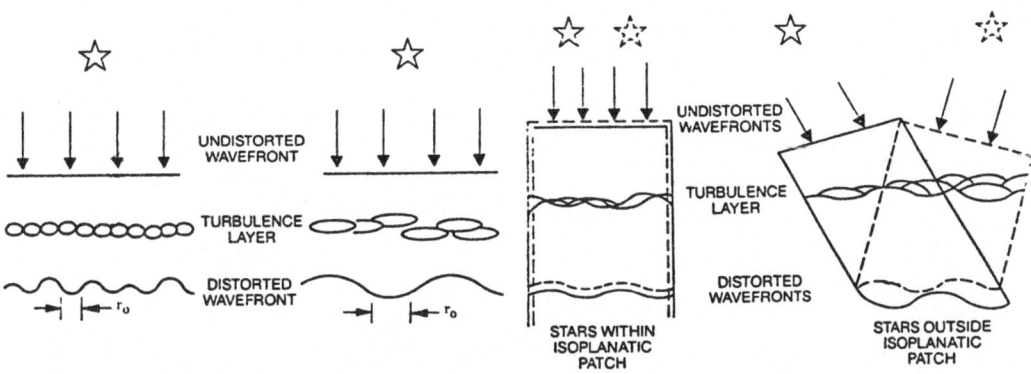

Figure 1a. The scale of r_o

Figure 1b. The size of the isoplanatic angle.

r_o (see figure 1a). This is a rough definition because flatness is, of course, a function of wavelength. The parameter r_o is of critical importance to the use of adaptive mirrors because the number of active segments of the mirror is roughly pi times the square of the ratio of the telescope diameter to r_o. Most daytime measurements yield r_o's in the range of 6 to 10 cm, but exceptional sites must sometimes have r_o's of 20 to 40 cm or more.

The second parameter of considerable importance is the isoplanatic angle. Roughly speaking, the isoplanatic angle is the limiting angle over which the wavefront from two sources at infinity suffer the same distortion through the atmosphere (see figure 1b). The size of the isoplanatic patch determines the angular size over which adaptive correction is effective. For solar sites this angle is in the range of 2 to 10 arc seconds.

Active mirrors.- We have used a high frequency response mirror for a number of years to remove the mean tilt of the incident wavefront and hence image displacement from our observations. The mirror and servo system has a unity gain cross over point at about one kilohertz. It is tilted by PZT actuators. Figure 2 shows an optical schematic of the tracker system. The high speed mirror is placed in an image of the objective formed by a reimaging lens that is just forward of the primary image plane of the telescope. Part of the light that reflects from the tracker mirror passes through a 96/04 beamsplitter and is

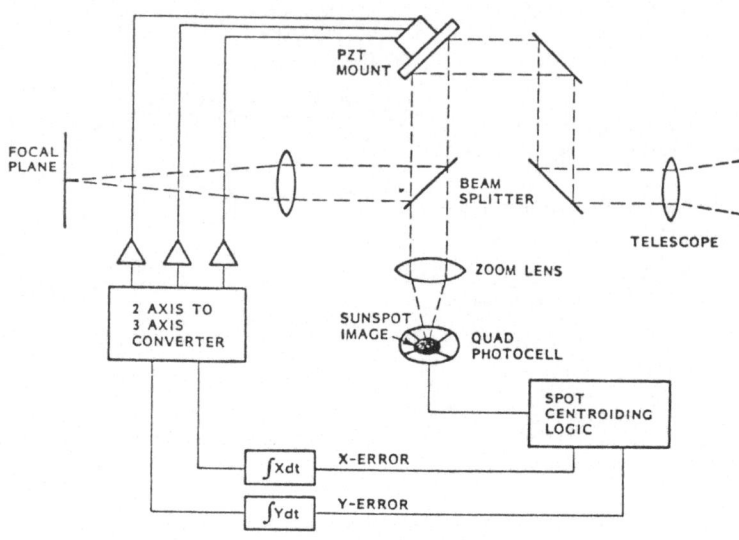

Figure 2. Schematic of active mirror system.

reimaged onto a quadrant photodetector, while the majority of the light
is reflected toward another reimaging lens which forms the experiment
image. In operation a small sunspot or pore is positioned onto the
quadrant photodetector. From the signals produced by the four quadrants,
centering logic calculates the displacement of the tracker mirror required to center the spot on the detector. The displacement signals
are sent to a type one servo amplifier which drives the mirror displacement electronics.

The amount of residual jitter in the image is difficult to measure on
the telescope because of the variable blurring and distortion caused
by seeing. But both films and videotapes have shown that the residual
jitter is always much less than the seeing diameter. In the laboratory
the tracker can remove jittering test signals to better than 0.01 arc
seconds.

Figure 3 shows a one axis signal from the servo drive in seeing of

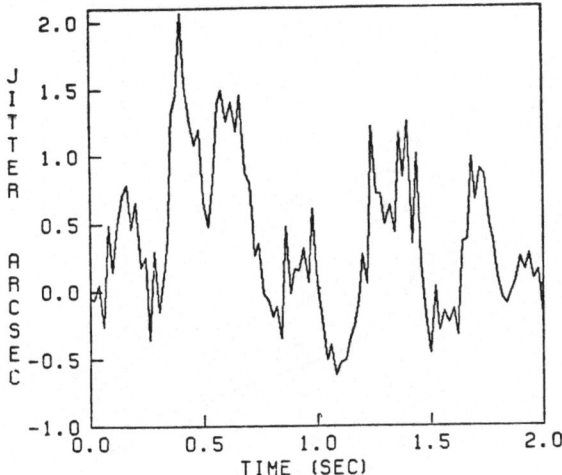

Figure 3. One axis servo signal.

better than one arc second. The waveform of the image displacement is nearly rectangular with steps of 0.5 to 2.0 arc seconds. The typical width of the basic rectangular step is about a half second of time. This motion can be clearly seen if the tracker is turned off. A higher frequence jitter a few tenths of an arc second in amplitude is super-posed on the large scale displacement. Figure 4 shows the power spectrum in bad seeing, and a composite spectrum. All of the spectra are quite similar in shape. The humps in the spectra arise because of resonances in the telescope and telescope guidance system. From the power spectra it is clear that the mirror response should be above 100 Hz.

It has been our experience that seeing conditions are often quite good at Sacramento Peak Observatory in the presence of significant amounts of image motion. This may be due to a combination of relatively low alti-tude, large scale disturbances and wind induced telescope shake. We also found occasional good frames in the middle of fundamentally bad seeing.

It is worth emphasizing that the tracker consists of two parts. One is the high speed mirror with its servo electronics and the other is the tracker sensor electronics. For the system discussed above a simple sunspot tracker is used as a sensor. This is limiting because it requires operations near spots or pores. It would be much more

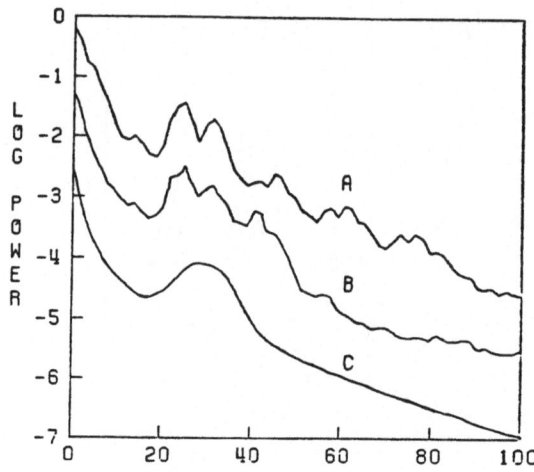

Figure 4. Power spectra of servo error signal in good
(A), poor (B), and averaged (C) conditions.
Graphs displaced for clarity.

desirable to stabilize anywhere on the solar disk. This can be done via
a correlation trackering sensor system which locks up on the solar
granulation pattern. Such systems are currently under development at
Freiburg, Sac Peak, and our lab. It is reasonable to expect correlation trackers to be operational in 12 or 18 months, but there is no
reason that high speed steering mirrors should not be standard observing
equipment now. At present tracker's of our design have been built at
High Altitude Observatory and Big Bear Observatory and similar systems
exist at the National Solar Observatory.

Adaptive mirrors.- An adaptive mirror system for atmospheric seeing
compensation consists of a sensor that determines the wavefront error
introduced by the atmosphere, a mirror with a suffcent number of segments to remove the atmospheric distortions and control electronics that
connect the two together. Essentialy our adaptive mirror concept is a
straightforward generalization of the active steering mirror. The
objective is imaged onto a segmented mirror where each segment corresponds to a size of order r_o on the objective. The control beam, reflected from a beam splitter after the adaptive mirror, is reimaged by a
set of lenses each one of which gathers light from a single segment.

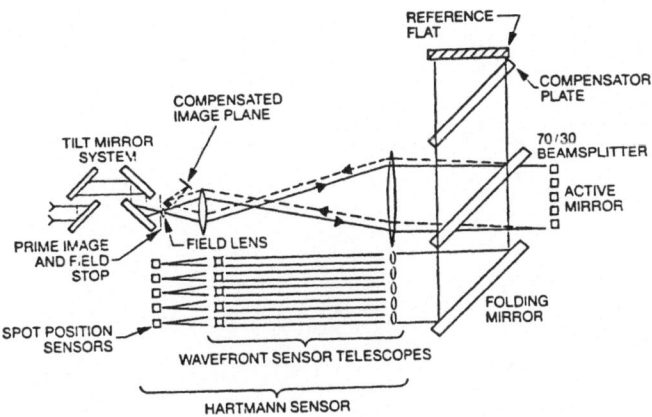

Figure 5. Schematic of adaptive mirror system.

The individual images of a small spot or pore are separately sensed
and a servo loop is closed on the proper individual mirror segment to
center it's image on it's sensor (see figure 5). This controls the tilt
of all the segments and serves to cause all of the subimages to be
coincident in the experiment beam. The method of sensing subimages is
equivalent to Hartmann measurement of the wavefront surface tilts.
Figure 6 shows a diagram of a segments of our adaptive mirror. Figure
7 is a photograph of the completed adaptive mirror.
In order to form diffraction limited images the relative phase or
piston of the individual mirrors must   also be controlled. If the
adaptive mirror is flat for a perfect input beam, controlling the
relative phase is equivalent to making the edges of the individual
segments contiguous. Edge fitting is accomplished with a amplifier
network that connects all the tilt signals together. A block diagram
of the electronic control is shown in figure 8.
At the present time the active mirror has not succeeded in producing
diffraction limited images. However, it has been successful in
significantly improving both poor and fair seeing. Six arc second
seeing has been improved to about 2.5 arc seconds. This has been done
without closing the phase loop. The segment sizes under these conditions
are poorly matched to r_o, but the tilt correction serves to overlap the
images from the separate subsegments. This is an advantage of separate
tilt and phase sensing.

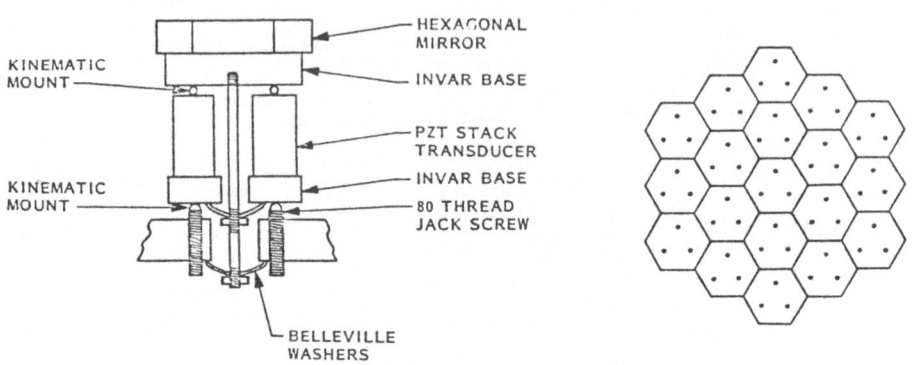

SINGLE SEGMENT

MIRROR CONFIGURATION

HEXAGONAL MIRROR

KINEMATIC MOUNT

INVAR BASE

PZT STACK TRANSDUCER

INVAR BASE

KINEMATIC MOUNT

80 THREAD JACK SCREW

BELLEVILLE WASHERS

Figure 6. Diagram of mirror segment.

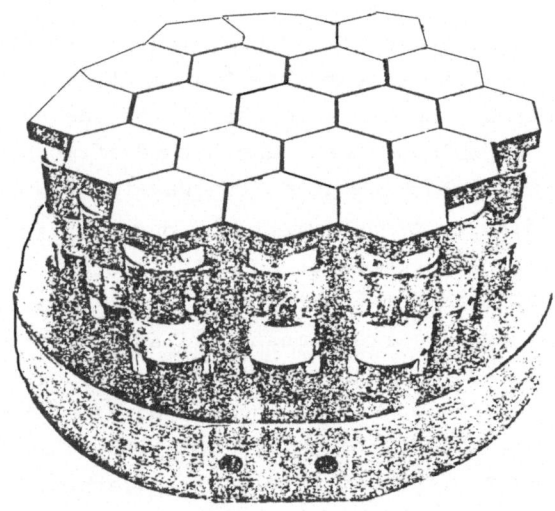

Figure 7. Photograph of 19 segment adaptive mirror.

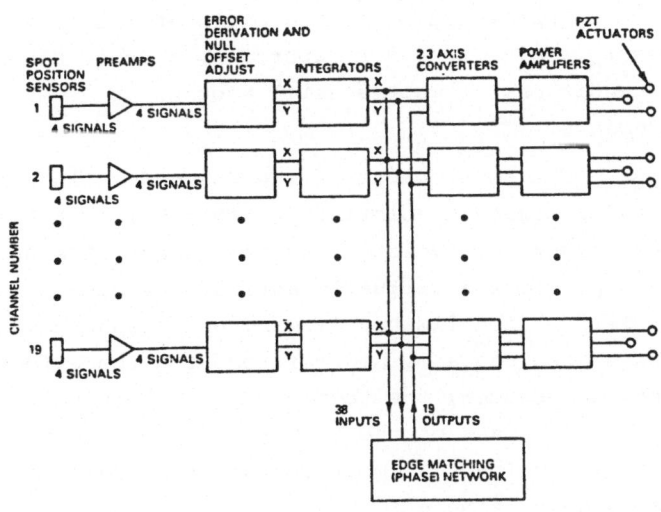

Figure 8. Electronics block diagram.

Two arc second seeing has been taken to under one arc second with the phase loop closed. At present there are still some optical problems in the alignment interferometer used for initial calibration of the mirror control system which introduces on the order of a wave of distortion across the aperture. We are presently correcting the appropriate optics and expect to achieve diffraction limited performance equivalent to that of a 50 to 70 cm telescope during this year.

When the adaptive mirror has been made operational it will suffer from some major limitations. The first is caused by the fact that this mirror has only 5 segments across a diameter. This means at best the effective aperture is 5 times r_o. When r_o is 10 cm this corresponds to to 50 cm telescope which at 5000 A can yield about a quarter arc second imagery. At a better site using a bigger telescope a tenth arc second might be achieved on occasion. However, the more optimistic one's assumptions about r_o the more seldom will be the diffraction limited images. Further, the improved images will probably be limited to 5 to 10 arc seconds in diameter because of the size of the isoplanatic patch.

The current mirror has 19 segments and 57 actuators. To add the next ring of segments requires 24 more segments for a total of 43 segments

and 129 actuators. This additional ring improves the diffraction aperture to 7 times r_o. To go to a 9 r_o aperture, a meter class telescope
in reasonable conditions, requires still another 48 segments for a
total of 91 segments.

Space Telescopes.- The development of the Space Shuttle by NASA and
the Space lab and Instrument Pointing System (IPS) by ESA allows
improving solar imagery by avoiding the atmosphere. The space environment, of course, provides a virtually automatic vacuum telescope system.
Achievement of diffraction limited imaging only requires a nearly
perfect optical system and a method for removing the effects of an
imperfect mounting. However, the technology for solving the mounting
proglem is essentially that of the high speed steering mirror. But as
stated in the introduction the construction of a nearly diffraction
limited system is never simple !

Space telescopes are further complicated by the requirement to satisfy
launch and landing loads. This constraint is serious because it makes
mountings the optics much more difficult. Perhaps the most important
concern in the design of a space experiment is the overwhelming requirement that it work without any new fine adjustments.

In April of 1985 Spacelab 2, which contains the Solar Optical Universal
Polarimeter (SOUP) should be launched. SOUP is a 30 cm Cassegrain
telescope with diffraction limited performance over the spectral range
4900-7000 A. It is mounted on adjustable legs so that it can point
independent of IPS or co-point with any of the other solar experiments.
Flying with SOUP is the High Resolution Telescope Spectrograph (HRTS)
of the Naval Research Laboratory and two other lower resolution solar
experiments. SOUP has two independent focal plane systems for scientific observations. The first records a 2 arc minute a white light image
on film. The second and primary system is a tunable filtergraph consisting of polarization analysers, blocking filters, a 40 mA tunable birefringent filter, film camera, CID camera, digital image processor
and scan converter. An optical schematic of the SOUP is shown in figure
9. The filter can be set to 5 mA absolute and 1 mA relative. The film
camera has a 1.5 by 2 arc minute field, while the CID has a 35 by
44 arc second field.

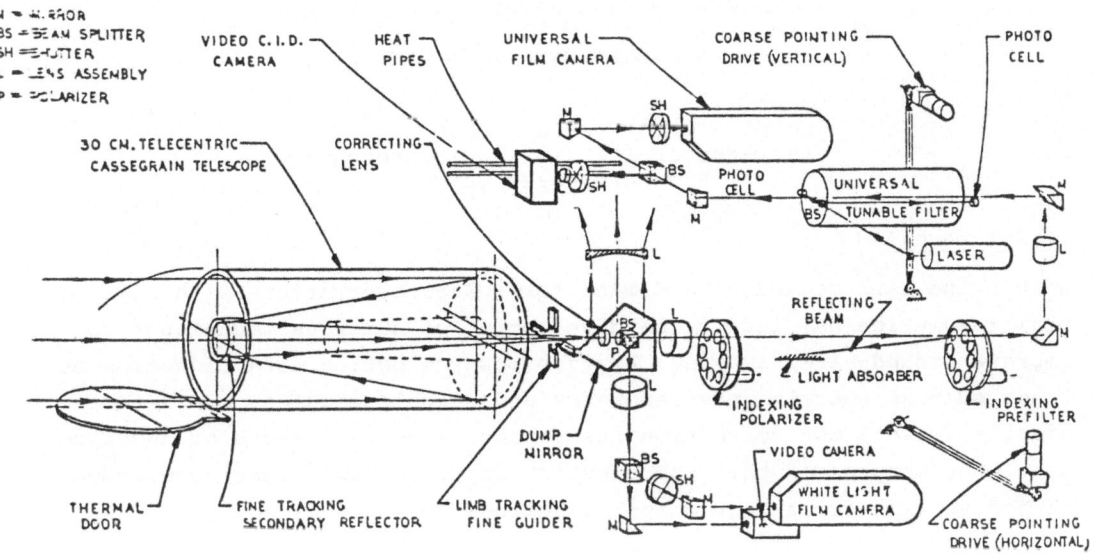

Figure 9. Schematic drawing of SOUP.

ADAPTIVE IMAGE STABILIZATION OF SOLAR

OBSERVATIONS: A REVIEW

O. von der Lühe
Kiepenheuer-Institut für Sonnenphysik
Freiburg, West Germany

I. Introduction

Within the last decade, the demand for solar observations with high spatial resolution has led to the development of numerous techniques to improve ground-based imaging. Besides post-detection data reduction methods such as speckle interferometry and speckle imaging, pre-detection adaptive wave front correction has reached a certain state of development. The purpose of this contribution is to review recent developments of active wave front correction techniques for solar imaging.

Generally speaking, a system that adapts to a given perturbation of the wave fronts emerging from an astronomical source consits of three major parts (see fig. 1):

- The wave front error detector or wave front sensor partially or fully detects the perturbations of the incoming wave front.
- The servo logic converts the error signals to drive the
- Active optical element, which applies the appropriate phase shifts to the incoming light such as to partially or fully compensate perturbations.

Usually, these three constitutes are combined in a closed-loop system, where the wave front sensor detects residual errors after compensation rather than the perturbations of the incoming light and tries to drive the residual errors to null. Table I identifies possible sources for wave front distortions. The quantities relevant for adaptive correction are mainly the expected dynamical range of the deformations and their temporal behaviour. In terms of bandwidth and complexity, atmospheric seeing may be most demanding to an adaptive system, so we will be concerned with seeing in the balance of this contribution.

II. Atmospheric effects

Consider a plane wave travelling through the earth's atmosphere, which imposes spatial and temporal variations on the light path. Let $\phi_\nu(\bar{x},t)$ describe the phase delay of the plane wave, at entrance pupil coordinate vector $\bar{x}$ and at time t, due to a point source at the angular location ϑ in the sky (see fig. 2). It might be advantageous to expand $\phi_\vartheta(\bar{x},t)$ into a series of orthogonal functions over the entrance pupil, such as Zernike-polynomials, especially if one distinguishes between first-order (image motion) and higher-order (blurring) wave front correction. But now, we will be more concerned with statistical parameters of the phase fluctuations. The resolution obtained when observing through the atmosphere depends on the phase structure function

$$\mathcal{D}_\phi(r) = \langle |\phi_\vartheta(\bar{x},t) - \phi_\vartheta(\bar{x}+\bar{r},t)|^2 \rangle \; ; \quad r = |\bar{r}| \tag{1}$$

When we assume a Kolmogorov law for the temperature inhomogenities (Tatarskii, 1971), we may take

$$\mathcal{D}_\phi(r) = 6.88 \left(r/r_0 \right)^{5/3} \tag{2}$$

following Fried (1966). In most cases, the validity of eqn. 2 is apparently confirmed to a satisfactory extent (Roddier, 1981). The parameter r_0 describes the quality of the seeing and the obtainable long-term resolution.

If we wish to fully compensate wave front perturbations in a situation with a given r_0, we may obtain a measure of the dynamic range Δz of the light path variations required in an adaptive system (v.d. Lühe, 1983):

$$\Delta z = 1.77 \left(\frac{D}{2r_0} \right)^{5/6} \cdot \lambda \tag{3}$$

where D is the telescope entrance pupil diameter. Also, we have to be able to apply phase delays in spatial scales as small as r_0. Therefore, the "number of actuators", the number N of independent parameters that control the deformation of an active element, has to be of the order of

$$N = \left(\frac{D}{r_0} \right)^2 \tag{4}$$

Another important quantity in connection with phase fluctuations would be the temporal mean-square difference:

$$\mathcal{J}_\phi(\tau) = \langle |\phi_\vartheta(\bar{x},t) - \phi_\vartheta(\bar{x},t+\tau)|^2 \rangle \tag{5}$$

Suppose the wavefront perturbations have been successfully compensated for at a given time t. If we keep the momentary shape of the active element, the correction would deteriorate as time goes on. We may define a parameter τ_0 such that $\mathcal{J}_\phi(\tau_0)$ describes just noticeable wavefront compensation mismatch of active element and momentary wavefront. e.g. $\mathcal{J}_\phi(\tau_0) = 1\,rad^2$. Therefore, τ_0 determines the required "speed" of the adaptive system and $1/\tau_0$ would specify a minimum bandwidth of the system.

The third important quantity to describe wavefront perturbations is the angular mean-square difference:

$$\mathcal{A}_\phi(\delta) \;=\; \langle \, | \, \phi_\vartheta(\bar{x},t) - \phi_{\vartheta+\delta}(\bar{x},t) \, |^2 \rangle \tag{6}$$

We may understand this quantity in a similar fashion as eqn. (5): Suppose that the perturbations of the wavefronts have been compensated for a particular line of sight ϑ. Since a slightly different line of sight eventually passes different portions of the atmosphere, the correction deteriorates and $A_\emptyset$ describes this amount of wave-front error variance. We may define a "radius of isoplanatism" δ_o for which $\mathcal{A}_\phi(\delta_o) = 1\ \text{rad}^2$.

Without pretension to completeness, let me compile some measurements to demonstrate the range of such parameters as defined above. In Fig. 3a, an example of the measurement of phase excursions under night-time conditions is presented (Breckindridge, 1976). Seeing conditions were inferior. This figure gives an idea on which dynamic range may be required. In Fig. 3b, the cumulative frequency of r_o under both nighttime and daytime conditions, as obtained with a rotational shearing interferometer and with image motion measurements, is presented (Bornigno and Brandt, 1982). The data were taken at La Palma, Canary Islands, and show the typical range of daytime r_o about 5...7 cm.

In Fig.s 4a and 4b, the temporal behaviour of the atmosphere under daytime conditions is demonstrated. In Fig. 4a, temporal fluctuations of the phase delay averaged over a patch 6 cm in diameter in the entrance pupil plane as well as the overall tilt of the wave front of a 30 cm telescope is shown. (Hardy, 1980, Brandt, 1969 and Tarbell and Smithson 1980). In Fig. 4b, amplitude spectra of unidirectional image motion and of the phase delay as measured in Hardy's (1980 a) experiment are presented. The bandwidth required for a stabilisation system clearly depends on the type of error to be compensated. For overall wave front tilt (image motion) compensation only, the bandwidth should be in excess of 200 Hz as shown in Fig. 4b.

The measurement of isoplanatism under daytime conditions is very difficult. Hardy (1980-a and b) claims that in his experiments, the size of the isoplanatic area sometimes was as small as $1''$. In Fig. 5, the results of analyses of distortion is presented. One curve (November, 1984) represents the rms value of the distortion observed in a time series of 50 frames covering a square of $14''$ side length of granulation data, whereas the second curve (Tarbell, 1984) shows distortion as obtained from solar limb data. Distortion is significant, since it reflects the "anisoplanatism" of atmospheric image motion. It is obvious that image motion is noticeably incoherent outside a field only a few arc seconds across. We may expect a similar size of the isoplanatic area for higher-order wave front aberrations.

How much improvement can we expect from adaptive systems? It is well known that pictures taken with short exposure show significantly more fine detail than long exposure photographs. A considerable part of long term blurring is caused by image motion alone. Consequently, image motion control alone significantly increases the spatial resolution of typical long exposure observations such as spectrograms and filtergrams.

Here, it would be convenient if we could express the gain in resolution obtained with an adaptive system in terms of average modulation transfer functions (MTF). Fried (1966) derived relatively simple theoretical expressions for average MTF's which may be used for the uncompensated and image-motion compensated case. In an improvement of Fried's theory, Wang (1977) published theoretical transfer functions for adaptive systems that compensate higher order terms up to astigmatism. His results indicate (Fig. 6), that, despite being far from diffraction limited, a system that compensates only a few low-order Zernike polynomial terms may significantly improve the modulation transfer at higher spatial frequencies. Thus, one could record data first and restore high frequency information during data reduction which would be entirely lost otherwise.

III. Image motion control

Image motion control (IM) means compensation of first-order wavefront errors only, thus the first adaptive control system in use were IM stabilizing devices. When characterizing such systems, it is advantageous to distinguish between the method of image displacement detection and the active element design.

III.1 Image displacement detection

There are basically two methods of displacement detection, spot tracking and correlation tracking.
Spot trackers are used in the simplest types of IM control systems and have been successfully employed by several groups for years. The principle (Fig. 7) is that of a four-point limb guider; a local intensity maximum or minimum is required for operation. The movement of the intensity extreme is measured with a quadrant or a position-sensitive solid-

state detector. The error signals are integrated to drive the servo. Typical targets for spot trackers are sunspots and pores. Occasionally, even multiple spots can be stabilized. The attempts to lock on individual granules were so far unsuccessful. Spot trackers operate satisfactory as long as there is any extreme to lock on, medium-sized sunspots are stabilized even during moments of severe blur. Coherent IM from telescope shake and guider jitter are fully compensated, seeing-induced IM is compensated only in the vicinity of the target spot, depending on the size of the isoplanatic patch. However, the stabilized region may be as large as one arc min to an accuracy of the size of the seeing disk, as has been observed by TARBELL and SMITHSON (1980). The accuracy also depends on the size of the target spot as compared to the size of the isoplanatic patch. Despite the limitation of possible targets, the application range of spot trackers is surprisingly large when no extreme accuracy is required (cf. fig. 11).

Correlation trackers are designed to stabilize arbitrary structure rather than intensity extrema, they are more complex than spot trackers. Their principle is to continuously scan an imaging detector such as a diode array and to compare the actual or "live" image with a previously remembered "reference" image of the target structure. The scan rate has to be high in order to resolve image displacements to the required accuracy. If the target structure, such as granulation, has a limited life time, periodic updating of the reference is required in order to stabilize the image for longer periods. Since the effective detector area can be chosen to be as small as the isoplanatic patch, a correlation tracker could guide to high accuracy better than the telescope resolution.

There are various algorithms to suitably compare reference and live images and to obtain error signals. If $I_R (x,y)$ is the reference and $I_L (x,y)$ is the live frame, x and y being focal plane coordinates, the "Mean-Square Residual Function", defined as:

$$ MSRF(\xi,\eta) = \iint [I_L(x,y) - I_R(x+\xi, y+\eta)]^2 \, dx \, dy \qquad (7) $$

and the Cross-Correlation function:

$$ CC(\xi,\eta) = \frac{\iint I_L(x,y) \cdot I_R(x+\xi, y+\eta) \, dx \, dy}{\left\{ \iint I_L^2(x,y) \, dx \, dy \cdot \iint I_R^2(x,y) \, dx \, dy \right\}^{1/2}} \qquad (8) $$

are suitable measures of equality. They have been examined in previous studies (SMITHSON and TARBELL, 1977 and VON DER LÜHE, 1983a and 1984). Image displacement is measured by detecting the location of the global

minimum in (7) and the global maximum in (8). When a pixel number larger
than 4 is used, it is more efficient to compute (8) by Fast Fourier
algorithms rather than by the ordinary "shift-and-multiply" scheme.
The computation of either (7) or (8) at a kilohertz rate requires complex modern hardware. The advantage would be that the tracker locks on
his target as long as there is sufficient overlap between live image
and reference. Another strategy, where only a few values of the MSRF or
CC function are computed, has been proposed by L. MERTZ (1984) for the
use on the SOT correlation tracker. As an example, the gradient of the
correlation function at the origin is taken for the error signal in both
directions:

$$\delta x = \sum_{i=2}^{N-1} \sum_{j=1}^{N} \left(I_R(i-1,j) - I_R(i+1,j) \right) \cdot I_L(i,j)$$

(9)

$$\delta y = \sum_{i=1}^{N} \sum_{j=2}^{N-1} \left(I_R(i,j-1) - I_R(i,j+1) \right) \cdot I_L(i,j)$$

where N is the pixel number in either direction of the detector and i,
j are pixel coordinates. The acquisition range of a tracker based on
this algorithm is target dependent (C. EDWARDS, 1984).
There is no correlation tracker operational today. A variety of studies
were carried out (see the previous 3 references and ANDREASSEN and ENG-
VOLD, this volume). A movie simulating the performance of correlation
trackers was made by R. PETROV from Nice. A brief summary of the major
conclusions from previous studies would contain:
- The field of view of a pattern detector should cover several granules
(4" x 4") when locking on granulation. It should not be considerably
larger than the isoplanatic patch for high accuracy guiding.
- A few hundred pixels (250) are sufficient for successful tracking.
The pixel size may be larger than the telescope resolution element.
- A two dimensional detector performs far better than two one dimensional, crossed detectors.
- For space-borne telescopes, a derivative strategy (eqn. 9) is efficient, for a ground-based tracker with occasional loss of lock due to
blurring a global scheme (eqn. 8) may be better.

Correlation trackers are now under development at LOCKHEED (derivative
strategy, eqn. 9) for the use on SOT and, in a joint effort, at the
National Solar Observatory, Sac Peak and KIS, Freiburg (global strategy,
eqn. 8).

III.2 Active elements for IM control

Recent developments of successful image motion control systems employ
agile high quality mirrors close to a pupil in the telescope's optical
path. Here, the dynamic range of deflection is smallest and system alignment is maintained. Although the frequency spectra in fig. 4b indicate
that the major part of image motion is caused by fluctuations below
approximately 50 Hz, it is advisable to allow for a much wider bandwidth
of the active element. Typical rms image motion caused by seeing of moderate quality is about 1 arc sec for a medium-sized telescope, but in
order to account for telescope shake, slow drifts due to possible guiding errors and solar rotation, a dynamic range of 10 arc sec ptp at
the image is practical. The effective dynamic range in the final focal
plane is the dynamic range of the agile mirror times the demagnification
of the entrance pupil image at the agile mirror. Thus, if the demagnification is 1/20, the device should have a range of 200 arc sec in order
to allow for the 10 arc sec control range in the final image. Since
small mirrors are used in order to decrease inertia and to increase
band-width, there is a lower limit of the mirror's dynamic range.

In the two principal concepts, the agile mirror is either piezoelectri-
cally or electrodynamically driven. The principle of a piezo-mirror, as
has been used by the LOCKHEED group for years, is shown in fig. 8. Three
piezoelectric actuators tilt the mirror such that the location of the
center of mass remains steady. The piezo-actuators have a range between
5 to 50 microns, and the typical tilt range of such a mirror is several
hundred arc seconds. High voltages in the range of 500 ... 1000 V are
needed to drive the piezos. The electric circuitry is simple and straightforward. The bandwidth is typically several hundred Hz. Fig. 9 shows an
impedance curve of a piezo actuator together with its HV amplifier and
amplitude and phase spectra of the KIS agile mirror prototype as obtained
by measuring the deflection of the mirror with a laser beam and a position-sensitive detector. The bandwidth is limited by resonances at 280
Hz and higher frequencies. Most probably, these resonances are caused
by internal vibrations in the piezos rather than in the mechanical design. The phase spectrum is sufficiently linear to allow complete IM
control in excess of 100 Hz. The cost of the entire system, including
spot tracker electronics but excluding work shop manpower, is less than
US $ 2000.-.
Fig. 10 gives an overview of the agile mirror that has been developed
by B. GRAVES (1984) at NSO - Kitt Peak. Here, a little mirror in a gim-
bal mount is driven by two off-the-shelf loudspeakers via a lever. The

system has a very large dynamic range of approximately 3 degrees. This allows a very small demagnification and thus the use of very small (1/2 inch) and light mirrors. Also, only low voltages are required by the actuators, which may make the system more reliable in a high altitude environment. The bandwidth is clearly less than the bandwidth of piezomirrors, and more complex circuitry is required to damp mirror overshoot. Along with the gain-vs.-frequency curve, a sample spectrum of image motion amplitudes observed at the McMath telescope is shown, which confirms the less stringent bandwidth requirements on a large solar telescope. The system is routineously used. Agile mirrors of the same design are now under construction for the NSO - Sac Peak facilities.

IV. Higher order wavefront correction

A variety of system designs were considered for higher order wave front compensation, some contributions to the meeting on Solar Instrumentation held at the Sacramento Peak in 1980 ("Solar Instrumentation: What's next?") were on such designs. But there have been only two serious attempts to use an adaptive system on solar targets.
HARDY adapted the ITEK "Real Time Atmospheric Compensation" (RTAC) system to the SPO tower telescope during two observing runs in 1980. A description of the RTAC system can be found by HARDY et al. (1975), results of the solar experiments are available (HARDY, 1980a and 1980b). The wavefront sensor uses a rotational shearing interferometer, which in principle can be understood as a time dependent Ronchi test. The active element was one of ITEK's Monolithic Piezoelectric Mirrors (MPM); a fast, analog logic generated the drive signals. Although substantial experience has been gained, the experiment failed to produce a stable improvement of image quality when working on solar targets. Hardy attributed this failure to the isoplanatic patch being smaller than the seeing disk, but it is possible that there are principal problems with his type of wave front sensor on the sun (v. d. Lühe, 1983b).
R. SMITHSON (1983) is presently working on a different approach. The reader is referred to Alan Title's contribution somewhere else in this volume for technical details and figures. The outstanding features of this design are its simplicity and its high degree of modularity. The wave front sensor is based on a combination of a Hartmann test principle and the spot tracker principle. The active element prototype has 19 hexagonal segments with 3 piezo actuators each, which is sufficient for a prototype, but more may be needed in a final design. Once the per-

formance of the system has been demonstrated, it should be easy to add
more elements and to acomodate higher complexity of wavefront perturbations.

The limitations of the system are mainly due to the principle of error
detection; again the types of possible targets are limited to local intensity extrema. Also, if a spot with a size larger than one isoplanatic patch is being locked on, the error signals would measure the average perturbations of several isoplanatic patches rather than the wavefront errors of one line of sight. It is not trivial to adopt the correlation tracker concept to a Hartmann-type sensor, since each sub-tele-
scope should cover an area of the diameter of r_o at most. If we try to
operate such a system under reasonably good seeing conditions ($r_o \sim 5cm$),
it would be difficult to detect the motion of granules with a resolving
power of 2 seconds of arc. Additionally, the segmented mirror is align-
ment-intensive and may cause straylight.

Some observing runs were made with this system operating, mostly under
bad seeing conditions. The system produces a noticeable and stable improvement of image quality, although far from what we would call diffraction-limited. However, since the system was operating with an entrance
pupil of 30 cm diameter, each segment corresponded to a segment of the
incoming wavefront with 5 cm diameter. When seeing is bad or moderate,
r_o is definitely smaller than 5 cm and we really could not expect better
performance   (cf. eqn. 4). It would be very interesting to see how this
system operates under better atmospheric conditions.

The simplicity of the system makes it an excellent device to learn to
understand the possibilities of full wavefront correction on a solar
telescope. It is presently the most promising approach in that direction,
and one should look forward to its further development and its future
results. If this system succeeds in producing stable and sufficient
image improvement under good seeing conditions, additional elements
should be added to gain performance under inferior conditions, making
it a very useful scientific tool.

V. Conclusions

Experience has shown, that image motion control significantly improves
the resolution of typical long exposure applications such as spectrograms and spectroheliograms. As I tried to point out, basic IM control
systems need only simple technology and, in most cases, are easily im-
plemented in most existing telescopes. Spot trackers are on their way
to being routineously used.

Simulations of correlation trackers are promising, however, they are
more complex. We will have to wait for   about another year for first
tests. In telescopes of the future, IM control systems based on spot
trackers as well as on correlation trackers should be included into the
design from the beginning.

As far as higher order correction is concerned, the systems still are
in an early stage of development. As far as we understand now, proper
wavefront error detection is the major diffiulty in those systems and
the Hartmann principle presently appears to be the most promising approach. Even with this principle, the application range may be limited.
With the existing system, bad seeing could be somewhat improved, but
diffraction-limited performance has not yet been demonstrated.

Perhaps we will have to change our attitude towards wavefront compensation systems. We should not expect them to do all the work under any
conditions, such systems, if at all feasible, would be significantly
more complex than the telescope itself including post-focus instrumentation. A realistic system would be able to turn extremely good seeing
to diffraction-limited performance and to improve medium and good seeing.
So, one should be able to significantly increase telescope time for
high resolution observations in the ordinary sense. Even if only partial wavefront error compensation can be achieved, one should consider
the possibility of post-detection correction of observations obtained
in this mode. Then, a thorough understanding of the combined action of
atmosphere and the adaptive system would be required. But the combination of pre-detection and post-detection correction techniques may
turn out to be the ultimate solution for ground-based high resolution
imaging.

Acknowledgements

Many persons contributed to this review by discussions and with partially
unpublished material. I wish to express my thanks to P.N. Brandt, R.B.
Dunn, C. Edwards, B. Graves, L.J. November, R. Radick, R. Smithson, R.
V. Stachnik, T.D. Tarbell, A. Title and A.L. Widener.

Literature

Many citations given below were published as contract or internal reports
or conference proceedings rather than in the open literature, or were
private communications. Whenever necessary and possible, the address of
the author or proceedings editor is given in order to facilitate access
to the publication.

-BORGNINO, J. and BRANDT, P.N. (1982): "Daytime and Nighttime measure-
ments at La Palma in June 1982", JOSO Ann. Rep. 1982, Ed. A.v. ALVENSLE-
BEN, Kiepenheuer-Institut f. Sonnenphysik, Schöneckstr. 6,7800 Freiburg
FRG
-BRECKINRIDGE, J.B. (1976): "Measurement of the amplitude of phase excur-
sions in the earth's atmosphere", Journ. Opt. Soc. Am. 66,2 143-144
-BRANDT, P.N. (1969): "Frequency spectra of Solar Image Motion", Solar
Physics 7, 187-203
-EDWARDS, C. (1984) private communication, LOCKHEED Palo Alto Res. Lab.,
3251 Hanover St., Palo Alto CA 94304, USA
-FRIED, D.L. (1966): "Optical Resolution Through a Radomly Inhomogeneous
Medium for Very Long and Very Short Exposures", Journ. Opt. Soc. Am. 56,
10, 1372-1379
-GRAVES, B. (1984) private communication, National Solar Observatories -
Kitt Peak, 950 N. Cherry Ave. Tucson AZ 85726, USA
-HARDY, J.W.; LEFEBVRE, J. E.; KOLIOPOULOS, C.L. (1977): "Real time at-
mospheric compensation", Journ. Opt. Soc. Am. 67,3 360-369
-HARDY, J.W. (1980a): "Solar Imaging Experiment", Contract final report,
ITEK corporation, Optical Systems division, 10 Maguire Rd., Lexington,
MA 02173, USA
-HARDY, J.W. (1980b): "Solar Isoplanatic Patch Measurements" in "Solar
Instrumentation: What's next?", see below
-VON DER LUEHE, O. (1983a): "A Study of a Correlation Tracking Method to
Improve Imaging Quality of Ground-based Solar Telescopes", Astron. As-
trophys. 119, 85-94
-VON DER LUEHE, O. (1983b): "Adaptive Optical Systems for LEST", LEST-
Foundation Technical Report II, Ed. O. ENGVOLD, Institute of Theoretical
Astrophysics, Oslo University, P.O. Box 1029, Blindern, N.-Oslo 3, Norway
-VON DER LUEHE, O. (1984): "A Study of Correlation Tracker Strategies for
Ground Based and Space Borne Solar Telescopes", KIS internal report,
Kiepenheuer-Institut f. Sonnenphysik, Schöneckstr. 6, 7800 Freiburg, FRG
-MERTZ, L. (1984) private communication, LOCKHEED Palo Alto Res. Lab.,
3251 Hanover St., Palo Alto CA 94304, USA
-NOVEMBER, L.J. (1984) private communication, National Solar Observatory
- Sacramento Peak, Sunspot, NM 88349, USA
-RODDIER, F. (1981): "The Effects of Atmospheric Turbulence in Optical
Astronomy", in "Progress in Optics XIX", Ed. E. Wolf, North Holland Pub-
lishing Company, ISBN 0 444 85444 4
-SMITHSON, R.C. and TARBELL, T.D. (1977): "Correlation-Tracking Study
for Meter-Class Solar Telescope on Space Shuttle", Contract Final Report,
LOCKHEED Palo Alto Res. Lab., 3251 Hanover St., Palo Alto CA 94304, USA
-SMITHSON, R.C. (1983): "Solar Magnetic Fields Study", Contract Final Re-
port, LOCKHEED Palo Alto Res. Lab., 3251 Hanover Str., Palo Alto CA 94304,
USA
-SOLAR INSTRUMENTATION: WHAT'S NEXT? (1980) SPO Conference Proceedings,
Ed. R. B. DUNN, National Solar Observatory, Sacramento Peak, Sunspot
NM 88349, USA
-TARBELL, T.D. (1980): "A Simple Image Motion Compensation System for
Solar Observations" in "Solar Instrumentation: What's next?", see above
-TARBELL, T.D. (1984) private communication, LOCKHEED Palo Alto Res. Lab.,
3251 Hanover St., Palo Alto CA 94304, USA
-TATARSKII, V.I. (1971): "The Effects of the Turbulent Atmosphere on Wave
Propagation", Israel Program for Scientific Translations, Jerusalem,
Israel, 1971
-WANG, J.Y. (1977): "Optical Resolution through a Turbulent Medium with
Adaptive Phase Compensation", Journ. Opt. Soc. Am. 67,3 383-390

TABLE I

Cause	Result	Dynamic range	Time scales (sec)
a) Vibrations of telescope and mount	isoplanatic (coherent) image motion	up to several arc seconds in image plane	~ 0.1 10
b) Guidance errors	see above	$\lesssim$ 1 arcsecond in image plane	~ 1
c) Poor optical figure, misalignment	image shift, distortion, spatially varying PSF	up to several wavelengths in exit pupil, rapid spatial variation for large E.P.	temporarily stable
d) Gravitational stresses and thermal load on optics and mounting	see above	up to several wavelengths in exit pupil, slowly varying	~ 2ooo (1/2 h)
e) External and internal turbulent fluctuations of refractive index of medium (seeing)	image motion (distortion) and image blurring, isoplanatic angle ~ 10	up to several arc seconds image motion, up to several wavelengths wave front distortion in exit pupil, dependent on entrance pupil diameter, site, time of day, ... etc.	$> 10^{-3}$

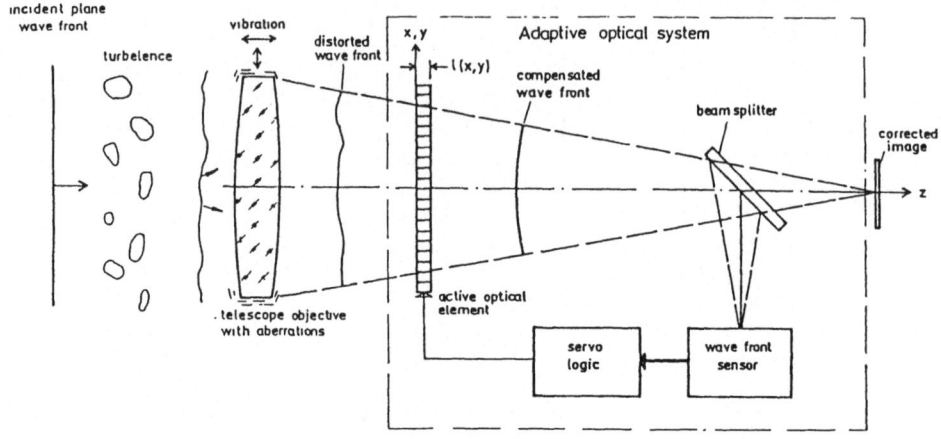

Fig. 1: Principle of adaptive optical system

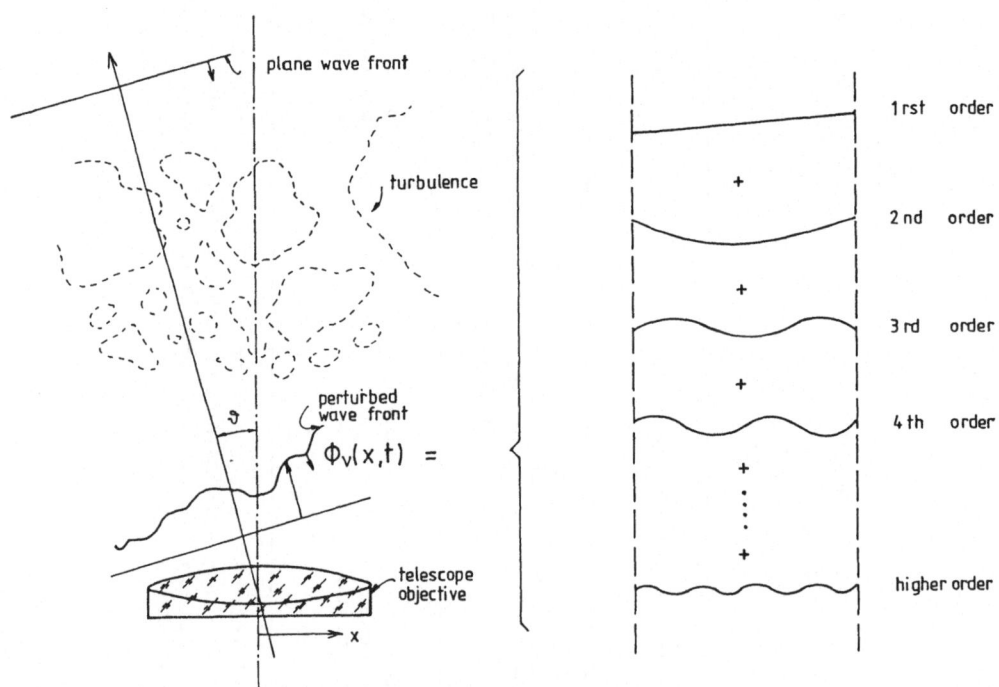

Fig. 2: wave front error definitions

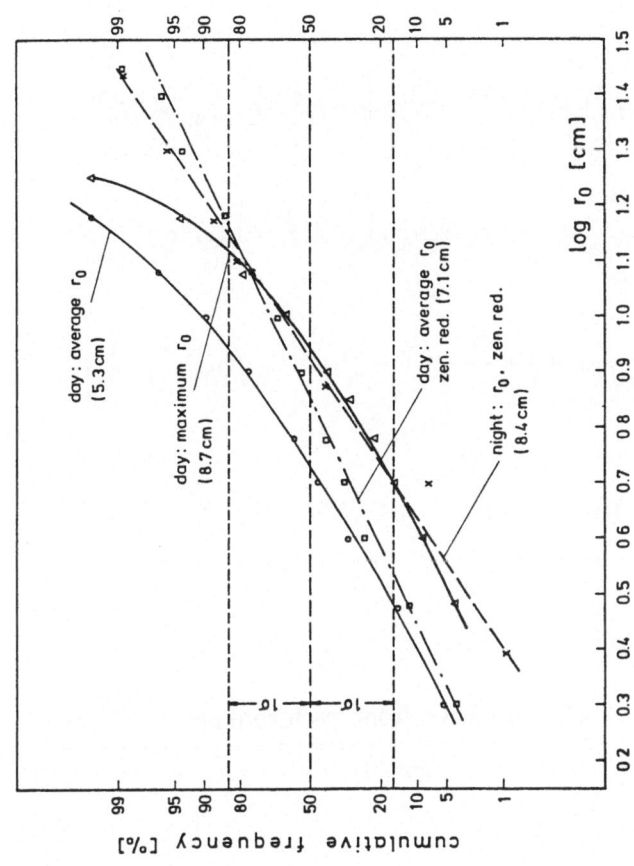

Fig. 3b: Cumulative frequency distributions of daytime and nighttime r_0
measurements ("zen. red." indicates that the values were reduced to
zero zenith distance).

Median values are given in parentheses.

after BORGNINO, I. and BRANDT, P.N. (1982)

Fig.3a: Plot of the visually measured fringe amplitude in
waves as a function of pupil point separation in meters across
the 1.5 m McMath Solar Telescope at Kitt Peak. Source was
a star near the zenith. Points on the graph are averages of
3 stars all observed on 4 different nights near the zenith.
Straight line shows a $\frac{5}{6}$ power dependence. Seeing was approximately 3 arc-sec, wavelength 600 nm.

after BRECKINRIDGE, J.B. (1976)

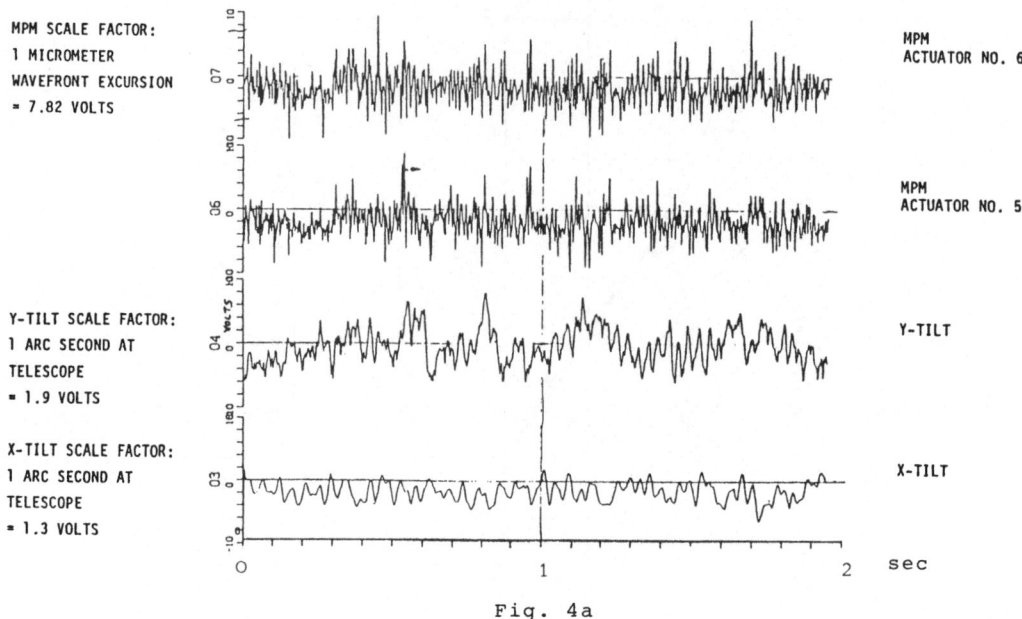

MPM SCALE FACTOR:
1 MICROMETER
WAVEFRONT EXCURSION
= 7.82 VOLTS

MPM
ACTUATOR NO. 6

MPM
ACTUATOR NO. 5

Y-TILT SCALE FACTOR:
1 ARC SECOND AT
TELESCOPE
= 1.9 VOLTS

Y-TILT

X-TILT SCALE FACTOR:
1 ARC SECOND AT
TELESCOPE
= 1.3 VOLTS

X-TILT

Fig. 4a

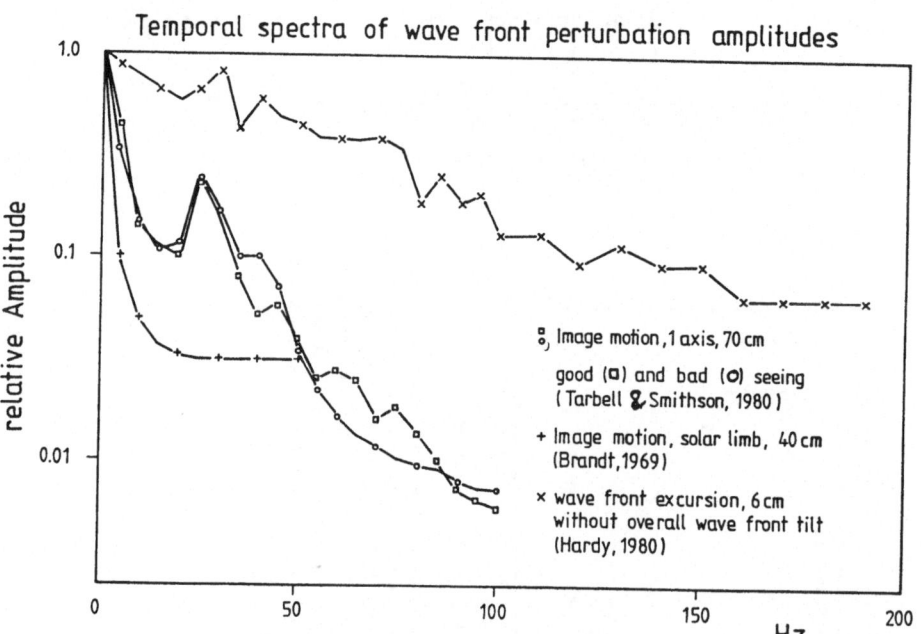

Temporal spectra of wave front perturbation amplitudes

relative Amplitude

□, Image motion,1 axis, 70 cm
  good (□) and bad (○) seeing
  (Tarbell & Smithson, 1980)

+ Image motion, solar limb, 40 cm
  (Brandt,1969)

× wave front excursion, 6 cm
  without overall wave front tilt
  (Hardy,1980)

Fig. 4a: Temporal variation of phase retardation (upper two curves)
and over-all wave front tilt (lower two curves), after HARDY (1980a).
Fig. 4b: Temporal spectra of phase retardation and wave front tilt;
curves are arbitrarily set to 1 at 0HZ.

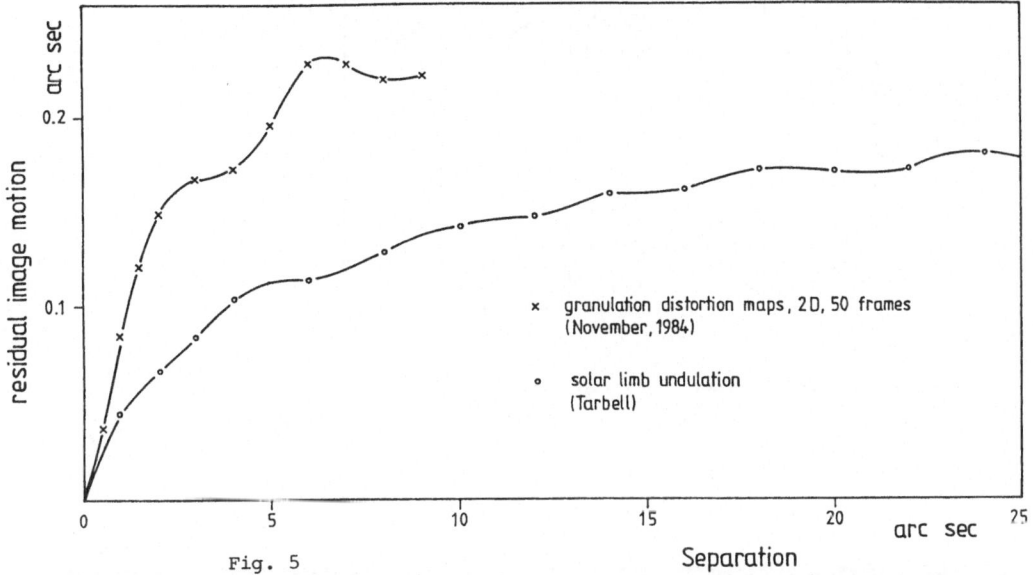

Anisoplanatism of image motion

Fig. 5

Differential rms image motion as a function of point separation

× granulation distortion maps, 2D, 50 frames
   (November, 1984)

○ solar limb undulation
   (Tarbell)

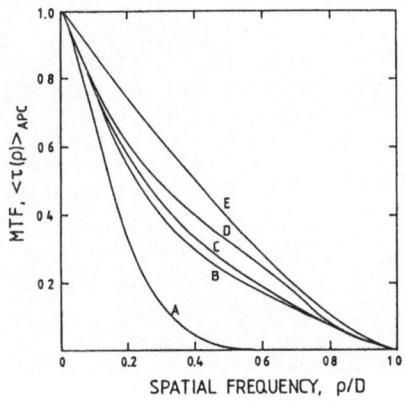

Fig.6a: Modulation transfer function with adaptive phase compensations for $D/r_0 = 2.0$. Curve A, no correction; curve B, tilt correction; curve C, tilt plus focus corrections; curve D, tilt plus focus plus astigmatism corrections; and curve E, ideal phase compensations and diffraction-limited performance.

Fig. 6b Same as Fig.a except that $D/r_0 = 7.0$.

from: WANG, J.Y. (1977)

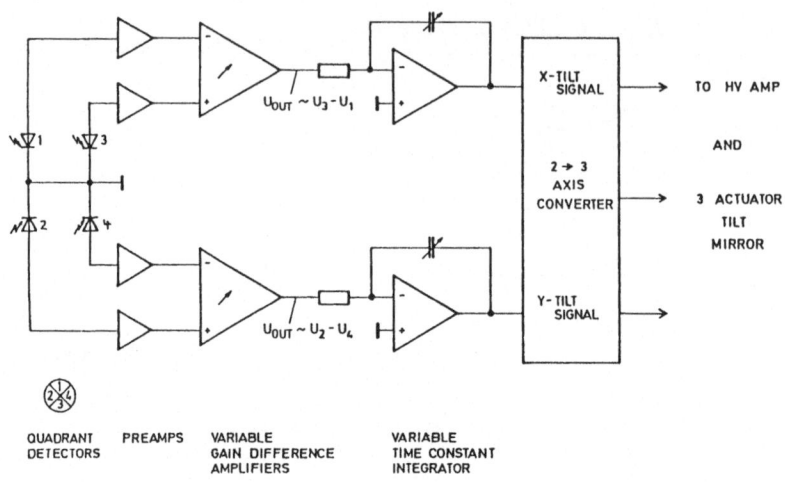

Fig. 7: Sunspot tracker electronic circuitry.

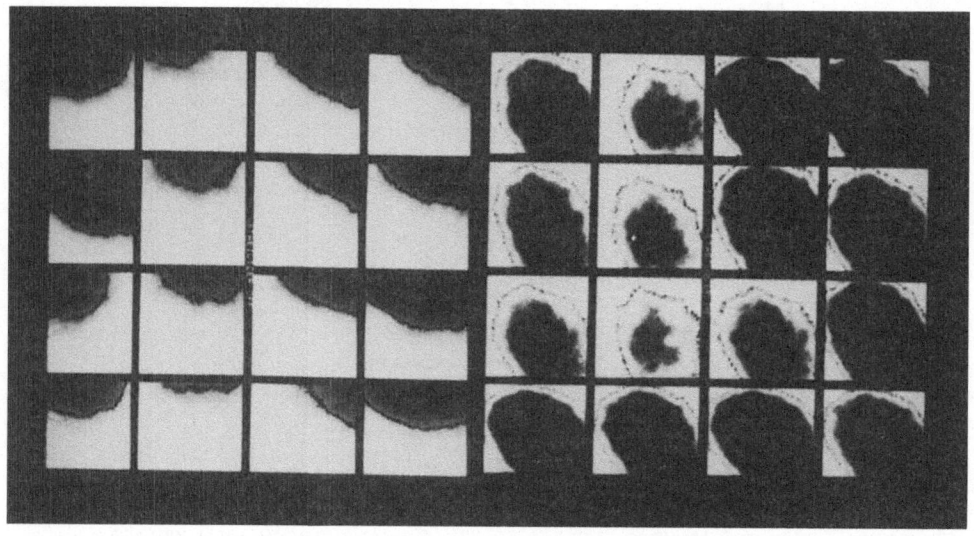

Fig. 11: An example of sunspot stabilisation with the KPNO system (cf. fig. 10). Left half: unstabilised, right half stabilised sunspot umbrae. Picture courtesy to R.V. STACHNIK.

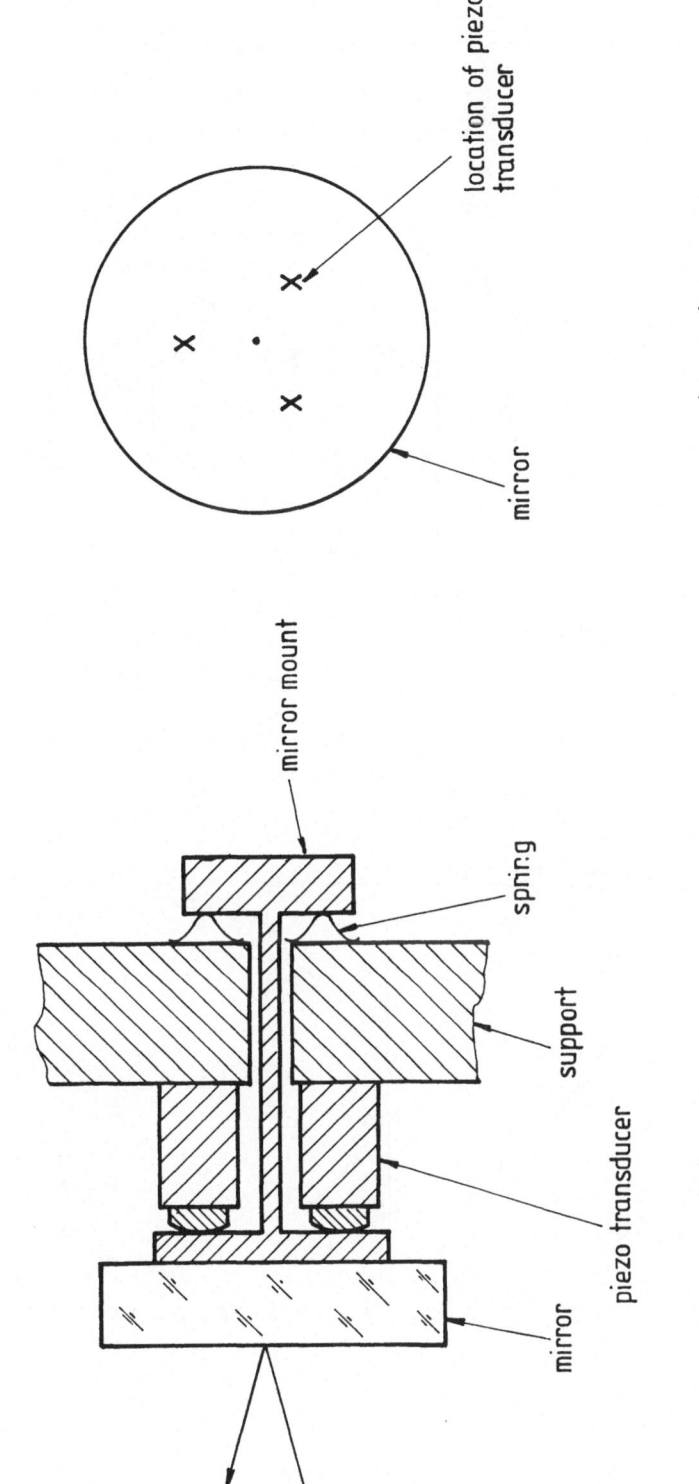

Piezo driven Agile Mirror

top view

location of piezo transducer

mirror

mirror mount

spring

support

piezo transducer

side view

mirror

Fig. 8

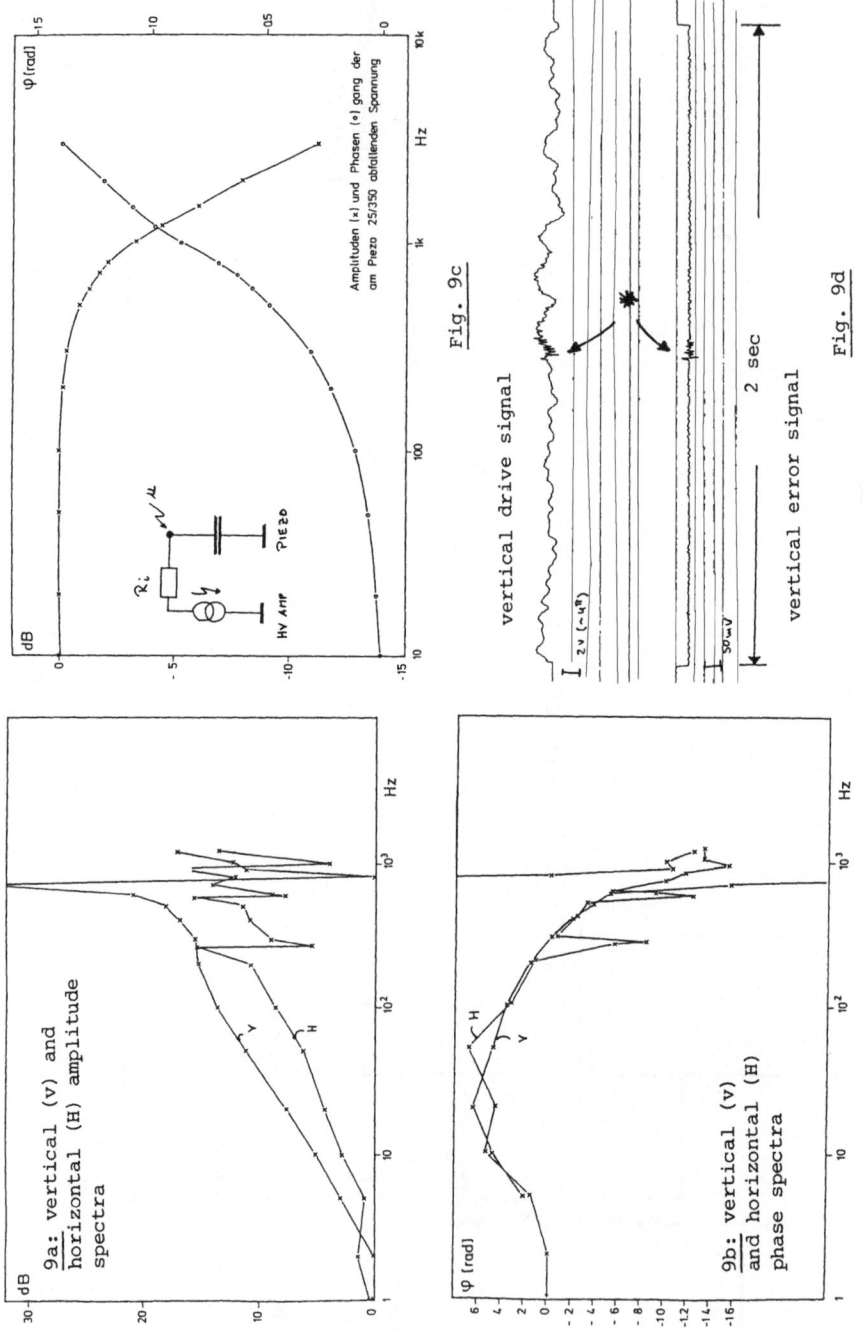

9a: vertical (v) and horizontal (H) amplitude spectra

9b: vertical (v) and horizontal (H) phase spectra

Fig. 9c

Fig. 9d

Figs. 9 a-d: A technical summary of the KIS agile piezo mirror. a and b: amplitude and phase spectra in either direction. c: impedance curve of single piezo. d: sample trace of drive and error signals when locking on a sunspot. The event at "*" is due to knocking on the optical bench.

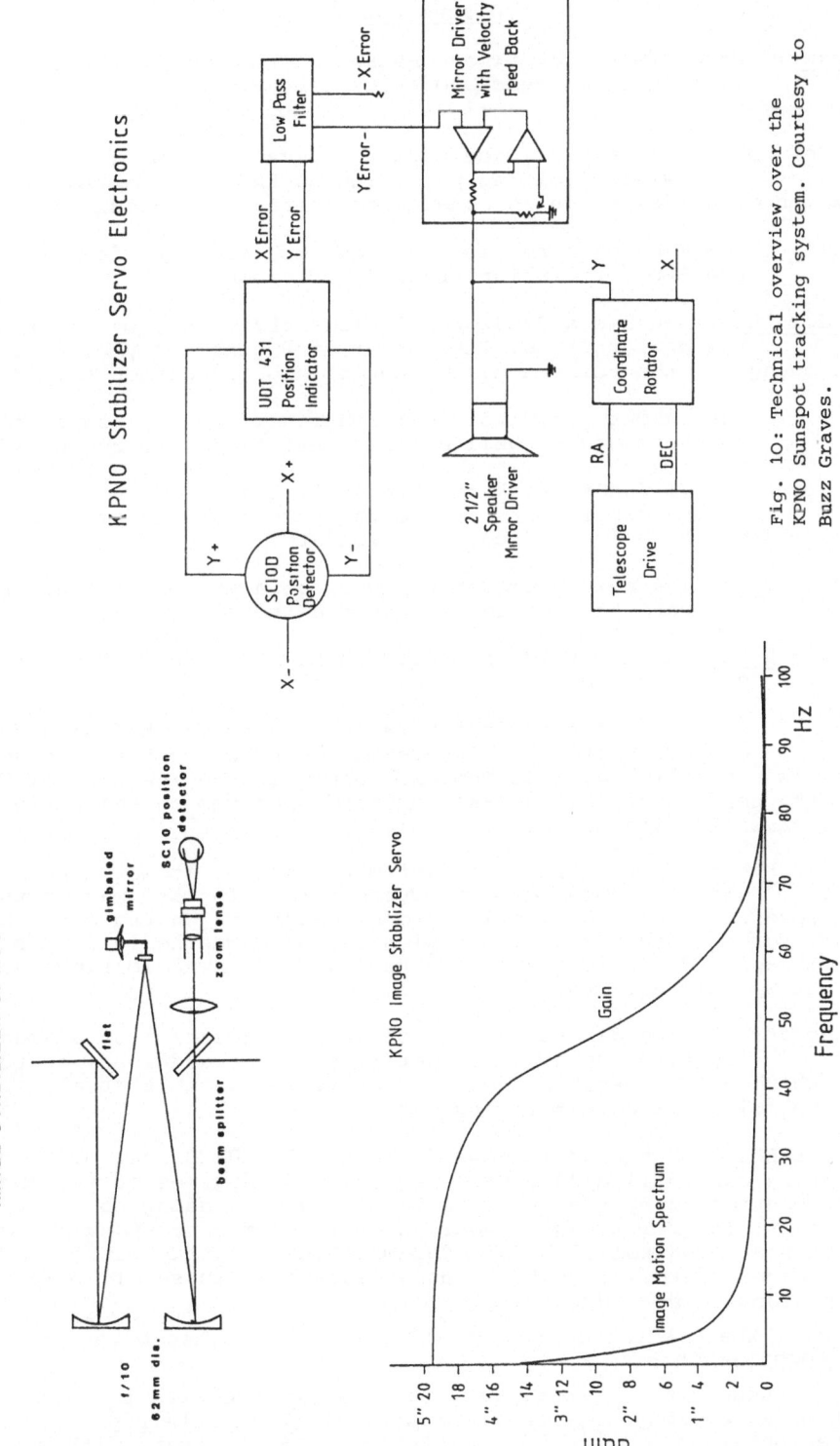

Fig. 10: Technical overview over the KPNO Sunspot tracking system. Courtesy to Buzz Graves.

Discussion

J.C.Pecker: What limits the time constant of the active optics response
(which should reach 10^{-4} sec to beat the effects of bad
seeing)?

O. von der Lühe: In general, the active optical element is the band-
width-limiting part. The LOCKHEED segmented mirror has
a bandwidth of 8 KHZ, which is considered plenty for typical day time use.

O. Engold: How many picture elements do the 1 millisec frame rate of
the SPO correlation tracker correspond to?

O. von der Lühe: We use a 32x32 pixel diode array, but we track only on
the center 16x16. The border of 8 pixels may be used for a
dynamic change of the reference, to support high precision scanning etc.

Artzner: What about laboratory calibration of the response of seeing
compensators to well defined, artificial turbulences?

O. von der Lühe: I don't think that we would learn too much from laboratory experiments, since the atmosphere behaves in a complicated manner which is difficult to simulate.

L.M.B.C. Campos: The next improvement would be a mirror with several
elements that could be adjusted independently?

O. von der Lühe: As a matter of fact, the LOCKHEED segmented mirror is
such a device.

A. Righini: We have observed in our radiosounding of the turbulence
over Tenerife that layers may occur where the Kolmogoroff
spectrum is not followed. This happens mainly in the lee side of Teide.
What is the bearing of this non-Kolmogoroff spectrum on the performances
of image stabilisation devices?

O. von der Lühe: This is an important observation. It may mean that pre-
dictions obtained with current theories of imaging through
the atmosphere may be occasionally wrong, depending on the contribution
of this layer as compared to other significant atmospheric layers. I
would not expect dramatic changes in bandwidth or dynamic range requirements, but we will have to gain experience.

M.V. Schmidt: In the discretization of an approximation of a smooth
function one can in general introduce addtional degrees of
freedom which allow instabilities in the system. Is this at all a problem in the multitelescope Hartmann test?

O. von der Lühe: Not that I know of. In the LOCKHEED system, there are
57 actuators controlling 19 mirror segments. 38 degrees of freedom are
entirely controlled by the Hartmann sensor corresponding to the tilt of
the segments. The remaining 19 degrees of freedom corresponding to segment piston are controlled by an edge-matching network, within which
the tilt signals are integrated. Thus, instabilities should only occur
in the piston of the segments.

R. Muller : What is the best resolution so far reached by an image
motion compensator ?

O. von der Lühe: There is not very much experience with the gain of
resolution under image motion compensation. Alan Title reports that his
group has obtained 0.5 arc sec resolution in filtergrams with several
seconds exposure under favourable conditions.

POSTERS : HIGH RESOLUTION TECHNIQUES OF
OBSERVATION AND DATA REDUCTION

COLLAGES OF GRANULATION PICTURES

Richard B. Dunn and Laurence J. November, AFGL Scholar
National Solar Observatory
Sunspot NM 88349

Introduction: One of the best techniques available to solar astronomers for obtaining a sharp picture is to take a burst of photographs
with short exposures during moments of good seeing. The astronomer
then selects the best frame within the burst by eye. If a movie is
desired, a series of bursts is taken and the astronomer selects those
frames that will form the movie. With the advent of the CCD array
with its digital output, one can do this selection process automatically with a computer. In addition, since the sharpness varies over
the scene, the computer could extend the selection process to small
areas and then, after a suitable time (the order of a few seconds),
could assemble all the best pieces into a final picture representing
the best image that occurred during the burst.

This paper describes two small-area selection schemes that we
have applied to CCD observations of solar granulation. The first
scheme, which we call the "mosaic," divides the 128x128 array into 64
subarrays each containing 16x16 pixels. On each picture in the burst
the RMS contrast of the fine structure is measured in each subarray
and compared to the corresponding value in a table that contains the
highest previous RMS values. If the new value is higher, then that
value is adopted and the corresponding subarray replaces the previous
best subarray. After the entire burst is processed, the best subarrays are reassembled into a mosaic by aligning each subarray by
comparison to a running time average of the original data. A border
of four extra pixels around each 16x16 subarray is saved so that each
subarray may be moved slightly when the mosaic is reassembled. The
final mosaic may contain subarrays from any of the frames within the
burst.

The second scheme, which we call a "collage," is similar except
the RMS value is calculated smoothly within a sliding Gaussian window
over the entire scene and the value of an individual pixel is gated
into the final collage whenever the RMS contrast at that pixel location exceeds that of all previous frames taken during the burst. The
size of the Gaussian window used for calculating each "RMS Map" sets

the size of the individual pieces of the final collage.

If the scene is differentially stretched, because of atmospheric seeing, as is often the case with solar pictures, then the edges of the parts of the final collage do not fit. One of us, November, has developed a "destretch" algorithm called the "Continuous Correlation Tracker" (November, 1984) that smoothly stretches the data frame with respect to the running average. This algorithm locally shifts the spatial coordinates of the image to maximize the correlation with the running average. It is very effective in removing the image stretch due to seeing.

Seeing Quality and "RMS Map": We tried different schemes of assessing the seeing quality. We found that there was no noticeable difference between histograms of brightness in scenes that looked good and bad to the eye. We found no difference in the overall statistical properties of images that were convolved with a high-pass spatial filter. We found that the calculation of RMS was successful in choosing between good and bad scenes only if it was done on a small part of a scene that had previously been high-pass filtered.

Two types of high-pass spatial filters were used successfully. The first is the simple high-pass defined as the operator $\Omega_c(j)$ acting upon the two-dimensional image j as follows:

$$\Omega_c[j(x,y)] = F^{-1}\left[F[j(x,y)] \cdot \begin{cases} 0 \text{ for } k<c-\delta \\ \text{ramp for } (c-\delta)<k<c \\ 1 \text{ for } c<k \end{cases}\right],$$

where c is the lower spatial cutoff frequency and j is the intensity in the original two-dimensional image. F is our notation for the two-dimensional Fourier transform and F^{-1} is its inverse. $k^2 = (k_x^2 + k_y^2)$ is the radial wavenumber in Fourier space and δ is small compared to c.

The other high-pass spatial filter is a well-known, edge-locating function (Hildreth, 1980; Marr, 1982; Poggio, 1984) used in robotics and artificial intelligence. It is the convolution of the image with the Laplacian ∇^2 of a Gaussian. It is very similar to the difference between two Gaussians (Ochs, 1979; Muller and Roudier, 1984) but rigorously cancels all DC and constant derivative signals. The scale ℓ defines the width of the Gaussian and gives the edge scale for which this operator is most sensitive. The operator we call $\Lambda_\ell[j]$ is defined:

$$\Lambda_\ell[j(x,y)] = j(x,y) * \nabla^2\left[\exp\left(-\frac{1}{2}\frac{(x^2+y^2)}{\ell^2}\right)\right],$$

where '*' is the two-dimensional convolution defined by the Fourier transform $F[\ ]$ as:

$$f * g = F^{-1}[F[f] \cdot \overline{F[g]}], \text{ where } \overline{G} \text{ is the complex conjugate of } G.$$

The Laplacian operator $\nabla^2 = [\frac{\partial^2}{\partial x^2} + \frac{\partial^2}{\partial y^2}]$.

We used this operator to process our data, because in addition to providing a high-pass filtered picture for the RMS map, it provided an edge-located image that we thought was interesting for studying the evolution of granules.

The RMS Map, $R_g[k(x,y)]$, is formed from the convolution of a Gaussian with either the high pass $m(x,y) = \Omega_c[j(x,y)]$ or the Laplacian $m(x,y) = \Lambda_\ell[j(x,y)]$ as follows:

$$R_g[m(x,y)] = \left[m^2(x,y) * \frac{\exp\left(-\frac{1}{2}\frac{(x^2+y^2)}{g^2}\right)}{2\pi g^2} - \right.$$

$$\left. m(x,y) * \frac{\exp\left(-\frac{1}{2}\frac{(x^2+y^2)}{g^2}\right)}{2\pi g^2} \right]^2 \right]^{1/2},$$

where g is the RMS window size in units of (x,y). This follows from the statistical formula given for standard deviation (Bevington 1969, Eq. 2-9, p. 19), but with summations replaced by the convolution with a normalized Gaussian. This is natural, since the idea is to compute the spatially-local rms here defined within the Gaussian window and continuously in space via the convolution.

The Data: The data were taken on July 17, 1983 with an RCA "thin" CCD device on the NSO/Sac Peak Vacuum Tower Telescope and recorded on magnetic tape. The exposure was 4ms. Optical filtering included a 5Å band-width interference filter centered on $\lambda5172\text{Å}$, a Schott Kg-3 filter to remove the IR, and an 0.8 neutral density filter. Each 30-micron pixel represented 0.11 arcsecond. A raster size of 132x130 pixels (about 14x14 arcseconds) was sampled. The interval between pictures is 0.55 seconds, so that 27 frames represents a time interval of 15 seconds, which is the burst interval used in our reduction scheme. About 7000 images were recorded, representing about 1 hour of observations.

The Reduction: A data image is read in from tape, corrected for dark current and gain (Figure 1a). The image is spatially-shifted to maximize the correlation with an ongoing average in the central 32x32 pixels to recenter each frame using the correlation algorithm studied by von der Lühe (1983).

The data is then filtered by the Laplacian of a Gaussian (Figure 1b) or by the high-pass filter (Figure 1c) so that the RMS values and the RMS maps may be calculated.

Figure 2 is the result of the mosaic process described in the introduction. The RMS value was calculated in each 16x16 subarray from the scene filtered by the Laplacian of a Gaussian. After the selection process each subarray was shifted to maximize the local cross-correlation with an ongoing average. The edges can be seen and there is some mismatch between the subarrays, but the overall mosaic is an improvement over the "best" scene that is shown in Figure 1.

In the second scheme, each scene is recentered and "destretched" with respect to an ongoing average. This scene is then filtered with either the Laplacian of a Gaussian or the high-pass filter and RMS maps are formed like those shown in Figure 3. The destretched data from the 27-frame burst are then gated by the amplitude of the RMS maps, pixel by pixel, to form two collages shown in Figure 4, which may be compared to Figures 1 and 2. Figure 5 shows the same result without the destretch. Clearly, the edges are badly mismatched, which shows that the individual scenes during the burst are differently stretched, and emphasizes the importance of removing the distortion from the original observation. (The "destretch" would also improve the mosaic process). We plan to process all 7000 frames with both schemes, stretched and destretched, and present the result as a movie for comparison. Small area selection appears to be a useful first step in image reconstruction of solar scenes where the object is extensive and bright.

References

[1] Bevington, P.R. Data Reduction and Error Analysis for the Physical Sciences, McGraw Hill, New York, 1969.
[2] Hildreth, E.C. Implementation of a Theory of Edge Detection, AI-TR-579, Artificial Intelligence Laboratory, Massachusetts Institute of Technology, 1980.

[3] Marr, D. Vision, W. H. Freeman and Company, 1982.
[4] Muller, R. and T. Roudier. The Structure of Solar Granulation. Proceedings of the Conference on the High Resolution in Solar Physics (1984) in Toulouse IAU, this volume.
[5] November, L.J. (in preparation), 1984.
[6] Ochs, A. L. J. Opt. Soc. Am., 69 (1979), 95.
[7] Poggio, T. Scientific American, 250 (1984), 107.
[8] von der Lühe, O. Astronomy and Astrophysics 119 (1983), 85.

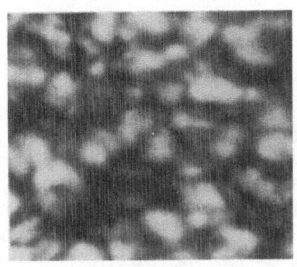
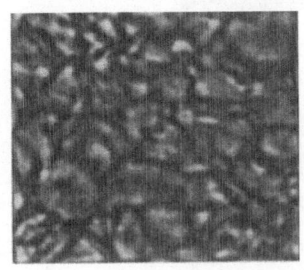
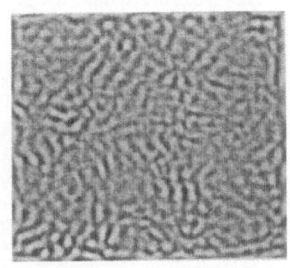

Fig 1a   Fig 1b   Fig 1c

Fig. 1a Data (14"x14") corrected for dark current and gain. This is the best overall scene found within the 27-frame burst, as judged from the RMS contrast in the central 32x32 pixels.

Fig. 1b Same as a, but filtered with the Laplacian of a Gaussian (ℓ=1).

Fig. 1c Same as a, but filtered with a high-pass filter (c=20).

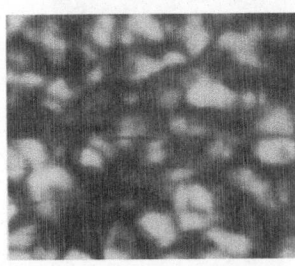
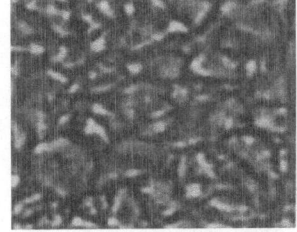

Fig. 2a   Fig. 2b

Fig. 2 Best mosaics assembled from the 27-frame burst. a = data, b = filtered by the Laplacian of a Gaussian (ℓ=1).

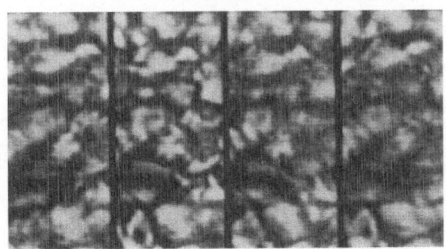

Fig. 3a Four scenes separated in time by 0.55 seconds and filtered by the Laplacian of a Gaussian (ℓ=1). The second from the left scene looks the best to the eye. The other scenes show small-scale distortions and blurring.

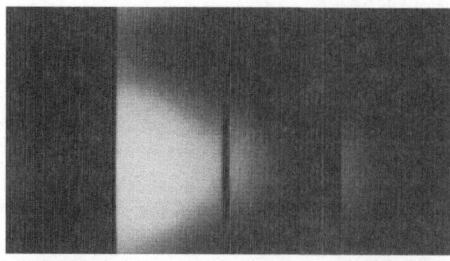

Fig. 3b RMS maps made from the above pictures with a Gaussian (g=30). Clearly the RMS map picks out the best scene.

Fig. 3c RMS maps made of the same four pictures, but with a high pass filter, (c=20), and a Gaussian (g=5). Most of the pixels in the second scene have a higher contrast than the corresponding point in the others.

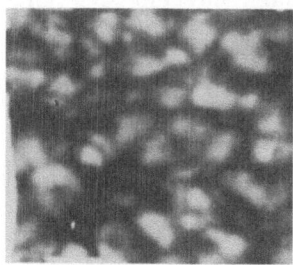 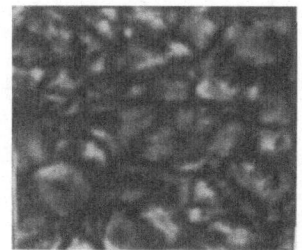

Fig. 4a   Fig. 4b

Fig. 4 Collages (destretched) made from the 27-frame burst. a=data, b = filtered by the Laplacian of a Gaussian (ℓ=1).

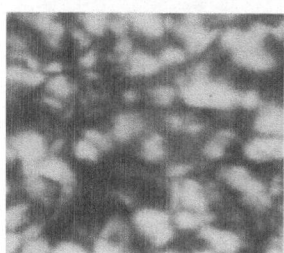 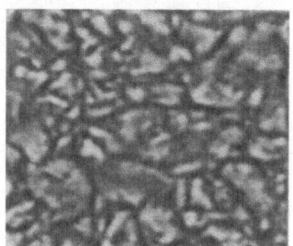

Fig. 5a   Fig. 5b

Fig. 5 Collages formed from the original data (not destretched) showing mismatch.
a = data   b = filtered by Laplacian of a Gaussian (ℓ=1).

SIMULATED CORRELATION TRACKING ON SOLAR GRANULATION

Ø.ANDREASSEN AND O.ENGVOLD
Institute of Theoretical Astrophysics
University of Oslo
Oslo, Norway

R.MULLER
Observatoire du Pic-du-Midi
Bagneres-de-Bigorre, France

The objective of the study is to test the suitability of low contrast granulation images of the Sun for use in telescope tracking over an extended period of time. The method is being developed by von der Lühe (1983) and collaborators. A possible limitation to using the correlation tracking technique in solar telescopes is the fact that the granulation pattern changes with time. The lifetime of photospheric granules are typically 7-15 min (Bray et al. 1984).

The reliability of correlation tracking may be tested by tracking simultaneously on individual and neighbouring regions on the Sun, and then intercompare the results over time periods longer than the lifetimes of the granules. This test operates on image shifts caused by telescope vibration and tracking errors, which are coherent over the entire field of view. (It is well known that the image motion produced by atmospheric seeing is coherent over small angular areas only; usually 2-5 arcsec (Greenaway 1981)). We have performed such a test by simulating correlation tracking in a time series of high resolution images of solar granulation.

The present series was observed on 16 May 1979 with a cinecamera of the 50 cm solar refractor at Pic-du-Midi. The time resolution of the observations is about 1 minute. The selected frames represent seeing of about ½ arcsec and better. The photographic images were digitized with the rapid scanning microphotometer of Institute of Theoretical Astrophysics, University of Oslo. The image scale is 5.45 arcsec/mm, and the total scanned areas correspond to 83.5×83.5 arcsec^2 (512×512 image elements).

Figure 1 illustrates the calculation of cross-correlation power. The relative displacement of two images is given from the location of the peak (Δx, Δy) in the cross-correlation power.

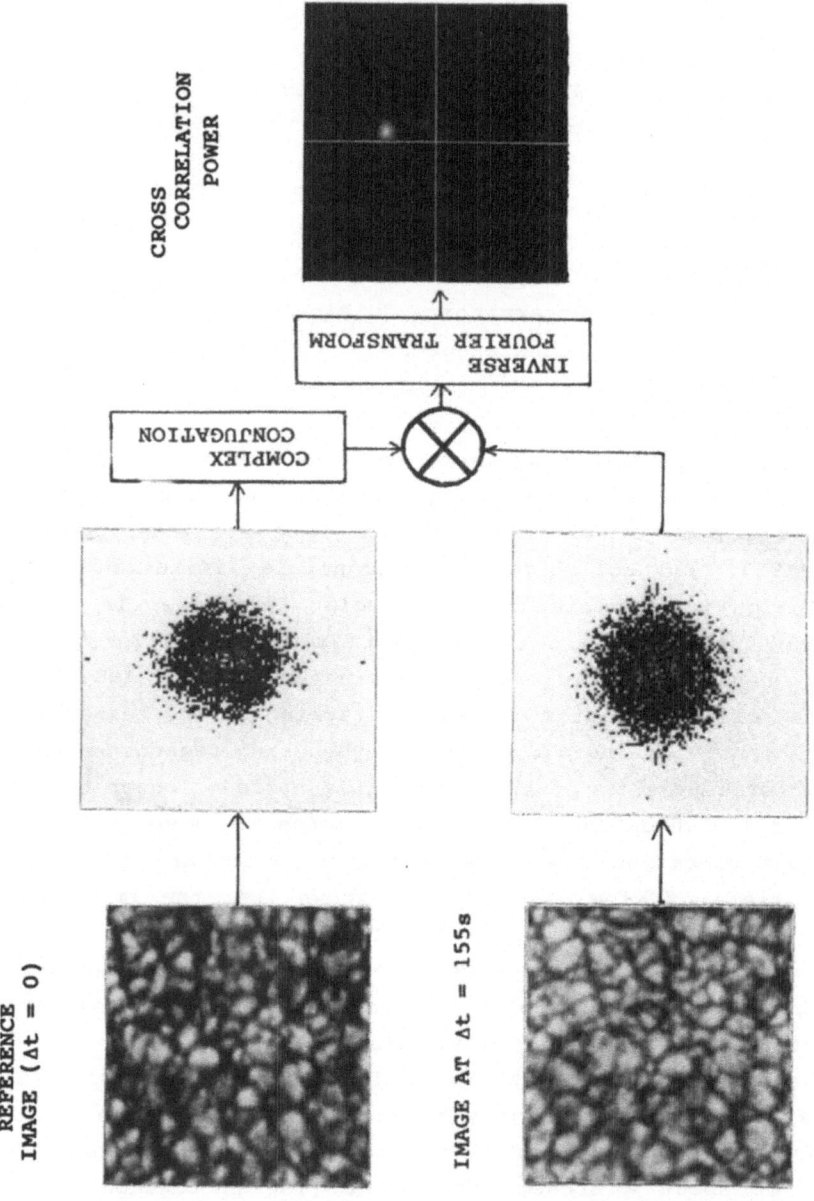

Figure 1. Illustration of cross-correlation power calculations.

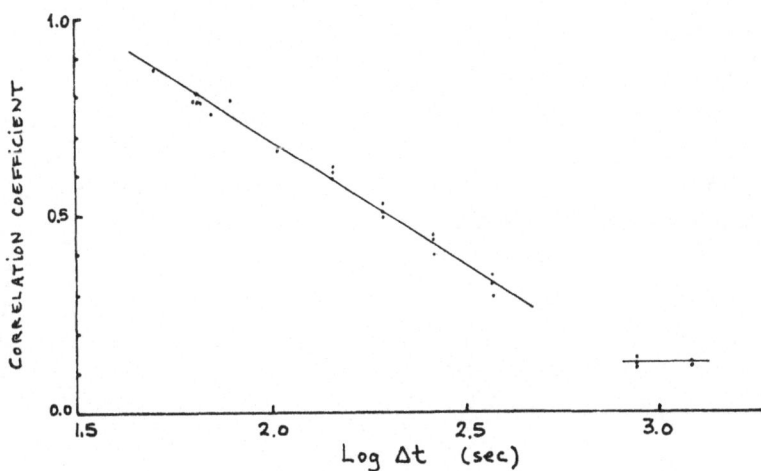

Figure 2.   Variation of correlation peak values as function time
differences of the images.

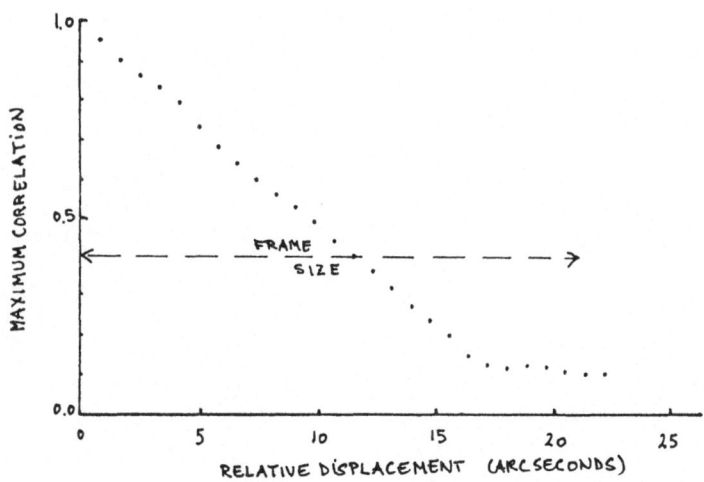

Figure 3.   Peak values of the cross-correlation as function of area
overlap of the two images.

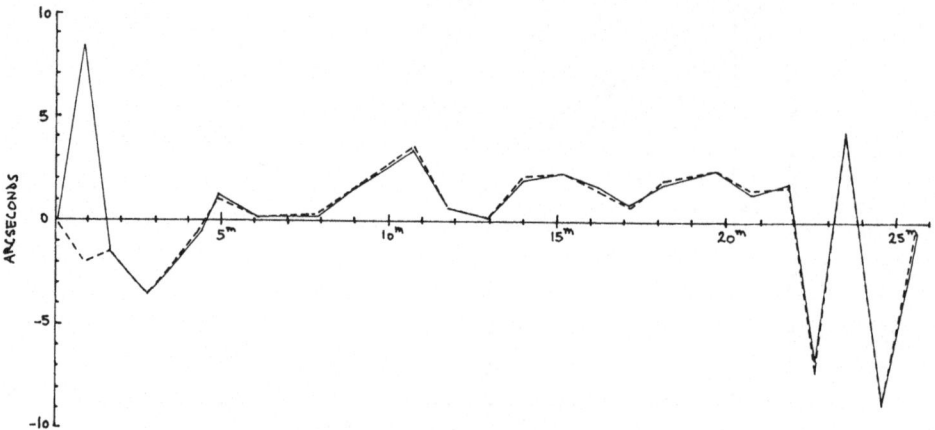

Figure 4. The calculated shift (Δy) of successive images taken about 1 minute apart. The two curves represent two different 21×21 arcsec^2 areas.

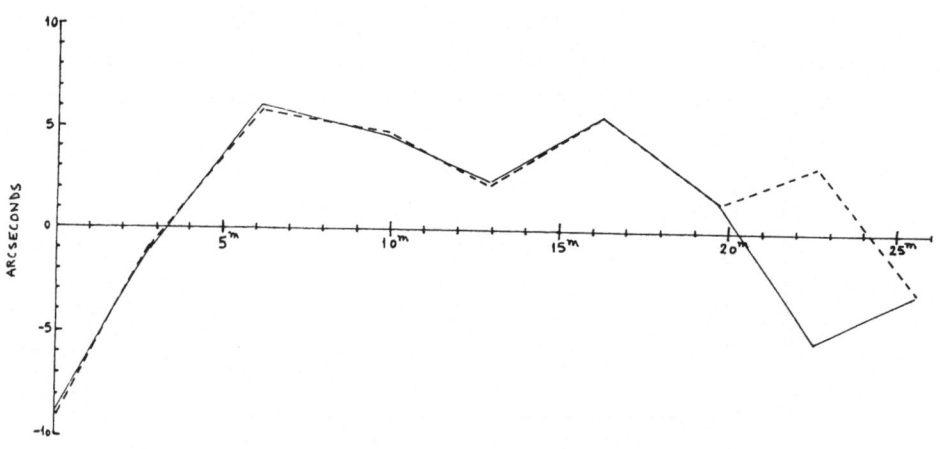

Figure 5. The same as above. The images are separated by about 3 minutes.

The typical cross-correlation peak value decrease with increasing time differences between the images. The time variation is in accordance with the noted lifetimes of individual granules (Figure 2). The area of the correlated images corresponds to 21×21 arcsec^2 represented by 128×128 picture elements. One notes that the cross-correlation falls off to a noise level value of 0.12, which arises from the presence of a typical scale and pattern of the granulation. A substantially higher value (0.32) was noted by von der Lühe (1983), which he ascribed to limitation in his image processing technique.

The maximum values of the cross-correlation is also a function of the overlap between two images. The peak value of the correlation power is halved for a linear shift of about half the image frame size (see Figure 3).

The correlation scheme was applied to areas of 21×21 arcsec^2 in our time series. Figure 4 shows the derived shift (Δy) of successive images in the series taken about 1 minute apart. The two superimposed curves represent the displacements derived for two separate areas within the camera frames. The 2-10 arcsec excursions arise from variations in telescope pointing. The differences of < ½ arcsec between the two curves are caused by seeing movements in the images. The "correlation tracker" lost lock for the second image of the time series because the overlap of the two images was less than 50%.

Figure 5 shows a similar plot for the case the reference image is refreshed about every 3 minutes, i.e. at a cross correlation peak value of about 0.5.

Our study shows that the granulation structure observed during good seeing shows identifiable peak values of the cross-correlation for time differences less than 8-9 minutes. The cross-correlation technique may be used to recognize a given area of the sun for the purpose of tracking over periods that are several times the lifetime of granules provided the reference frame is updated every few minutes.

References.

Bray, R.J., Loughhead, R.E., and Durrant, C.J.: 1984, "The solar Granulation", Cambridge University Press.

Greenaway, A.H.: 1981, "Solar Instrumentation: What's Next?" SPO, Ed. R.B.Dunn, 403.

von der Lühe, O.: 1983, Astron. Astrophys. 119, 85.

HIGH RESOLUTION SPECKLE IMAGING OF SOLAR SMALL-SCALE STRUCTURE: THE INFLUENCE OF ANISOPLANATISM

O. von der Lühe
Kiepenheuer-Institut für Sonnenphysik
Freiburg, West Germany

I. Introduction

Speckle imaging is a generalization of LABEYRIE's (1) speckle interferometry technique. A series of atmospherically degraded short exposure
pictures of an astronomical object is taken in rapid succesion, the
series is then analyzed to recover both Fourier amplitudes and Fourier
phases of the astronomical source at spatial frequencies larger than
the seeing limit. There is a variety of such techniques available, an
overview is given in DAINTY's (2) review. The number of useful techniques
is very limited in the case of solar observations, since the sun's surface is practically unlimited, its small scale structure has low contrast
(only 10% of the photons contribute to the information) and the lifetime of the structure is of the order of a few minutes. Only the most
general speckle imaging techniques, like KNOX' and THOMPSON's (3) pro-
mise successful reconstructions. The KNOX-THOMPSON technique was applied
to solar observations by a group at Harvard (4, 5, 6, 7).

When an extended object is observed from the ground, the light emerging
from two separate object points pass different portions of the atmosphere and the point spread functions of the two image points may differ.
There is a controversity whether this anisoplanatic effect affects the
reconstruction or not. FRIED (8) suspected, that, if the observed object
is larger than the isoplanatic patch, i.e. the area over which the
point spread function is correlated, a KNOX-THOMPSON reconstruction of
the object may be distorted or even completely meaningless. BATES (9)
claims that anisoplanatism would not seriously influence the reconstructed image.

I report on a method to evaluate the reliability of KNOX-THOMPSON speckle
image reconstructions. This method was developed to monitor the reliability of reconstructions obtained under various conditions; here, particular emphasis is put on the effects of anisoplanatism.

II. Methods

The principle of the KNOX-THOMPSON technique is described elsewhere
(e.g. 3, 5). Also, the technique to correct the atmospheric attenuation
of the Fourier emplitudes is to be found elsewhere (10,11).

Let us consider the digital representation F_i (I,J) of the two-dimensional Fourier transform of the i-th picture in the series of K Pictures.
Let I and J be the pixel coordinates in the Fourier plane and let N be
the number of pixles in either direction. F can be expressed as:

$$F_i(I,J) = C_i(I,J) \exp[j\phi_i(I,J)] \tag{1}$$

where C_i is the modulus, ϕ_i is the phase function and $j = \sqrt{-1}$. The
major task of the KNOX-THOMPSON technique is to recover the object phase
ϕ_0 from a set of bispectra D_u and D_v :

$$D_u(I,J) = \frac{1}{K}\sum_{i=1}^{K} F_i(I,J) \cdot F_i^*(I-1,J)$$

$$= \frac{1}{K}\sum_{i=1}^{K} C_i(I,J) \cdot C_i(I-1,J) \exp[j(\phi_i(I,J) - \phi_i(I-1,J))] \tag{2}$$

$$D_v(I,J) = \frac{1}{K}\sum_{i=1}^{K} F_i(I,J) \cdot F_i^*(I,J-1)$$

$$= \frac{1}{K}\sum_{i=1}^{K} C_i(I,J) \cdot C_i(I,J-1) \exp[j(\phi_i(I,J) - \phi_i(I,J-1))]$$

If K is sufficiently large, the phase differences in eqn. 2 are an estimate of the phase differences of the object phase ϕ_0 (I,J). Thus, we
dispose of two NxN phase difference arrays to obtain one NxN phase
solution. Successful techniques to integrate the phase differences are
iterative least-squares integration algotithms like the Gauss-Seidel
algorithm (12). The two bispectra are not independent if they describe
one solution $C_0(I,J) \exp [j.\phi_0 (I,J)]$, and we may take advantage of
their dependency to estimate errors in the reconstruction. There are several schems one could dream up; here is one that evaluates the second
derivative of the phase:

$$D_{uv}(I,J) = D_u(I,J) D_u^*(I,J-1) \; ; \; D_{vu}(I,J) = D_v(I,J) D_v^*(I-1,J) \tag{3}$$

By inserting eqn. 2 into eqn. 3, it is easily seen that for an error-
free reconstruction D_{uv} has to equal D_{vu} everywhere. I will evaluate
the ratio D_{uv}/D_{vu}, I will call this check a "consistency test" of the
bispectra and I will assume that the test is successful if D_{uv}/D_{vu} equals
the real number one. I attribute deviations from this result to unreco-
verable errors in the bispectra. This test works fairly local, it involves a 4 pixel square in the Fourier plane.

III. Data material and results

A time series of 30 solar granulation pictures covering 14 by 14 arc
sec and 15 s of time were analyzed (for more information on the data
used see NOVEMBER and DUNN, this volume). Granulation is a useful target structure because of the isotropy of it's statistics. Fig. 1 shows
a sample frame from the series and an edge-enhanced version of this
frame to bring out the fine detail. I present four different reconstructions obtained from the same series. Firstly, the whole field (14"),
128 by 128 pixels large, was analyzed. All frames were brought into registration over an area covering 32 by 32 pixels in the center. When
displaying this data set as a movie, considerable distortion was visible;
the set appeared stabilized only over an area of approx. 4 arc sec in
the center. For the second reconstruction, the series was subject to a
newly invented treatment by L. NOVEMBER (13), which effectively removes
anisoplanatic image motion (distortion). Thirdly, the field was restricted to the center 64 by 64 pixels covering approx. 7 by 7 arc sec and,
at last, only the center 40 by 40 pixels corresponding to 4.4 arc sec
square were analyzed. A KNOX-THOMPSON reconstruction was obtained from
each of these cases and the consistency test was performed. The isotropy
of the granular pattern makes it possible to calculate azimuthal averages
in the Fourier plane; the errors inferred from eqn. 3 exhibit radial
symmetry. Fig. 2 a-d are plots of the quantities:

$$\text{"average phase"} = \frac{1}{2\pi} \int_0^{2\pi} PHA \left(D_{uv}(s,\varphi) / D_{vu}(s,\varphi) \right) d\varphi$$

$$\text{"absolute phase"} = \frac{1}{2\pi} \int_0^{2\pi} | PHA \left(D_{uv}(s,\varphi) / D_{vu}(s,\varphi) \right) | \, d\varphi \tag{4}$$

$$\text{"rms phase"} = \left\{ \frac{1}{2\pi} \int_0^{2\pi} PHA^2 \left(D_{uv}(s,\varphi) / D_{vu}(s,\varphi) \right) d\varphi - (aver.\ phase)^2 \right\}^{1/2}$$

$$\text{"abs. lg amplitude"} = \frac{1}{2\pi} \int_0^{2\pi} | lg_{10} AMP \left(D_{uv}(s,\varphi) / D_{vu}(s,\varphi) \right) | \, d\varphi$$

where the operators AMP and PHA return modulus and phase of a complex
number, respectively and s,φ are polar coordinates in the Fourier plane.
Figs. 3 a-d show the corresponding reconstructions, uncorrected as well
as corrected for atmospheric attenuation of Fourier amplitudes. The
brightness range is scaled by the maximum and the minimum in each pic-

ture. In fact, the uncorrected pictures have a rms intensity contrast
of 4% while the corrected pictures have an rms contrast of 14%.

IV. Conclusions

The comparison of the sample frame with the reconstructions reveal the
failure of the KNOX-THOMPSON technique when the field is considerably
larger than the isoplanatic patch. It is striking that in the first case
the reconstruction is unsuccessful even for frequencies smaller than
the seeing limit. The removal of distortion apparently fixes the problem
for the large structures, but the errors for high frequencies are still
unacceptably large. This diagnosis is also reflected in the radial behaviour of the phase and amplitude errors. Successive reduction of the
field of view leads to successive reduction of the errors in the region
between seeing cutoff and the limit of useful signal. Beyond this region, the average amplitude error is approx. a factor of 3 and the phase
error is consistent with a uniform distribution of the phase between π
and $-\pi$. Except for edge artifacts, most of the fine detail seen in the
reconstruction from the smallest field appears to be of solar origin.

The results presented here confirm FRIED's suspicion that the KNOX-
THOMPSON technique would have problems under anisoplanatic conditions.
Apparently, error monitoring is unrenouncable in order to produce results with a certain photometric reliability, and earlier results should
be carefully reexamined before drawing conclusions from them.

Acknowledgements: It is a pleasure to thank R.B. Dunn, L.J. November
and R. Radick for the data, many discussions and the substantial soft-
ware support. Pictures courtesy to R. Faller, NSO Sacramento Peak.

Literature
(1) LABEYRIE, A. (1970) Astron. Astrophys 6, 85-87
(2) DAINTY, C. (1984) "Stellar Speckle Interferometry", in Vol. 9 of
"Topics in Applied Physics: Laser Speckle and Related Phenomena",
Springer, 2nd edition
(3) KNOX, K.T. and THOMPSON, B. (1974), Astrophys. J. (Letters) 193, L45ff
(4) STACHNIK, R.V.; NISENSON, P.; EHN, D.C.; HUDGIN, R.H.; SCHIRF, V.E.
(1977), Nature 266, 10 149-151
(5) STACHNIK, R.V.; NISENSON, P.; PAPALIOLIOS, C. (1980): "Solar Speckle
Imaging" in "Solar Instrumentation: What's next?" SPO conference proc.,

Ed. R.B. Dunn, 502-509
(6) NOYES, R.W.; STACHNIK, R.V.; NISENSON, P. (1981): "Speckle Image Reconstruction of Solar Features", AFGL Final Report, AFSC USAF Hanscom
AFB, MA 01731, USA
(7) STACHNIK, R.V.; NISENSON, P.; NOYES, R.W. (1983), Astrophys. J.
(Letters) 271, L37-40
(8) FRIED, D.L. (1979), Optica Acta 26, 5 597-613
(9) BATES, R.H.T. (1981), Phys. Rep. Reviews 90,4 203-297
(10) VON DER LUEHE, O. (1984), J. Opt. Soc. Am. A 1, 510-519
(11) VON DER LUEHE, O. (1984) in: Proceedings of the IAU Colloquium 79
on "Very Large Telescopes", appears in Nov. 84
(12) SOUTHWELL, W.H. (1980), J. Opt. Soc. Am. 70,8 998-1006
(13) NOVEMBER, L.J. (1984) submitted to Applied Optics

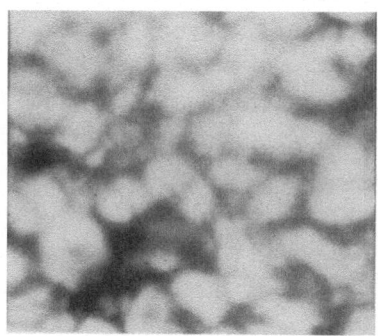   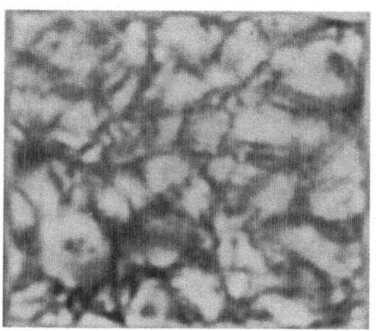

1a   1b

Figs 1a and b: A sample frame (fr. no. 1) from the analysed time series
of granulation pictures 1a: original frame, 1b: edge-enhanced version
of 1a. The field of view is 14 arc sec side length. Solar granulation
is observed through a 517 nm interference filter 8 nm wide. Exposure
time was 4 msec, pictures were taken 550 msec apart. Pictures courtesy
to R.B. DUNN and L.J. NOVEMBER.

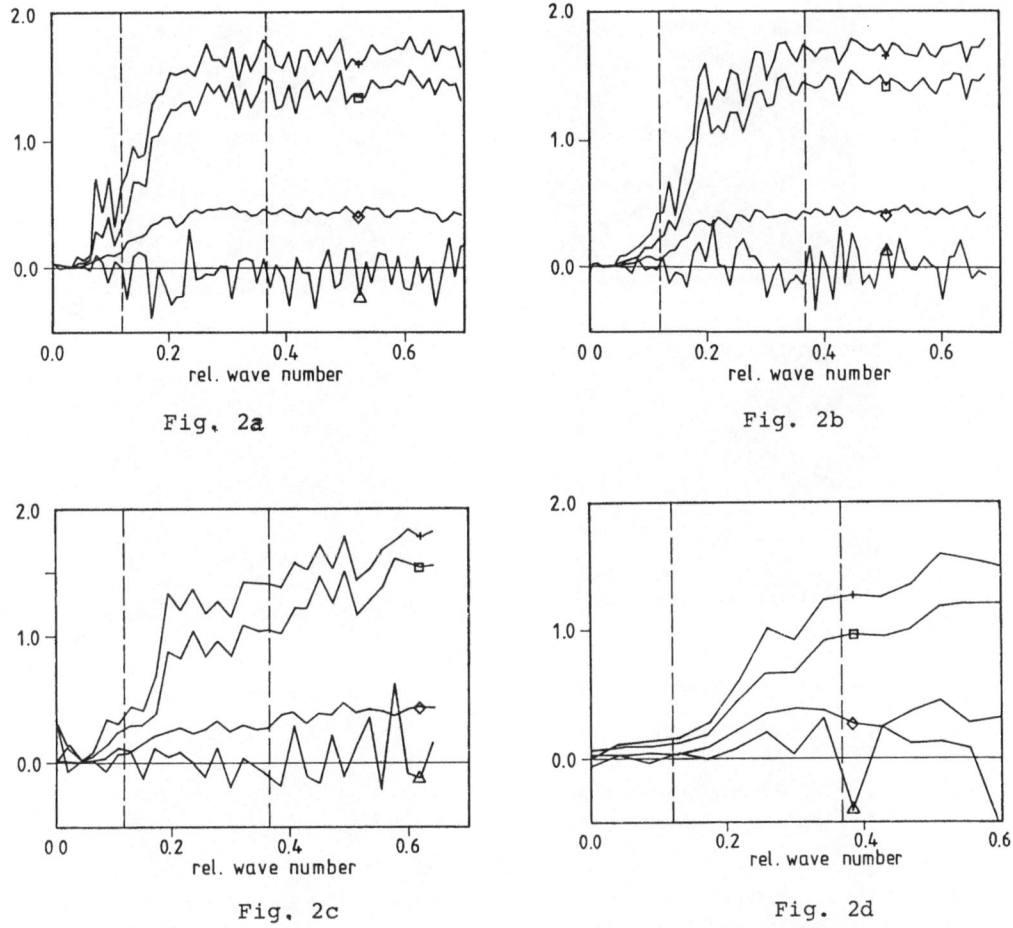

Fig. 2a

Fig. 2b

Fig. 2c

Fig. 2d

Figs. 2a-d: Radial plots of the quantities in eqn. (4) for all four reconstructions △: average phase, □ : absolute phase, + : rms phase, ◊ : absolute lg amplitude of D_{uv}/D_{vu}

The units are radians for the phase plots. The radial coordinate is expressed in units of the cutoff wave number of the telescope ("relative wave numbers"). The vertical dashed lines correspond to the seeing limit at 0.12 ($\hat{=}$ 6.9 Mm^{-1} on the sun) and the limit of usable signal (SNR > 2) at 0.36 ($\hat{=}$ 20.6 Mm^{-1}).

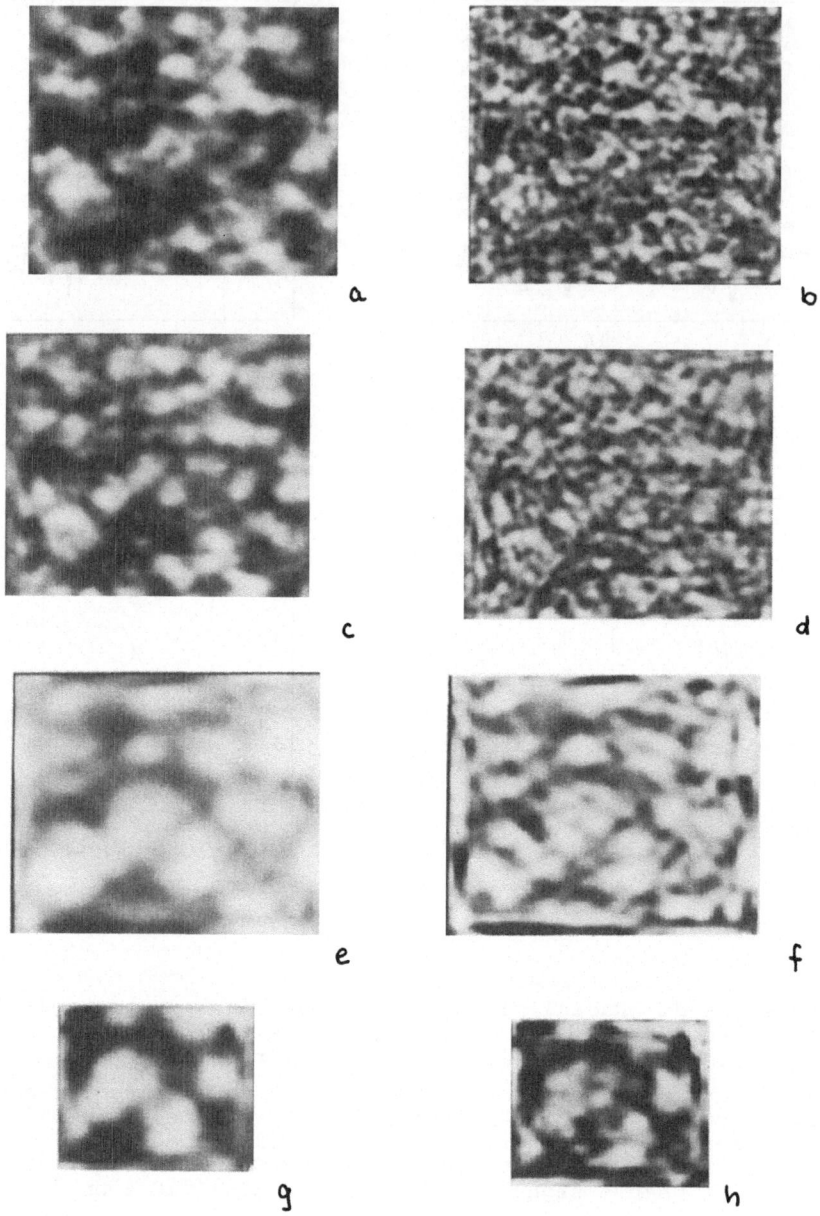

<u>Figs. 3 a-h:</u> KNOX-THOMPSON reconstructions obtained from a 30 frame series. Left column: "raw" reconstructions, right: seeing corrected reconstructions. First row: 14" field, centered only, second: 14" field, distortion removed, third: 7" field, fourth row: 4" field.

SPECKLE INTERFEROMETRY TECHNIQUE
APPLIED TO
THE STUDY OF GRANULAR VELOCITIES

Aime C., Borgnino J., Druesne P., Harvey[+] J.W., Martin F., and Ricort G.

Département d'Astrophysique de l'Université de Nice, E.R.A. 669 du CNRS

Parc Valrose - 06034 Nice Cedex-France.

(+) Kitt Peak National Observatory

950 North Cherry Avenue, P.O. Box 26732 Tucson, Arizona-U.S.A.

I - INTRODUCTION

Following the pioneering stellar measurements of Labeyrie[1] (1970), several authors[2,3] have shown that the speckle interferometry technique applies also to the study of brightness fluctuations in the solar photosphere. We report here the preliminary results obtained by the speckle technique on the correlation between intensity and velocity in the solar granulation.

II - OBSERVATIONS AND DATA PROCESSING.

The speckle experiment was set up at the McMath main telescope of K.P.N.O. on July 1980. It consisted in recording the signals collected by two photomultipliers set 0.25 Å apart on the wings of the Fe 5233 Å line whilst the solar image was rapidly moved across the entrance slit of the spectrograph (velocity of 500"/s). Every 40 seconds, the same region of about 50" x 150" at the centre of the solar disk was scanned with a 0.1" x 5" slit in 100 lines at 1.5" intervals, each line being 50" long and running in the same direction. A total number of 6800 lines was processed. The brightness b(r) was defined as the sum of the two outputs of the photomultipliers, and the Doppler velocity v(r) as their normalized difference, assuming linearity for the wings of the line.

The data processing consisted first in computing the cross-spectrum matrix $W(\nu)$ between $v(r)$ and $b(r)$, defined by :

$$W(\nu)= \begin{vmatrix} <|\hat{b}(\nu)|^2> & <\hat{b}(\nu).\hat{v}^*(\nu)> \\ <\hat{b}^*(\nu).\hat{v}(\nu)> & <|\hat{v}(\nu)|^2> \end{vmatrix}$$

where the symbol ^ stands for a Fourier transform, * for the complex conjuguate, and < > for the ensemble average. The one-dimensional analysis, combined with the use

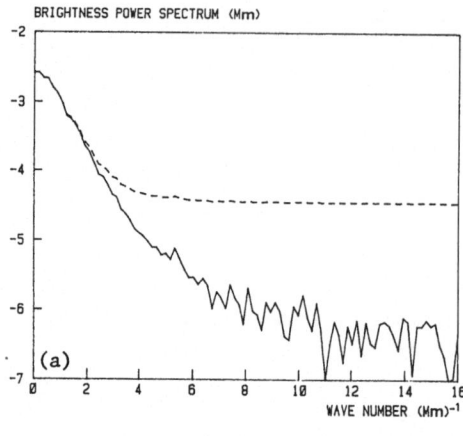

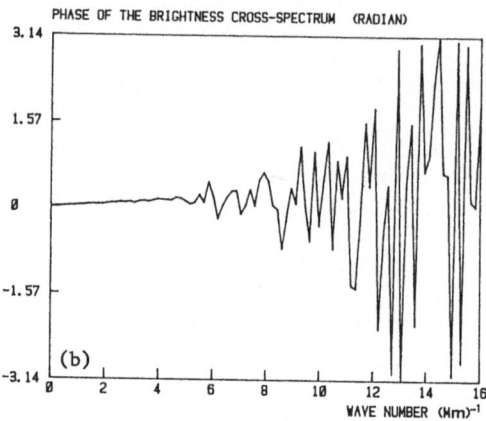

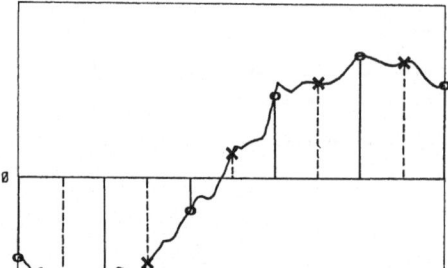

Figure 1. Example of the cross-analysis
technique applied to the determination
of the brightness power spectrum.
(a) Result of the standard single channel
technique (-----)with its background noise, compared with the unbiased cross
analysis estimate (————)
(b) phase of the cross analysis between
the odd and even points shown by (x)
and (O) in the figure (c). Note the
linear phase variation corresponding to
the shift between the two sets of data.

of a slit, give us access to a radial cut of the two-dimensional spectrum.

The raw computation of W (ν) made visible a high level of background photon
noise. The level of this noise was higher than the signal for wavenumbers greater
than 2.5 Mm^{-1}. In order to avoid the delicate problem of the subtraction of this
bias [4], we undertook a cross-analysis of interlaced samples of the signal [5]. The
data of each line scaned were splitted in odd and even samples of points (figure 1)
and the cross spectra between them computed. Provided that the band pass of the
electronic be large enough, the successive samples are free of correlation at the
level of the photon noise. The cross spectrum analysis retains the part of the
signal relevant to solar structures and makes the background noise bias vanish[6].
This is shown in figure 1 for a partial estimate of the brightness power spectrum.
In dashed line, is the result of the standard technique for the power spectrum, and
in continuous line, the result of the cross spectrum technique. Note that the phase
of the cross spectrum shows in the 0 to 5 Mm^{-1} region the expected linear variation
due to the shift between the two set of data. Beyond 11 Mm^{-1}, the photon noise makes
the phase to be uniformely scattered between $-\pi$ and $+\pi$, which forbids further

exploitation of the results in this range.

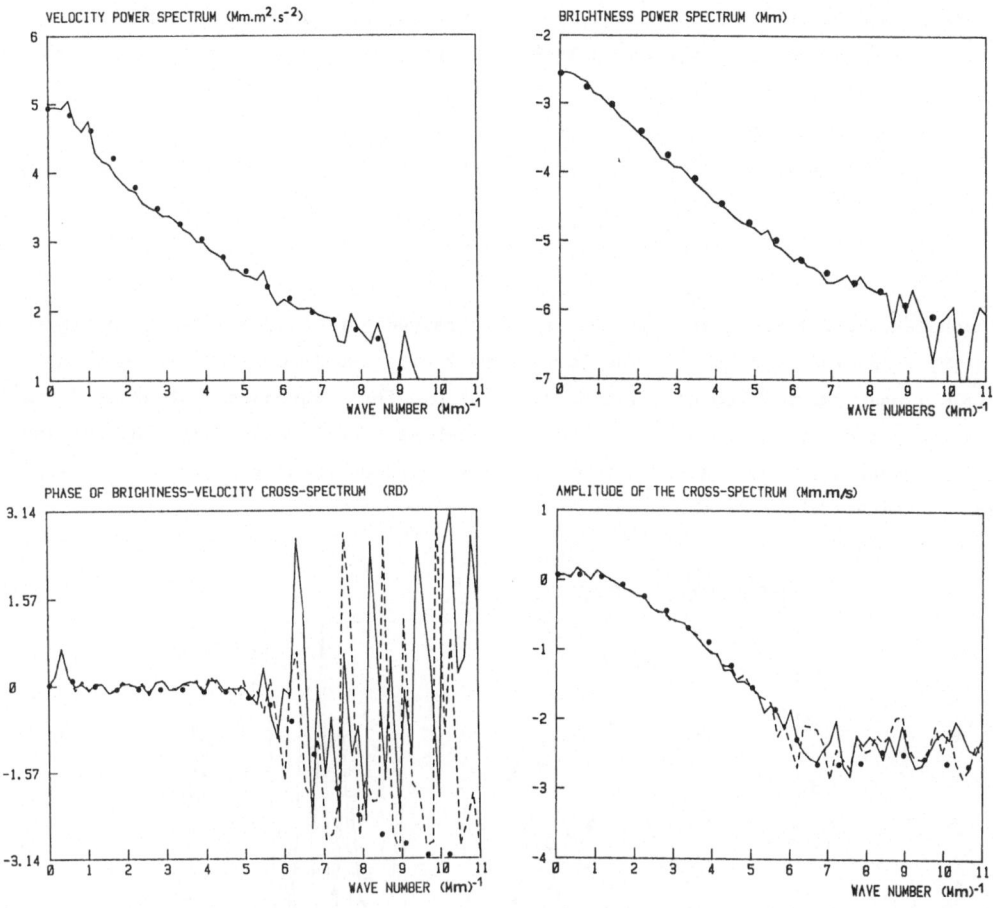

Figure 2 : Brightness velocity cross spectrum matrix W (ν).

III - <u>RESULTS AND DISCUSSION.</u>

The brightness velocity cross spectrum matrix is shown in figure 2. The curves drawn in continuous lines have the basic resolution of the experiment (0.17 Mm^{-1}), corresponding to the inverse of 50", while the values indicated by ● correspond to a lowered resolution of 1.5 Mm^{-1} and display a better signal to noise ratio (S.N.R.). The data processing technique allowed two estimates of the cross spectrum to be computed from the same set of scans (continuous and dashed lines). Note that the difference between them is fully due to photon statistics. The phase of the cross spectrum clearly shows where the likeness between velocity and brightness is accurately measured. Between 0.5 Mm^{-1} and 6 Mm^{-1}, the correlation is positive since the excursion of the phase around zero is less than 0.08 radian r.m.s., while the variance of the photon noise makes it become uniformly scattered between $-\pi$ and $+\pi$ for wave-numbers greater than 7 Mm^{-1}. A trend to $-\pi$ variation, i.e. a negative correlation, appears in the smoothed curve shown by ● . However, the weakness of the signal makes this surprising result subject to caution.

It is quite difficult to interpret the amplitude of W (ν), since the data are still contaminated by atmospheric seeing effects. However, the coherence function $\gamma^2(\nu)$, defined by :

$$\gamma^2(\nu) = \frac{|<\hat{b}(\nu).\hat{v}(\nu)^{\star}>|^2}{<|\hat{b}(\nu)|^2><|\hat{v}(\nu)|^2>}$$

is much less sensitive to seeing. The values obtained for γ (ν) are shown in figure 3 with the same drawing convention as for figure 2. The amplitude of $\gamma^2(\nu)$ shows a maximum linear correlation coefficient of about 0.6 for a wavenumber of about $3.5Mm^{-1}$ and a steep decrease in lower and higher wavenumbers. These results are in substantial agreement with those derived from spectrostratoskop data by Durrant and Nesis[7] (1982), indicated by X in the figure.

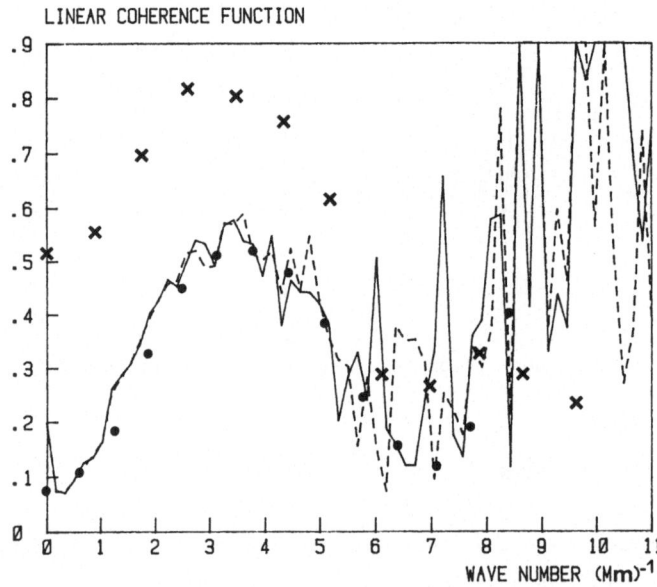

Figure 3. Linear coherence function $\gamma(\nu)$. The curves drawn in continuous and dashed lines have the basic resolution of the experiment (0.17 Mm^{-1}). The differences between them are fully due to photon statistics. The values indicated by x are from Durrant and Nesis[7]. The value indicated by ● correspond to γ (ν) computed from the smoothed curves of figure 2.

Between 0.5 and 6 Mm^{-1}, the curves are very similar in shape. The lower amplitude of our coherence function is probably due, as suggested to us by Nesis[8] at this meeting, to the fact that we measure the correlation between velocity and line intensity (instead of continuum intensity for Durrant and Nesis). The lower value of our curve agrees with their assumption that the velocity structure of the ascending gas better maintains its shape than the temperature fluctuations, when traveling through the photosphere [9].

The trend to a negative correlation, together with a hypothetic increase of the coherence in the high wavenumber, if not an artefact, would be a relevant information It stands to reason that better S.N.R. data are needed to conclude in this sense. A light increase of the S.N.R. could be obtained running the same experiment under better seeing conditions. A significant improvement probably requires the application of the speckle technique to the study of spectrograms with CCD-type detectors.

REFERENCES

1. Labeyrie, A., Astron. Astrophys. *6* , 85, 1970
2. Harvey, J.W. and Breckinridge, J.B., Astrophys. J. *182*, L 137, 1973
3. Ricort, G. and Aime, C., Astron. Astrophys. *76*, 324, 1979
4. Harvey, J.W. and Schwarzschild, M., Astrophys. J. *196*, 221, 1975
5. Mein, P., private communication, 1983
6. Aime, C., Kadiri, S., Martin, F., Petrov, R. and Ricort, G., Astron. Astrophys. 134 , 354, 1984
7. Durrant, C.I. and Nesis, A., Astron. Astrophys., *111*, 272, 1982
8. Nesis, A., private communication, 1984
9. Nesis, A., Durrant, C.J. and Mattig, W., publication of Sacramento Peak Observatory, July-August 1983.

THE INTEREST OF SIMULTANEOUS SPECTRAL AND
SPATIAL HIGH RESOLUTION SPECTROSCOPY
IN THE INFRARED

L. Delbouille[1], N. Grevesse[1] and A.J. Sauval[2]

(1) Institut d'Astrophysique, Université de Liège
    B-4200 Ougrée-Liège (Belgium)
(2) Observatoire Royal de Belgique, 3, Avenue Circulaire,
    B-1180 Bruxelles (Belgium)

1. Introduction

Up to the present, small scale structures on the sun have mostly been studied through observations made in the visible and ultraviolet. After having recalled some of the main advantages of infrared observations, we give a few samples of high spectral resolution/low noise solar spectra now routinely obtained in this spectral range by using Fourier transform spectrometers (FTS). We then show that such high spectral resolution spectra could be obtained in short period of time and for small scale solar structures, using a LEST type collector.

2. Main advantages of infrared observations

The number of lines per unit wavelength interval decreasing when λ increases, the infrared solar spectrum is much "cleaner" than the visible spectrum and the line profiles can be measured with much higher accuracy.

The continuous opacity is very well known.

Near 1.6 μm we can observe the deepest photospheric layers.

Planck's law is much more objective in the infrared i.e. it tends to equally weights the hottest and coldest area of an inhomogeneous surface.

Atomic lines show larger Zeeman splitting in the infrared; suitable atomic lines have already been proposed or used (Harvey, 1973; Harvey and Hall, 1975). Furthermore many atomic lines show large hyperfine structure (HFS); such lines should also be very sensitive to magnetic fields.

There are also in the infrared a very large number of suitable molecular lines [CN (red system), C_2 (Phillips and Ballik-Ramsay), CO (vibration-rotation, Δv = 1,2), OH (vibration-rotation and pure rotation)]. These molecular lines are formed in local thermodynamic equilibrium;

their depths of formation cover from the deepest photospheric layers
to the temperature minimum allowing to scan the whole photosphere (Figure 1). Furthermore, these lines are very sensitive to the temperature, some of them [CO, OH(pure rotation), CN] being two to three times
more sensitive than atomic lines (Figure 2).

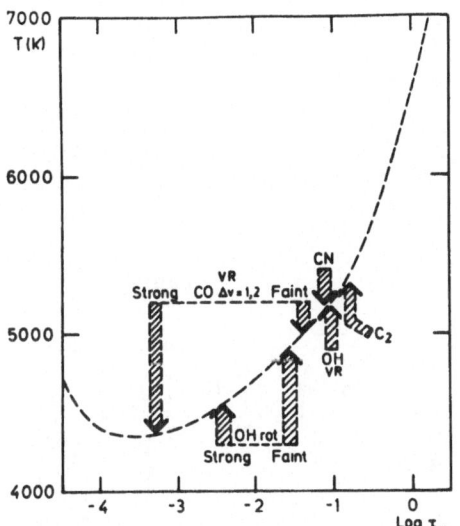

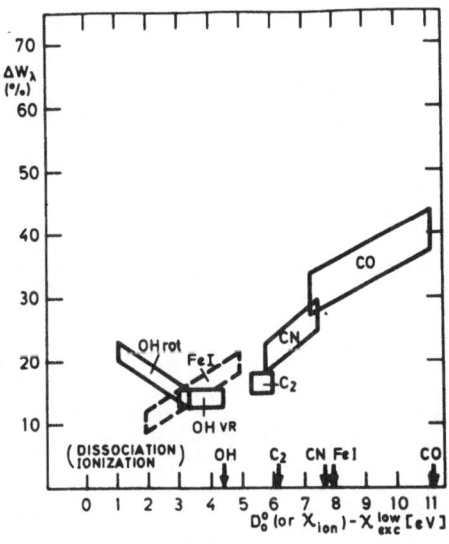

Fig.1 - Photospheric layers
where the infrared molecular
lines are formed.

Fig.2 - Relative variation of the
equivalent widths of the infrared
lines as a function of the ionization (dissociation) energy (from
the lower level) for a $\Delta T = 100K$
in the layers where the lines are
formed.

(The so-called infrared emission lines between 10 and 13 µm (800-1000
cm^{-1} ; Brault and Noyes, 1983; Chang and Noyes, 1983) should also be
kept in mind as very sensitive indicators of magnetic field and temperature).

3. High spectral resolution in the infrared

Using Fourier transform spectrometers (FTS) it is now possible to obtain very high resolution solar spectra (resolution limit < 0.01 cm^{-1})
together with very high signal over noise (S/N > 1000). We show hereafter a few samples (Figures 3 to 6) of infrared solar spectra (for
the center of the solar disk) recorded at Kitt Peak National Observatory (solar atlas between 2000 and 10000 cm^{-1} (1-5µm) by Delbouille,
Roland, Brault and Testerman, 1981) and at the Jungfraujoch Solar Physics Laboratory (Delbouille, Roland and Demoulin, 1984).

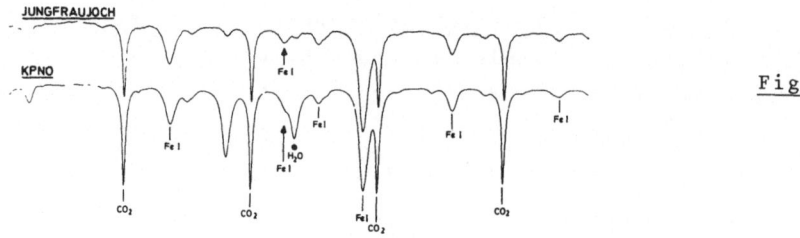

JUNGFRAUJOCH

KPNO

Figure 3

THE Fe I LINE AT 6514.485 (ARROW) IS BLENDED
WITH A WATER VAPOR LINE; AND MUCH MORE EASILY
MEASURED FROM THE JUNGFRAUJOCH.

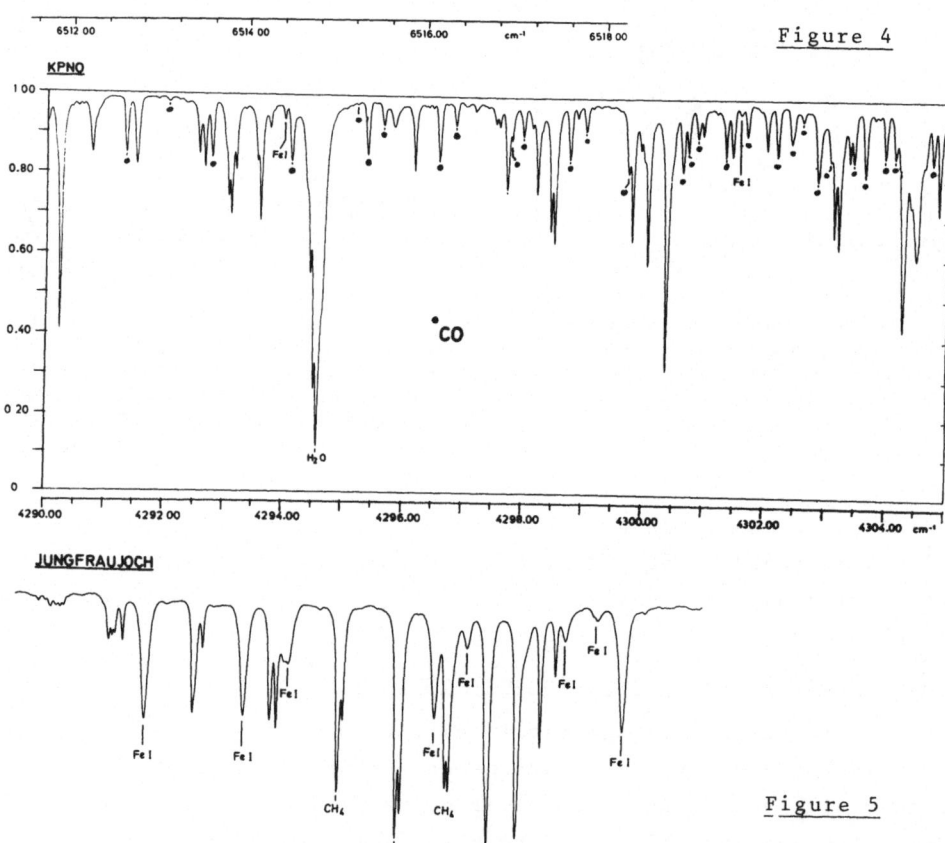

Figure 4

KPNO

JUNGFRAUJOCH

Figure 5

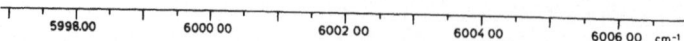

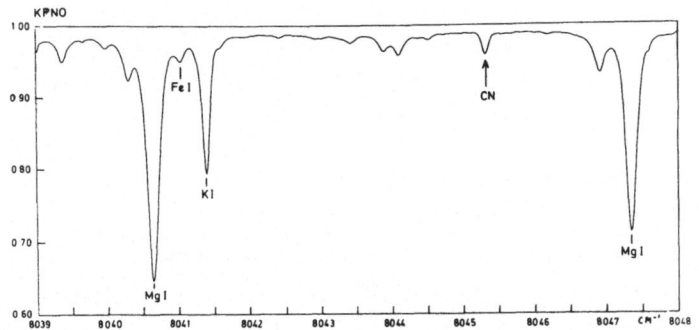

Figure 6

4. Simultaneous high spatial and spectral resolution spectroscopy

Figure 7 gives the performances around 6000 cm^{-1} of a Fourier transform spectrometer used as a post-focus instrument of a 2.4 meter solar telescope (LEST). They have been extrapolated from actual results obtained with the N.S.O. Kitt Peak Solar F.T.S. (Delbouille, Roland, Brault and Testerman, 1981) and gives the time needed to reach a given spectral resolution with a single detector "seeing" a small area on the solar disk (expressed by its angular diameter). Replacing it by an array of N similar elements, it will be possible to obtain simultaneously N spectra of adjacent small domains of the solar surface.

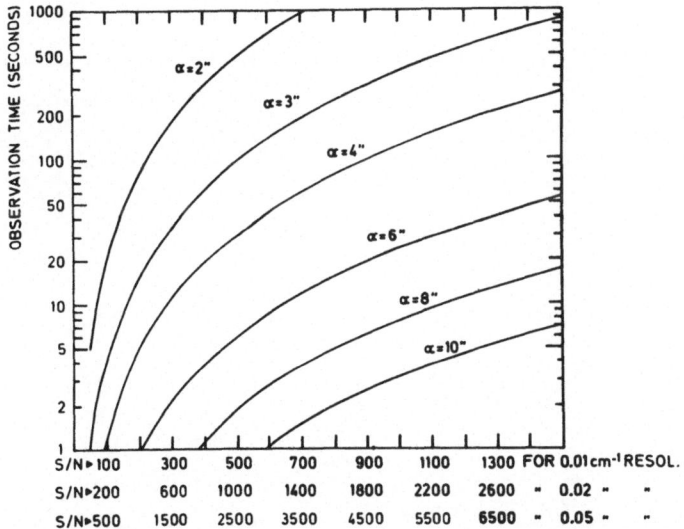

Figure 7 - Predicted performances of a Fourier transform spectrometer installed in the focal plane of a 2.4 meter diameter solar telescope. An optical bandwith of 200 cm^{-1} around 6000 cm^{-1} (1.6 μm) is assumed. α is the angular diameter of the field of view for one detector.

5. Conclusions

Post focus instrumentation for the LEST telescope should include a multiple detector Fourier transform spectrometer for infrared observations.

References

Brault, J.W., Noyes, R.W., 1983, Astrophys. J. 269, L61.

Chang, E.S., Noyes, R.W., 1983, Astrophys. J. 275, L11.

Delbouille, L., Roland, G., Brault, J.W., Testerman, L., 1981, Photometric Atlas of the Solar Spectrum from 1850 to 10000 cm^{-1}, Preliminary Data, Kitt Peak National Observatory, Tucson.

Delbouille, L., Roland, G., Demoulin, P., 1984, unpublished results.

Harvey, J., 1973, Solar Phys. 28, 9.

Harvey, J., Hall, D.N.W., 1975, Bull. Am. Astron. Soc. 7, 459.

PRELIMINARY RESULTS OBTAINED WITH A NEW EXPERIMENTAL APPARATUS FOR SOLAR SPECTROPOLARIMETRY

W. Scholiers
Institut d'Astronomie,d'Astrophysique et de Géophysique
Université libre de Bruxelles
50,av.F.D. Roosevelt - B-1050 Bruxelles

A new experimental apparatus has been built in order to investigate the possibilities of two-dimensional photoelectric detectors for solar spectropolarimetry. The use of such detectors allows the immediate observation of spatial variations in one direction on the solar disc of whole Stokes profiles simultaneously obtained in different spectral lines (without any scanning).

Apparatus description :

The functional diagram of the Stokes-spectro-polarimeter is given in Fig.1. The photo-electric sensor of the camera is a 100 x 100 pixel RETICON array. A minification optics adapts the geometric scale of the spectograph to the dimensions of the array so that spectral lines can be observed in a wavelength range of 1.3 to 2.6 Å (at 6300 Å) along one spatial direction on the solar disc covering 50 to 100 arcsec. The polarization modulator for the measurement of the 4 Stokes parameters consists into two rotating achromatic λ/4 plates positionned with stepper motors and a fixed polarizer.

The camera and the modulator are under control of the electronic equipment which proces-ses the camera video signal in order to display immediately on a screen the Stokes profiles corresponding to the full spatial dimension covered by the detector. To meet this aim, the equipment uses two fast arithmetic processors and five memories, each one having a capacity of 10^4 16 bit words.

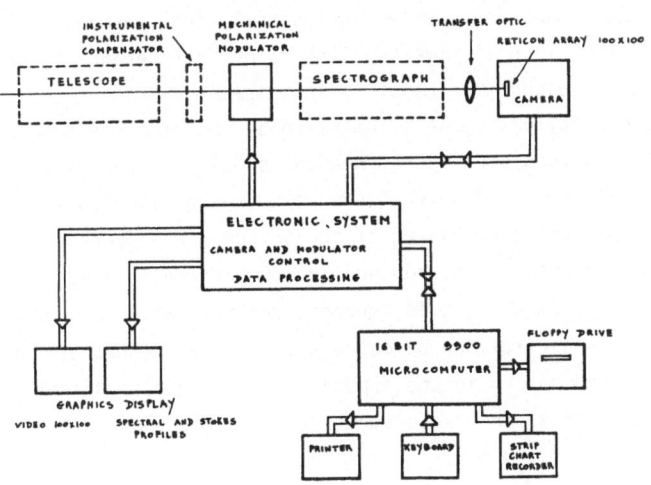

Fig.1 A block diagram of the spectropolari-meter optical and elec-tronic systems adapted to the Gregory coudé telescope of the Locar-no Solar Station.

The first memory is devoted to the continuous acquisition of new raw data. Two other memories are look-up tables for the real time correction of :

a) the dark currents of the array pixels

b) the differing responsivities of the array pixels and the fringing modulation due to interferences occuring mainly in the array window

The two last memories store the calculated Stokes profiles with the possibility to add a selected number of successive measurements in order to increase the signal to noise ratio.

A 16 bit microcomputer controls the acquisition and the recording of valuable data on floppy disks. Several programs stored in the microcomputer allow the observer to process further the acquired data in order to investigate immediately the properties of the observed magnetic structures. Left and right Zeeman components displacement, continuum and line core intensity, values of integrated V profiles can be visualized for inter-active observations.

Instrumental polarization

The telescopic birefringence mainly resulting from reflections on two plane mirrors induces cross-talk between U,Q and V Stokes parameters. This effect can be almost completely canceled with a Bowen compensator placed in front of the modulator. The possibility to visualize directly the Stokes profiles allows an easy fine adjustment of the compensator. For this purpose, structures with almost purely longitudinal magne-tic fields are observed near the center of the disc with a linear polarization modulator. The compensator is adjusted until the characteristic signature of the V profile disap-pears leaving a zero signal or a very small residual U or Q profile. Correction of other errors resulting from the telescopic polarization will be discussed later.

Real time correction of interference fringes :

Before starting a sequence of measurements in a selected spectral range, a calibration test is performed in order to correct in real time the effects of interference fringes (and at the same time the differing responsivities of the pixels).

During this test, each row of the array has to be illuminated by the same spectrum. For this purpose, the telescope is defocussed (and if needed, slowly drifted) and the alignment of the array is accurately adjusted so that the rows are parallel to the dispersion. The 100 spectra obtained at each reading of the detector are stored in the microcomputer. The averaged spectrum obtained by adding the 100 spectra is almost free from fringes and gain variations. By comparing this averaged spectrum with each stored spectrum, a correction factor for each pixel can be calculated. These factors are sent by the microcomputer to the dedicated memory in the apparatus allowing a real time correction of the video signal as long as the measurements are performed in the same spectral range.

A more accurate averaged spectrum can be obtained by correcting, before adding the 100 spectra with dividing factors derived from a flat field measurement. Such a flat field operation is performed by adding several successive spectral measurements obtained during a fast rotation of the grating. This method gives accurate results so long as the swept spectral range is free from strong lines.

Presentation of preliminary results :

We present some results obtained during an observation campaign performed in June-July 84 at the Locarno Solar Station of the Universitäts-Sternwarte Göttingen (with the Gregory coudé telescope). These results demonstrate the good functioning of the apparatus and show his many possibilities for the study of solar magnetic structures.

1. Asymetry of V profiles in network and plages.

All our measurements of Stokes V profiles in network and plages at good spatial and temporal resolution show pronounced asymetries. In order to quantify this asymetry we propose two parameters :

1. the peak ratio : $P.R. = \dfrac{\text{maximum amplitude blue polarization}}{\text{maximum amplitude red polarization}}$

2. the excess polarization: $E.P. = \dfrac{\text{area blue polarization} - \text{area red polarization}}{\text{area blue polarization} + \text{area red polarization}}$

The precise measurement of these two parameters is particularly difficult (in strong plages, the maximum amplitude of V is lower than 5% of the continuum intensity) and requires a careful evaluation of all instrumental errors affecting the V profiles.

So :

$$- V_{observed}(\lambda) = A.V_{true}(\lambda) + B.I(\lambda) + C + F(\lambda)$$

- the term $B.I(\lambda)$ originates from three different sources : the lower left term of the telescope's Muller matrix, the errors introduced by the polarization modulator and the brightness variations occuring during the observations.
- the term C mainly originates from ripple signal and reset noise on the detector and may therefore be different from row to row.
- the last term $F(\lambda)$ results from the different interference pattern for both positions of the $\lambda/4$ plate.

The evaluation of the two first error terms can be done by using the fact of vanishing Zeeman polarization in the continuum and by selecting the observed structures so that they always contain an at least small non magnetic region. The term $F(\lambda)$ constitutes the residual error which limits the accuracy of the measurements, its effect can only be reduced by averaging several rows.

After correction our measurements of P.R. and E.P. in enhanced network and plages for Fe 6301.5 (g=1.67) and Fe 6302.5 (g=2.5) show strong spatial variations : $1.2 < PR < 1.5$ and $0\% < E.P. < 15\%$

Fig.2a) represents the variation of the total polarization

$$\int_{\lambda_o}^{\lambda_o - \Delta\lambda} V d\lambda - \int_{\lambda_o + \Delta\lambda}^{\lambda_o} V d\lambda$$

λ_o : wavelength of V zero-crossing point

$\Delta\lambda$: chosen in order to evaluate the whole V area.

for lines 6301.5 and 6302.5. This measurement represents an one dimension spatial extension of an enhanced network. Such kind of observations performed simultaneously in different lines can bring information about spatial extension variations of magnetic structures with height. Such investigation has been performed with the Mgb2 line and the two neighbouring photospheric lines : no spatial extension with height has been noticed within the uncertainty range resulting from seeing and diode quantification (1.2 arcsec). The absence of telluric line traces on Fig.2b constitutes a check of the correct application of all correction terms on the observed V profiles.

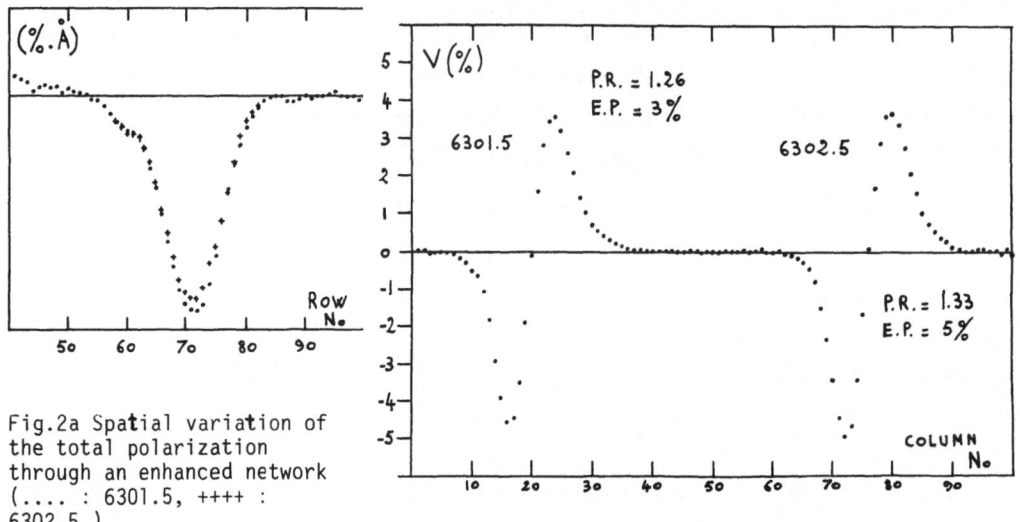

Fig.2a Spatial variation of the total polarization through an enhanced network (.... : 6301.5, ++++ : 6302.5)
1 diode row corresponds to 0.67 arcsec

Fig.2b Mean V profiles averaged over 4 rows (2.8 arcsec) corresponding to the peak on Fig.2a.P.R. and E.P. are indicated for both Fe lines. Distance between diode columns corresponds to 17.3 m Å

2. Mass motions inside fluxtubes

In order to investigate the important problem of velocity and velocity gradient within fluxtubes, we have recorded the wavelength position variations of opposite circular polarized profiles I+V, I-V and V zero crossing point through several magnetic structures. The positions of the profiles were determined using two moving windows of 70 mÅ width and 70 mÅ separation. Fig.3a and 3b concerning Fe 6302.5 show typical cases. It is worth noticing the following points : 1) in several regions (A,B,C on Fig.3a and E,F on Fig.3b), one polarized profile undergoes important position variations while the other remains practically at the same position 2) on Fig.3b, the mean position of the profiles within a strong plage is redshifted compared to the position of the profiles in the surrounding regions with no or small field. Such a shift, as well as the very typical pattern shown from rows 10 to 30 can be observed on practically all our measurements made at the border of strong magnetic regions. Fig.4 indicates the redshift of the V zero-crossing point referred to the mean position of the bisectrices for the two polarized profiles corresponding to region 3 of Fig.3a.

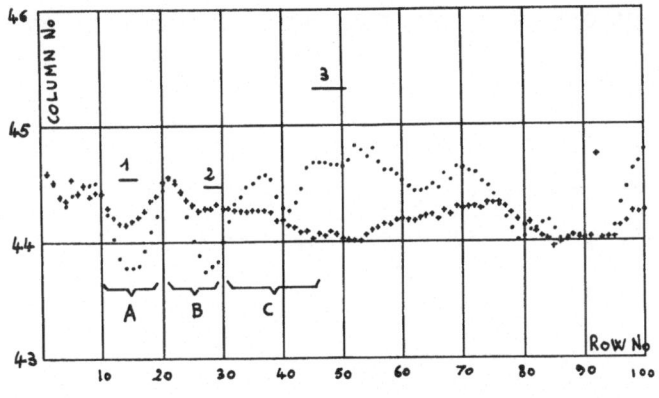

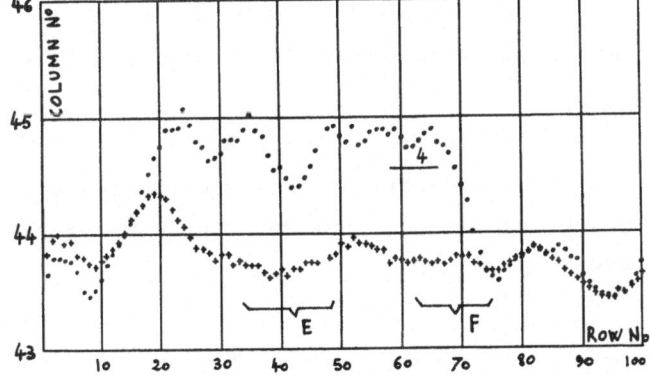

Fig.3a and Fig.3b

Spatial variation of the
wavelength position of profiles I + V (++++) and
I -V (....) of Fe 6302.5
The zero-crossing point of
profile V (averaged over
4 regions 1,2,3,4) is
indicated. The distance
between diode columns corresponds to 17.3 mÅ or
825 m/sec or 370 G. The
distance between rows corresponds to 0.67 arcsec
so that both observed
structures have a length
equal to 67 arcsec.

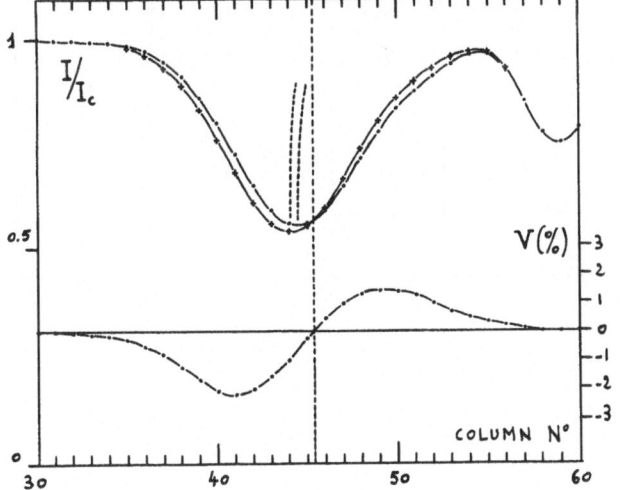

Fig.4 Mean profiles I + V
(++++), I - V (....) and V
corresponding to region 3
of Fig.3a (rows 45 to 51)

Acknowledgements

This work was supported by the
Fonds de la Recherche Fonda-
mentale Collective under
contracts 2.4518.75 and
2.4536.83. I would like to
thank the authorities of the
Locarno Solar Station (Universitäts-Sternwarte,
Göttingen) who made their
observatory's facilities available for operating this apparatus. During the observations campaigns, the help of Dr. E. Wiehr was particularly appreciated.

SOLAR TWO-DIMENSIONAL SPECTROSCOPY WITH UNIVERSAL BIREFRINGENT FILTERS AND FABRY-PEROT INTERFEROMETERS

D. Bonaccini, F. Cavallini, G. Ceppatelli, A. Righini.
Osservatorio Astrofisico di Arcetri, I-50125 Firenze, Italy

1.0 INTRODUCTION

The utility of high resolution monochromatic images of the solar atmosphere obtained with narrow bandpass filters (~ 20 mA FWHM) has been already well stated and demonstrated.

In this paper we briefly describe the project of a bidimensional solar spectrometer, working in the visible part of the spectrum (4200 - 7000 A), which might be a good approximation of the SUF (Super Universal Filter, Righini and Rutten, 1978).

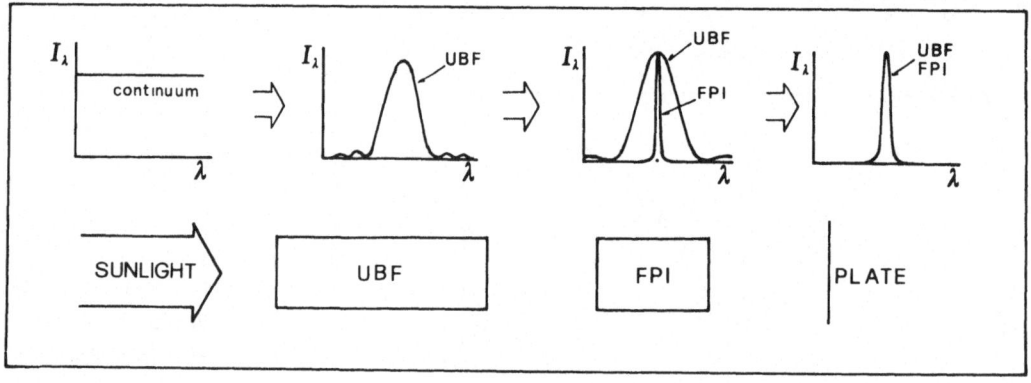

Fig. 1: basic idea underlying the project

The basic idea underlying the project is to use a Fabry Perot interferometer (FPI) in cascade with a tunable Universal Birefringent Filter (UBF) which acts as an order sorter of the FPI's transmission orders.

The advantage of this mounting is its compactness and its high spatial and spectral resolution. This scheme is sketched in Fig. 1.

We propose a telecentric optical configuration for both UBF and FPI. Thus the image field is spectrally homogeneous (i.e. bandpass shape and position does not change in the field), but the original instrumental profile of the interferometer is broadened. However a double passing of the light in the interferometer might increase the spectral purity of the images, lowering the characteristics "high wings" of the FPI passband.

The greatest problem which has to be solved is to maintain the passbands of both UBF and FPI tuned at the same wavelength.

2.0 THE UBF-ZEISS FILTER

The Zeiss Universal Birefringent Filter (UBF) is essentially a chain of 9 Lyot elements, each composed by a calcite or quartz retarder plate, a $\lambda/4$ achromatic waveplate, an input and an output linear polarizer. For a more detailed description see e.g. Beckers (1975).

Each Lyot element is independently tunable by rotating its input polarizer. The physical dimensions of the whole filter are 50x50.5x48cm. The input angular field is 2 degrees, while the output aperture linear dimension is 2.9 cm. in diameter. UBF is continuously tunable in the spectral range 4200 - 7000 A. The FWHM passband ranges from 90 mA at 4200 A to 250 mA at 7000A. The transmissions range from 4% to 16% in the same spectral interval. The wavelength setting time is less than 1 sec., and with a proper algorithm (November, 1983) an absolute spectral positioning error of 10 mA may be obtained, with a reproducibility of about 1mA.

3.0 THE INTERFEROMETER

Optically contacted piezoscanned interferometers are suitable for our purposes, provided that an appropriate plates gap control is provided. Such interferometers have already been successfully used in solar spectroscopy, see e.g. Cavallini et al. (1980) and Cavallini et al. (1982).

The parallelism of the plates may be monitored by optical and photometric systems like those based on "Brewster fringes" (Ramsey et al., 1966). It must be noted that the telecentric mounting of FPI's requires a high flatness of the interferometer; with flatness of $\lambda/200$-$\lambda/300$ it is possible to obtain a homogeneity of 3 - 2 mA of the bandpass position along the image field.

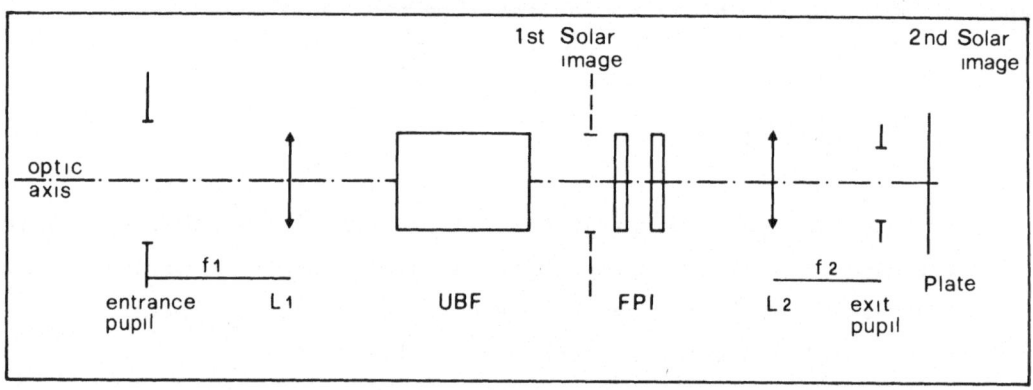

Fig. 2: Schematic layout of the optical interface

3.1 THE OPTICAL PATH

The telecentric optical interface of the system is shown schematically in Fig. 2. A lens (L1) collimates the pupil image, and gives an image of the sun between UBF and FPI. The field of view must be chosen in order to avoid vignetting. L2 transfers the image of the sun onto the revelator (a photographic plate or a photoelectric dice).

It is well known that the acceptance angle of FPI is lower than that of wide field birefringent filter. However the telecentric interface transfers this limitation in the image field, thus allowing to preserve the luminosity and the spatial resolution. For our system the following relation may be written: $\beta = \theta \cdot \Phi/\Psi$ where β is the angular field of the observed image θ is the angular beam aperture incident on the FPI Φ and Ψ are the linear diameters of the FPI and of the entrance pupil of the optical system.

Actual values calculated for the Kiepenheuer Institute Vacuum Tower Telescope on the Canary Islands give figures of β ~ 47 arcsec with Ψ = 60 cm., Φ = 2.8 cm. and θ = 17 arcmin.

3.2 THE WINGS OF THE FPI PASSBAND

It is known that the "wings" of the FPI bandpass are rather high, when compared with other spectral devices as Lyot filters or grating spectrometers.

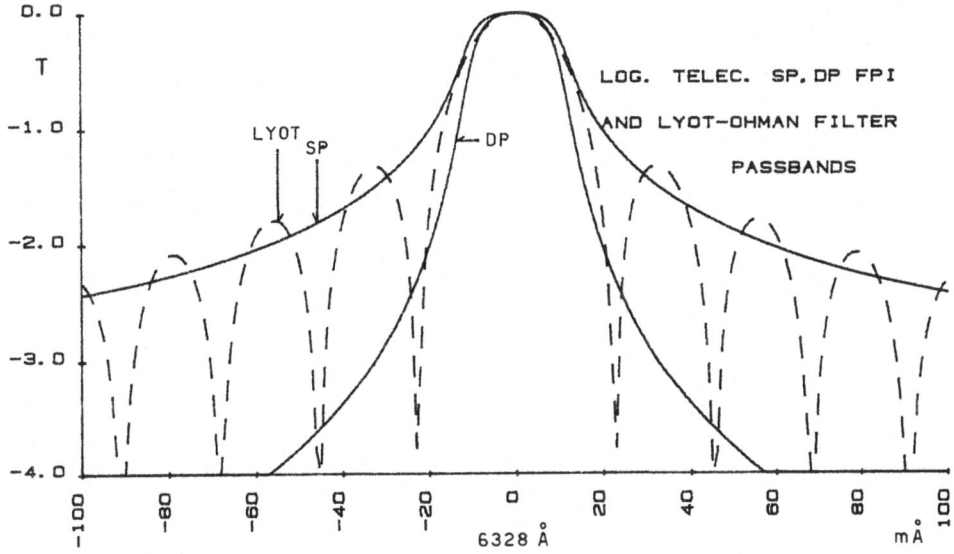

Fig. 3. Logarithm of the passband of a telecentric-single pass FPI, telecentric-double pass FPI and a Lyot Filter of the same FWHM of the S.P. FPI (20 mÅ)

To improve the instrument we propose the use of an FPI in double-pass configuration. Fig. 3. shows the logarithm of the passband of the proposed FPI in single pass mounting (highest wings curve), of a Lyot

Filter with the same FWHM (20 mA), and of the same FPI in double pass mounting (lowest wings curve). From this computer simulation it is clear that a double passage of the light in the interferometer lowers the FPI passband wings, improving its contrast and making it better, also with respect to an equivalent Lyot filter.

REFERENCES

Title, A.: 1979, Lunar and Planetary programs, NASA LMSC-D674593.
Falciani, R.: 1983, Solar Physics 89, 323.
Hernandez, G., Mills, O.A.: 1973, Appl. Optics 12, 126.
Cavallini, F., Ceppatelli, G., Righini, A.: 1980, Astron.&Astrophys 85, 255.
Cavallini, F., Ceppatelli, G., Righini, A.: 1982, Astron.&Astrophys. 109, 233.
Ramsay, J.V., Smartt, R.N.: 1966, Appl. Optics 5, 1297.
Righini, A. and Rutten, R.J.: 1978, Proc. of the JOSO Workshop, Florence
November, L.: 1983, Appl. Optics 23, 2333
Beckers, J.M., Dickson, L., Joyce, R.S.: 1975, AFCRL-TR-75-0090, Sac. Peak Instrum. Papers, n.27.

SOLAR HIGH RESOLUTION BALLOON SPECTRA
OBTAINED IN THE 190-300 NM WAVELENGTH BAND

D. SAMAIN, P. LEMAIRE

Laboratoire de Physique Stellaire et Planétaire
B.P. 10 Route des Gatines 91371 VERRIERES LE BUISSON CEDEX

ABSTRACT

A balloon gondola with a solar telescope spectrograph mounted on an equatorial platform has been launched from Aire/Adour (France) October 1, 1982. High quality spectra with a spectral resolution better than 1.5 pm and with a 4-5. arcseconds angular resolution have been recorded in the atmospheric window domains : 195-212 nm and 272-292 nm. Results from this first flight are discussed.

I   INTRODUCTION

Since many years, the increase in temperature of the outer atmosphere of the Sun remains unexplained and the failure in representing the heating mechanism as a global phenomenon has shown the interest of high spatial, spectral and temporal resolution. The objective of this experiment is the observation of the fine structures of the Sun in the earth atmosphere ultraviolet windows : wavelengths of this spectral range allow a simultaneous analysis of the solar layers (in continuum and in lines) from the top of the photosphere to the chromospheric plateau through the temperature minimum.
The LPSP has a long record of near ultraviolet solar spectroscopy using balloon gondola. During the 1968-1973 years several balloon flights have been performed with a two axis pointing system and with more increasing spectral and angular resolutions, (SAMAIN, 1971 ; LEMAIRE, 1969, 1971 ; LEMAIRE and SKUMANICH, 1973). In order to improve these results (2.5 pm and 5 arcseconds resolution in the last flights), the development of a new solar instrumentation and an equatorial mounting platform has been engaged few years ago. Hereafter the description of the instrumentation is given and results obtained during the first flight are presented. Some improvements of the instrumentation for future flights are now in progress.

II   INSTRUMENTATION AND PLATFORM

a - Instrumentation

The solar instrumentation is a telescope - spectrograph combination.
The cassegrainian type telescope has a 30 cm diameter aperture and a 6 meters equivalent focal length. A small portion of the field (6 arcminutes) goes to the spectrograph slit at the telescope focal plane. An other of the 32 arcminutes solar field is reflected by a plane mirror onto a motorized two axis translator with limb cells which are used as fine sensors to give the appropriated signal to the secondary

mirror actuators. The performance of such a system, as tested in laboratory, is able to get down to 0.1 arcsecond stability from a slow (below 5 Hz) ± 5 arcseconds jitter of the whole telescope.
In the focal plane the reflecting slit jaws send the slit image through lenses and Hα filter combination on a 100 x 100 CCD.
The spectrometer mounting is a cross dispersion echelle-Wadsworth mounting where the Wadsworth concave grating is used as the camera objective. This combination is stigmatic along the 30 arcseconds slit with .3 arcsecond angular resolution and along the 190-300 nm spectral range with 1.5 pm resolution. The beam issued from the concave grating is reflected on two multidielectric coated plane mirrors and collected on two photographic emulsions in the 195-212 nm and 272-293 nm wavelength ranges. The instrumental parameters are given in table I.

TABLE I

INSTRUMENTAL CHARACTERISTICS

Cassegrainian telescope	Theoretical	First flight
Primary mirror diameter	30 cm	22 cm
Diffraction limited resolution		
(500 nm - arcsec)	.4	.7

Spectrograph		
Collimator ≃ off axis parabola focal length	100 cm	100 cm
Echelle : Grooves/nm	79	79
Wadsworth grating		
Grooves/nm	600	600
Radius	400 cm	400 cm
Slit width	10 µm (3arcsec)	10 µm
Spectral resolution	1.5 pm	1.5, 2.0 pm
Spatial resolution	.5 arcsec	5. arcsec
Photographic Emulsions		SAI 101-01 KODAK

Slit jaw camera		100 x 100 CCD

b - The platform

The platform uses an equatorial mount. The gondola is stabilized along the local meridian with a magnetometer detector and an inertial wheel actuator. Then the two axis cardan system brings the instrumentation in right ascension and declination along the Sun-telescope axis using solar sensors. This stabilization system has, on the ground, a pointing capability of 2 to 3 arcseconds peak-to-peak jitter.

III FLIGHT AND DATA ACQUISION

The instrumentation has been flown from Aire/Adour (France) on October 1, 1982. The balloon has reached an altitude of 38 km (~ 3 mb pressure) where we made observations. During this first flight, through an aerodynamic process, a coupling appeared between the stabilization system and the swinging of the gondola at such a frequency that it cannot be filtered by the electronic and so limited the pointing jitter to a 4-5 arcseconds peak-to-peak amplitude.

Two sequences of spectra were taken near Sun center, one on a supergranular cell, the other one over an active network (or small facula) with exposure time ranging from 3 to 100 s in the 280 nm camera (SAI Kodak emulsion) and 30 to 900 s in the 200 nm camera (101-01 Kodak emulsion).

IV   FIRST RESULTS

Spectra recorded during the flight show the intensity variation related to the wavelength dependance of atmospheric transmission. In the 280 nm range the atmospheric transmission varies from 70 percent at 293 nm down to 0.01 percent at 272 nm whilst in the 200 nm range, outside the molecular oxygen absorption band, the atmospheric transmission is about 50%.

The spectral resolution is better than 2. pm in the 200 nm range and better than 1.5 pm in the 280 nm range. Calibration in intensity is made by comparison to rocket spectra. In the 200 nm range the spectra recorded by SAMAIN (1979) with 40 pm is used. Around 280 nm the rocket spectra obtained by KOHL et al. (1978) with 2.5 pm is the reference and the comparison is made in several wavelengths after adjustment to the same spectral resolution.

a. A sample of the results obtained in the 195-213 nm channel is shown in figure 1. It represents a portion of the Sun spectrum taken in a weak facula around the 208 nm Al I discontinuity. The continuum jump produced by ionization is higher than has been established before (≃ 4). The high spectral resolution permits to assign a higher value of the continuum intensity above 208 nm and we measure a discontinuity factor of ≃ 5. This result is in a good agreement with the theoretical computations (Vernazza et al., 1976) which predict a higher step that has been given by previous observations.

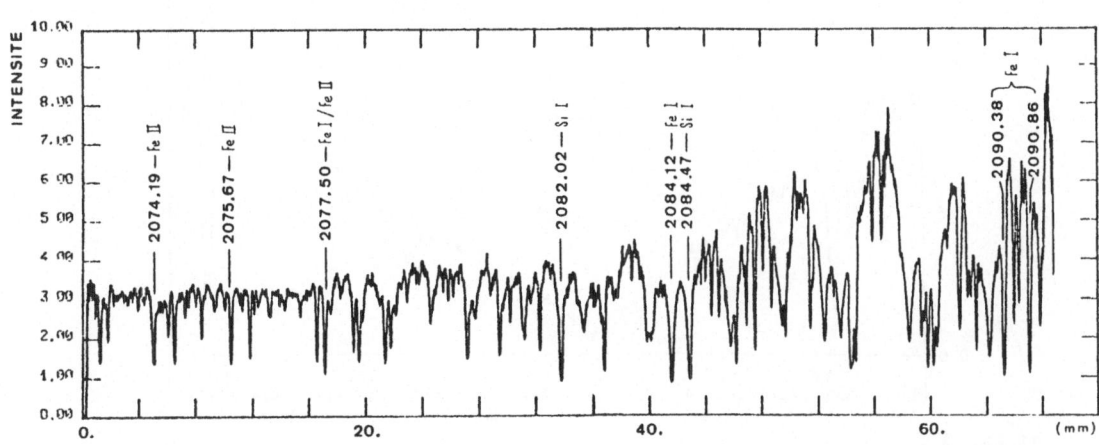

FIGURE 1
Solar intensity spectrum obtained in the 207.4 - 209.1 nm spectral range. The horizontal scale is the dispersion and the intensity is given in $erg/cm^2/s/$ Å $/sr$.

b. The 272-293 nm wavelength range.

In this part of the spectrum three strong depressions produced by strong absorption lines are detected :

- the 288.1 nm SiI line with a full width at half maximum (FWHM) of 0.6 nm and with wings contribution extending over more than 1.5 nm.

- the 285.2 nm MgI resonance line with a FWHM of 1.2 nm and wings extending over more than 3 nm.

- the 279.6-280.3 nm MgII doublet resonance lines with a FWHM of 1.9 nm and wings extending over more than 6 nm.

Figure 2 shows a comparison between the rocket spectrum obtained by Allen et al. (1978) and the balloon spectra in the vicinity of the 285.2 nm MgI line. The two spectra have the same spectral resolution, but because the balloon spectrograph has a greater dispersion and collects the data with a small granulary emulsion we obtain a better spectral quality. From the first data reduction that we have performed with the data collected over the supergranular cell at sun center we have not detected any evident variation of the shape and intensity of the MgI line across the slit whilst strong variations are seen in the MgII doublet lines. Higher angular resolution must give higher contrast within the fine scale structures.

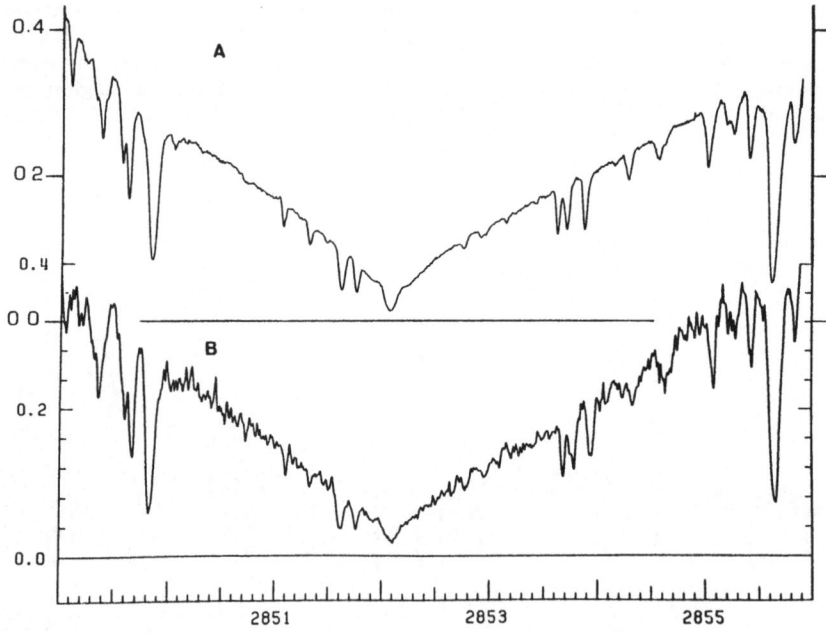

FIGURE 2
The MgI resonance line profile :
A. Observed by the balloon instrumentation
B. Observed by the Hawaï University balloon instrumentation (Allen et al., 1978). Wavelengths are in angstroms and the intensities are in arbitrary units.

V   CONCLUSION

From this first balloon flight we have obtained spectra with high spectral resolution and moderate angular resolution in the 190 nm–300 nm wavelength range outside the strong absorption ozone band. The data processing is in progress and some data calibration have been performed and shows the improve quality of spectra over ones previously obtained.

The main improvement which needs to be performed, is the angular resolution which depends on the quality of the pointing system. Study of such improvement is now in progress and we expect to go below the one arcsecond angular resolution during the next flight planned in 1985.

ACKNOWLEDGMENTS

The building of the instrumentation and the balloon equatorial platform has involved many people of the laboratory under the leadership of A. LABEQUE. Thanks for all the important contributions. The development of the experience has been performed under contract CNES 202-76-82 and the launch was made at the balloon CNES (Centre National d'Etudes Spatiales) facility of Aire/Adour. Data reduction is made using the PDS/CDSI microdensitometer and with the help of J. SALM, on the CNES CDC/Cyber 750 and the LPSP SEMS solar computer.

BIBLIOGRAPHY

ALLEN, M.S., Mc ALLISTER, H.C., JEFFERIES, J.T. : 1978,
"High resolution atlas of the solar spectrum 2678-2931 Å ", Institute for Astronomy-University of Hawaï

KOHL, J.L., PARKINSON, W.H., KURUCZ, R.L. : 1978,
"Center to limb solar spectrum in high spectral resolution 225.2 nm to 391.6 nm" HARVARD SMITHSONIAN Center for Astrophysics

LEMAIRE, P. : 1969, Astrophysical Letters, $\underline{3}$, 43

LEMAIRE, P. : 1971, "High resolution balloon borne spectrograph for the near solar ultraviolet in "New techniques in space astronomy", LABUHN and LUST eds., IAU symp. 41, 231 D. REIDEL, DORDRECHT

LEMAIRE, P., and SKUMANICH, A. : 1973, Astron. Astrophys., $\underline{22}$, 61

SAMAIN, D. : 1971, Compt. Rend. Acad. Sci., $\underline{273B}$, 1133

SAMAIN, D., : 1979, Astron. Astrophys., $\underline{74}$, 225

VERNAZZA, J.E., AVRETT, E.H., LOESER, R., : 1976, Astrophys. J. suppl., $\underline{30}$, 1

HUNDREDTHS OF ARCSEC RESOLUTIONS WITH NEW OPTICAL CORRECTORS ON DEEP U.V. PHOTORESIST

L. DAME and M. DECAUDIN
Laboratoire de Physique Stellaire et Planétaire
B. P. n° 10, Verrières le Buisson Cedex, France

ABSTRACT - This paper describes new optical correctors which are phase conjugated to a mirror and allow, in most cases, a residual distorsion of the outgoing wavefront as low as $\lambda/40$ (16 nm) peak to peak. Such a high resolution system is possible with Deep U.V. photoresist correctors registered at 257 nm in a diverging Michelson-Twyman interferometer <u>directly by</u> the mirror to be corrected itself, without any intermediary steps. Some results obtained with this technique are given. They show the potential interest of this method for optical imaging system in space involving large mirror and/or UltraViolet imaging.

1. THE ULTRAVIOLET RESOLUTION OF LARGE MIRRORS

When polishing a large mirror (diameter larger than 40 cm) there is always more or less residual deviations in between the final mirror surface and its desired initial shape. Deviations from the perfect surface can be divided according to the correlation length of the deviations, i.e. the lateral extension of the defects. Large size defects of frequency 1 or 2 cycles per diameter are mainly like classical geometrical aberrations, while medium size deviations in between 2 and 10 cycles per diameter (commonly called "ripples") introduce a more grating like effect. Very small deviations are the classical "roughness" which plays a diffusion role in the far UltraViolet.

For high resolutions, i.e. diffraction limited performances, it becomes important to introduce the mirror surface defects in the design analysis as soon as U.V. or even blue wavelengths have to be considered. In this wavelength range the degradation introduced by the mirror surface defects, mainly the ripples, is for most applications far larger than image motion of geometrical aberrations of the telescope combinaison. Examples of this influence can be found in WHETHERELL, W.B. (1982), DAME, L. and VAKILI, F. (1984) and DAME, L. (1983).

It is of paramount interest to remark that the same defect height (e.g. 16 nm peak to peak on the mirror surface, a surface quality of $\lambda/40$ at λ 632.8 nm) is much more dommageable to the resolution when its correlation length (number of defects per radius) is smaller. In other words ripples or zonal corrections degrade more the resolution than extended deviations. Figure 1 illustrate this point. The resolution (Normalized Spatial Frequency for a modulation transfer value of 0.25) is ploted as a function of the peak to peak defect height ω in λ units on the mirror surface for different correlation lengths, the defect topology beeing represented by concentric ripples simulated by a sinus function. Such a representation is suitable to illustrate most of the revolution defects generated by zonal polishing. We remark that for a correlation length of 0.25 (4 defects per radius), which is a common value, the degradation of the resolution between mirrors of quality $\lambda/8$ and $\lambda/4$ <u>is a factor 10.</u>

The Space Telescope mirror defect amplitude is $\lambda/20$ peak to peak at λ 633 nm (+ 15 nm) and its correlation length is in between 0.2 and 0.25 (ROBINSON, 1982 ; BÄCHALL and SPITZER , 1982). At λ 121 nm where it is intended to work its peak to peak quality is $\lambda/4$, and its resolution is more than 10 times less than its diffraction limited value. The reduction of a factor 2 of the mirror surface defect amplitude would improve the resolution by a factor 10 allowing a nearly diffraction limited resolution even in the far UltraViolet. This amelioration is however not possible by classical polishing techniques.

2. THE INTERFEROMETRIC CORRECTOR TECHNIQUE

We propose an interferometric technique which allow to register very high resolution correctors. These correctors are fixed correctors conjugated to a given mirror which allow diffraction limited performances in the far UltraViolet. Firstly intended for work in Space they might also be used in some ground application to accelerate the polishing, lower the cost and ameliorate the performance of large mirrors. The interferometric technique of registering could also be used to generate very special optics (astigmatism correctors, cylindrical lenses, etc) with an adequate choice of mirrors in the interferometer.

Interferences are obtained between a given large mirror with aberrations and a reference mirror in a Michelson-Twyman Interferometer with diverging light beams. The two mirrors are self collimated to their equivalent center of curvature. The form of the reference mirror is chosen in order to ensure a nearly perfect superimposition of the two wavefronts (cf. Figure 2). The interference pattern is a "flat tint" where the dominant intensity variations are the deviations of the large mirror surface from its perfect theoretical shape. The diverging Michelson-Twyman interferometer allow the use of a small reference mirror which can be realized to a far better quality than the large one. The phase delay between the two arms of the interferometer imply the use of coherent light. A good sensitivity and contrast of the interferogram pattern and the tremendous advantage of Deep U.V. Photoresists for the corrector medium imply the use of U.V. light. This particular U.V. coherent source is obtained through frequency doubling of an Argon ionized laser. The green line λ 514.5 nm is doubled in a cooled ADP Cristal. The U.V. light obtained at λ 257.2 nm is coherent on nearly one hundred meters and possess a power of several milliwatts.

The principal originality of the system is the corrector by itself. It is engraved by the incoming U.V. light in a deep U.V. Photoresist thin film coated on a high quality optical flat. Advantages are :

1 thin films (800 nm or less) are coated using a spinner with optical quality better than $\lambda/60$;
2 the photoresist response is linear for removed thicknesses less than 60 nm ;
3 roughness (micro irregularities) are less than 2 nm high ;
4 after treatment (i.e. development, rincing and drying) the photoresist relief is stable and ready for an U.V. coating (e.g. Al + MgF$_2$).

These 4 properties allow the thin film coating of deep U.V. photoresist to, directly, in a 1 step process, transform incoming intensity variations in thrue phase differences, i.e. in a physical relief representative of the mirror surface irregularities.

3. RESULTS

In a first attempt to show the feasibility of this correction technique, we have studied the optical quality of the photoresist films and the linearity of the photoresist. We have also realized the coherent U.V. source and a first interferometer to register correctors for a 10 centimeters spherical mirror of peak to peak surface defect of 46 nm. Finally, numerical simulations of the interferometer recording system were carried out in order to evaluate the intrinsic aberrations introduced on the wavefront by the interferometer in the astronomically more interesting case of parabolic mirrors.

3.1 PMIPK Deep U.V. photoresist properties

Sensitivity and linearity of the photoresist were discussed in DAME, L. et al. (1983). In that paper we mentionned that it could be suitable to use the photoresist 200 nm - 600 nm plage of removed thickness which is linear and more sensitive than the 0-60 nm plage. However the gain of sensitivity is payed by a strongly degraded resolution. This is mainly due to the fact that the solvent, evaporated in the near surface by the centrifugation and precooking of the photoresist, is still present in the deeper layers more fluid, more sensitive but much more inhomogeneous. If repeatability with more than 1 nm precision is possible in the range 0-60 nm, only 10 to 15 nm in the best cases can be achieved in the 200-600 nm range. This range is then clearly not suitable for high resolution optical correctors.

Optical flats of λ/60 peak to peak surface quality at λ 633 nm and of 31.5 mm diameter were coated using a spinner. The surface quality of the resists films were tested in a Fabry Perot by comparison to a reference flat. The fringe analysis shows that when the flat is well center and strictly parallel to the rotation axis, the surface quality is better than λ/60 . In this case, the photoresist deposit slightly compensate the residual aberration of the flat.

3.2 Interferometer tests and correctors

A self test of the interferometer was done using 2 spherical mirrors of same curvature radius to test the quality of the reference mirror and beam splitter assembly. The aberration of the system is 8 ± 4 nm peak to peak on the output wavefront. This is equivalent to, in the worst case, a λ/100 peak to peak surface quality for our reference system.

The mirror to correct was analyzed both by Foucault and by the Michelson-Twyman Interferometer fringes (Figure 3). The two results are in good agreement and the residual differences between the two can be attributed to the differential aberration introduced by the presence of the tilted beam splitter and the difference of curvature radius of the mirrors (cf. DAME, L., 1983).

A serie of corrector was registered with a limited success. Half amplitude correctors were done and studied in double reflection in the Michelson-Twyman interferometer with the mirror to correct. The surface of the corrector was also studied in the Fabry Perot. Both analysis showed effective compensation of the defect by the corrector but limited however to a factor of 2. The λ/14 mirror (46nm defect) was effectively ameliorate to λ/26 in the best case. This limit of λ/26 is imposed in our actual system by the vibrations of our interferometric bench and the slowly varying thermal gradients coupled with the weakness of the U.V. source whose output slowly decay with time (the ADP Cristal must stay in a perfectly dry ambiance otherwise it slowly melt). Dry Nitrogen permanent flow in the Cristal Assembly and hydrolic support of the bench are under development to overcome those non-fundamental limitations.

3.3 Numerical simulations of parabolic mirrors corrections

The reference mirror conjugated to a parabolic is an hyperbolic. This is true when the f ratio of the mirror is small. In the case of a 10 cm parabola at f/10 (focal of 1 m) the best reference hyperbola (of 3 cm only) determined by optimisation of the flat tint amplitude has (with a 2 x 10 mm beam splitter at 45°) an excentricity of 1.88 and curvature radius a = 236.4 mm and b = 376.3 nm. The nearest sphere (center-side fit) is at 25 nm. The residual aberration is an annulus of peak to peak amplitude on the output wave of 4 nm only. Slightly different curvature radius give a nearly as good defect amplitude but allow a deviation sphere-hyperbola much more important and which could be positive or negative. For a difference sphere-hyperbola of ± 0.5 μm, the additionnal aberration is less than 2 nanometers. This reference hyperbola can easily be done both by metallisation or elastic relaxation starting with a small and perfect spherical mirror.

In the case of a larger mirror (e.g. the Solar Optical Telescope primary) of diameter 1.25 m and of focal 4.5 m, the best reference is still an hyperbola but the larger f ratio (f/3.6) introduces the need to reduce a Schmidt like annulus aberration by an adequate metallisation of the hyperbola. This allow to use a still small hyperbola of 20–25 cm only but with an annular deformation of about 200 nm (with the 2 x 10 mm beam splitter a 15° incidence to obtain a perfect circular symetry for the deformation). The residual aberration on the wavefront can reach 16 nm (8 nm equivalent on the mirror surface) but this is perfectly acceptable as the correlation length is high (cf. curves of Figure 1).

4. CONCLUSION

The photoresist material is highly suitable for corrective optics ; it can be coated with surface qualities better than λ/60 peak to peak and is stabled during processing. A first serie of correctors was only a half success (amelioration of a factor 2 of a 10 cm mirror : λ/14 to λ/26) but not for fundamental reasons (blurring due to temperature variations and vibrations during a too long exposure time). Very high resolution optical correctors engraved in photoresists are however possible and numerical simulations show that even large parabolic mirrors can be adequatly corrected.

REFERENCES

BAHCALL, J.N. and SPITZER Jr., L. : 1982, The Space Telescope, Sci. Am. 247, 38
DAME, L., BONNET, R.M. and ARTZNER, G.E. : On the possible use of Deep U.V.
Photoresists correctors to obtain the ultimate U.V. Resolution of Space Borne
Telescopes, Proc. SPIE 446, to be published
DAME, L. : Correcteurs pour télescopes spatiaux obtenus par interférométrie sur
résines photosensibles au rayonnement Ultra Violet, D. ING. Thesis, Univ.
Paul Sabatier, TOULOUSE, France, October 27th 1983
DAME, L. and VAKILI, F. : 1984, The Ultra Violet Resolution of Large Mirrors via
Hartmann Tests and Two Dimensionnal Fast Fourrier Transform analysis, Opt. Eng.,
to be published (Nov/Dec 1984 issue)
ROBINSON, L.J. : An eye for tomorrow, Sky and Telescope 63, 1982
WETHERELL, W.B. : 1982, The effects of mirror surface ripple on image quality,
Proc. SPIE 332

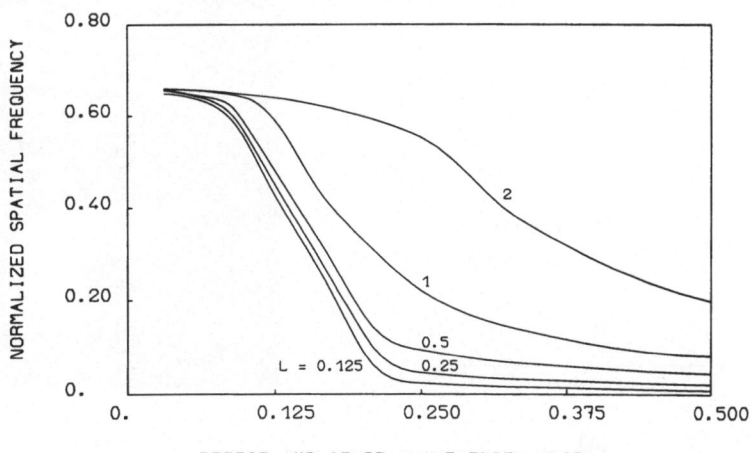

FIGURE 1 – Normalized Spatial Frequency in function of ω, the peak to peak defects amplitude on the mirror surface in λ units, and for different values of the correlation length (L, cycle frequency per radius)

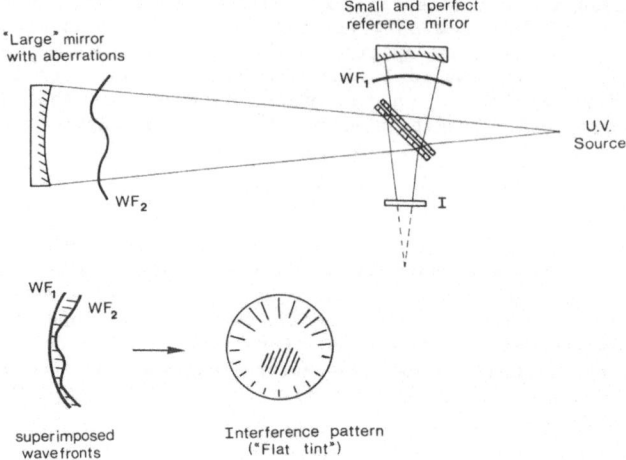

FIGURE 2 - Illustration of the corrector registering principle. Note that the large mirror with aberrations directly records its own corrector without any need for sophisticated measurements of the wavefront, the right adjustment of the interferometer being however ensure by the use of a U.V. λ 257.2 nm source. Note also that the corrector (interferogram plan I) is, by no need, of a given size or orientation. It can be tilted at any desired angle and be of any size (simply by moving it in the output converging/diverging beam) depending of its projected position in the final telescope or system where it is intended to work

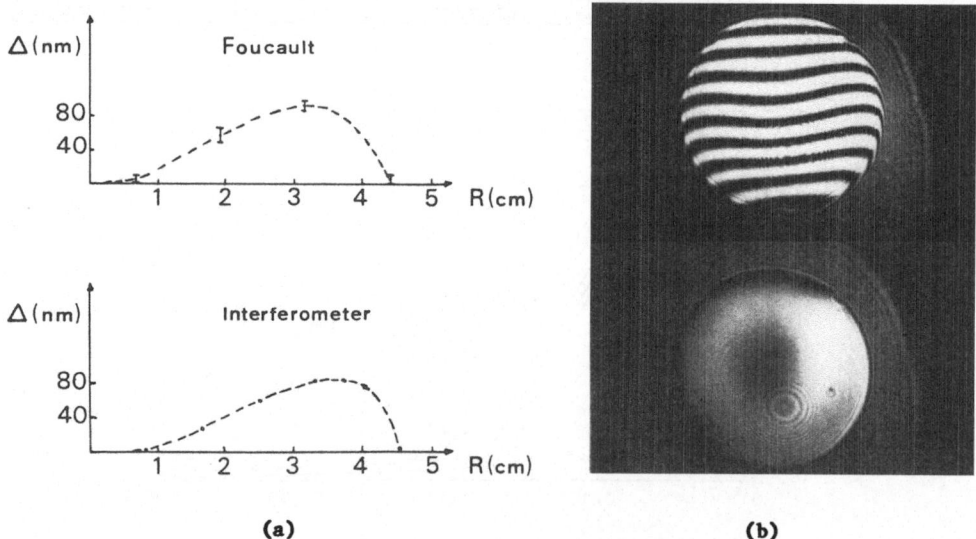

(a)   (b)

FIGURE 3 - a) Radial deformation of the test spherical mirror by Foucault and through the interferometer - b) Fringes and flat tint of the same mirror at λ 257 nm

OBSERVATIONS OF THE BIRTH AND FINE STRUCTURE OF SUNSPOT PENUMBRAE

M. Collados; J.I. García de la Rosa; F. Moreno-Insertis; M. Vázquez.
Instituto de Astrofísica de Canarias,
Tenerife,
Spain.

ABSTRACT.

High resolution white-light pictures of sunspot penumbrae are presented.
These include pictures showing details of their filamentary structure
and some instances of birth of a penumbra. The observations are discussed
in the framework of current penumbra theories.

INTRODUCTION.

After the flight of the Stratoscope (Bahng and Schwarzchild, 1961) the
well known theory of the structure of the sunspot penumbra as consisting
of convective rolls was put forward by Danielson (1961); this view was
in some aspects refined by Müller (1973), for whom the penumbra consisted of a pattern of bright granules moving towards the umbra and arranged
in elongated filaments on a dark background. In contrast with that view,
Moore (1981) has recently pointed out the possibility of individual dark
filaments emerging from the umbra and lying above the normal granulation.
Already Casanovas (1973) had arrived to similar conclusions based on observations carried out in his unpublished doctoral thesis. Additional
suppport to this theory of elevated dark filaments came from Giovanelli's
(1982) measurements on magnetic canopies extending from sunspots, and
partly, by the photometric study of Grossmann-Doerth and Schmidt (1981).
The Evershed and Wilson effects can be explained in the frame of this
theory, yet there remain important difficulties as to the mechanical
and radiative equilibrium of the elevated structures (Thomas, 1981;
Cram et al, 1981),

We present some pictures showing the evolution of the penumbra from the
first stages and two old spots having a peculiar penumbra in one part.
We think that they can help understand the penumbra structure. Some of
the frames were obtained during an observational program carried out in
the summer of 1980 to study the birth of sunspots (García de la Rosa,
1981)

ESSENTIAL FEATURES DERIVED FROM THE PHOTOGRAPHIC EXAMPLES:

I) The filamentary structure is clear from the first moment (see cases A, F and G, Fig.1). This is at variance with earlier descriptions of birth of the penumbra (Bray and Loughhead, 1964).

II) In the very first stages of its evolution the just born penumbra may not be stable. It may decrease or even disappear (see cases A and F)

III) In the majority of the cases observed, the penumbral filaments do not appear uniformly around the umbra, but rather by associations in the form of patches (see cases F, A and G). However some isolated filaments can sometimes be observed (see cases A and F).

IV) When a dark filament is first observed in our series it has about half its final length. The evolutionary stages prior to this first observation cannot be clearly made out in our pictures because of lack of resolution, yet we can set an upper limit of about one hour to the duration of this initial period. The further evolution of the filament occurs at a slower rate (see cases A, F and G).

V) A few dark filaments are seen in the pictures crossing each other (see case B, Fig.2). This can be taken as a strong indication that the filaments are in fact individual structures at different heights, a view which would give support to the theory of the elevated filament.

VI) In one case (see C and H, Fig.2), dark lanes on a light bridge in a spot are seen to have the same orientation as the dark filaments of the neighbouring penumbra. This can be interpreted as individual filaments crossing over the light bridge.

CONCLUSIONS.

A series of pictures have been presented, which give additional evidence of the existence of dark penumbral filaments as individual structures.

With respect to the birth of the penumbra some new observational aspects can be seen. The existence of the filamentary penumbra even in the first moments, its non uniformity and its short length are the major aspects derived from the pictures.

ACKNOWLEDGEMENTS.

We thank Monica Murphy and Miguel Briganti for their help in the edition of this contribution, and Prof. E. H. Schröter for providing very useful comments.

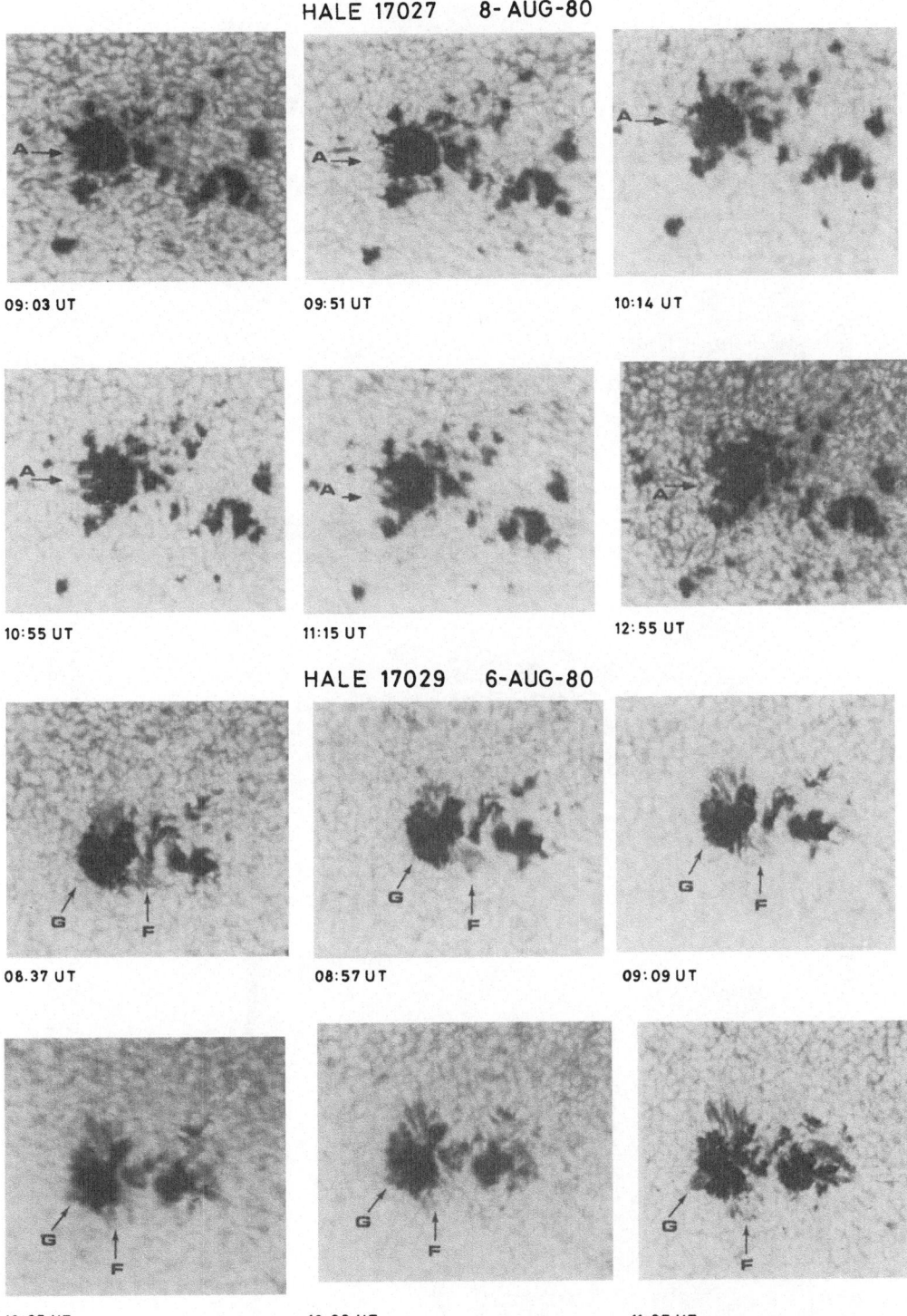

Fig.1. High resolution white-light pictures showing the birth of penumbrae

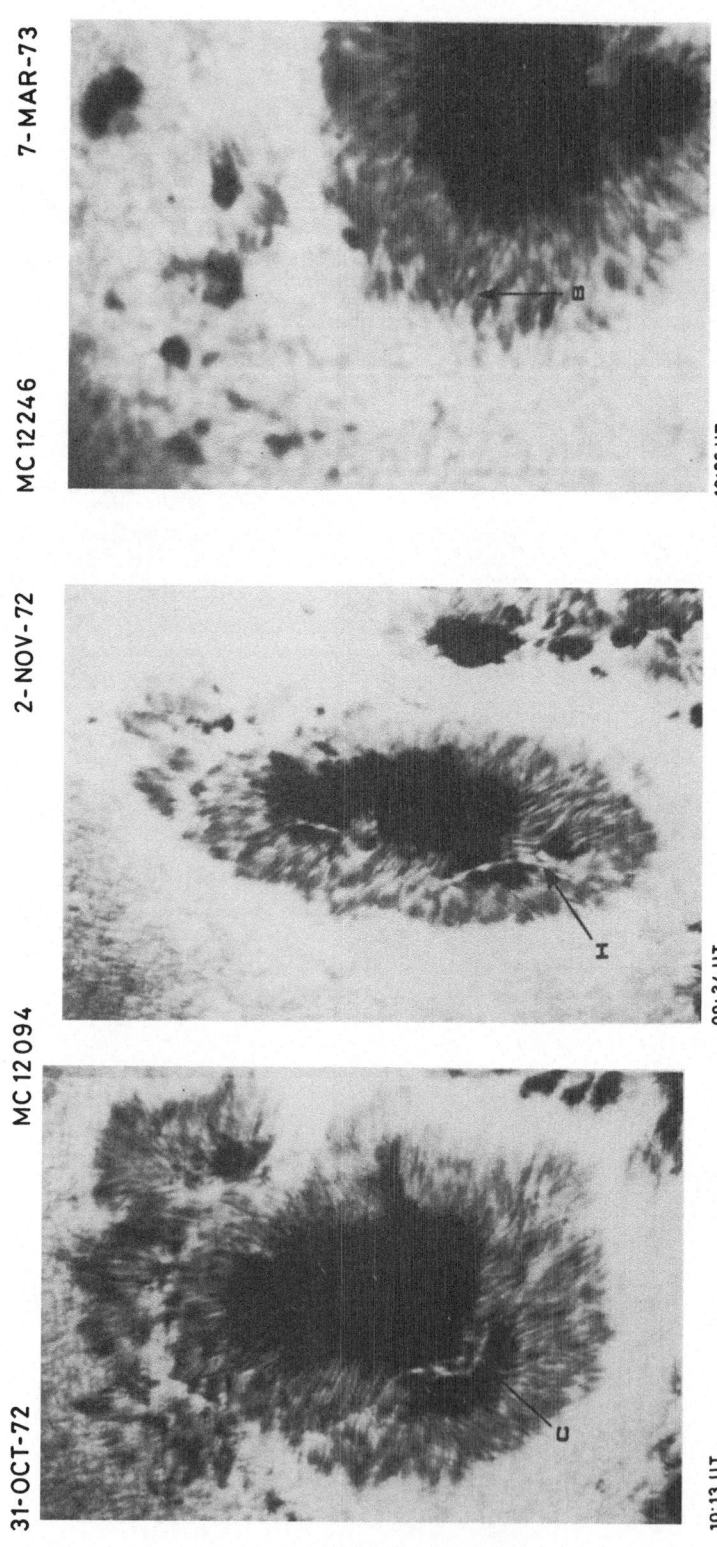

Fig. 2. Images of sunspot penumbrae containing indications (see arrows) of individual dark filaments crossing each other (B) or across a light bridge (C, H)

REFERENCES

Bray, R.J.; Loughhead, R.E.;:1964, "Sunspots", Edit. Chapman and Hall Ltd.
Bahng, J.; Schwarzschild, M.: 1961, Ap. J., 134, 337.
Casanovas, J.: 1973, PhD Thesis, Universidad de Barcelona
Cram, L.E.; Nye, A. H. ; Thomas, J.H.: 1981, "The Physics of Sunspots",
L.E. Cram, J.H. Thomas (eds)
Danielson, R.:1961, Ap. J., 134, 275.
Giovanelli, F.G.: 1982, Solar Phys., 80, 21
García de la Rosa, J.I.: 1981, Solar Phys., 74, 117
Grossmann-Doerth, U.; Schmidt, W.: 1981, Astron. Astrophys., 95, 366.
Moore, R.L.: 1981, Ap. J., 249, 390
Müller, R.: 1973, Solar Phys., 29, 55
Thomas, J.H.: 1981, "The Physics of Sunspots", L.E. Cram, J.H. Thomas (eds)

2. THE HIGH RESOLUTION STRUCTURE OF THE SUN

THE HIGH RESOLUTION STRUCTURE OF THE SUN

W. Mattig

Kiepenheuer-Institut für Sonnenphysik
Schöneckstr. 6, 7800 Freiburg, FRG

INTRODUCTION: In discussing the fine structure in the solar atmosphere
we open up an extremely wide field in solar research. Thus, it is im-
possible to review all the observations and their related problems in
any detail here. I can only touch upon some aspects of solar fine structure and have to restrict myself to a few problems. Fortunately, review
papers on most of the special aspects can be found in the literature.

Amongst other things, this means that I am unable to honour Prof. Rösch.
We all know the important contribution of observations made by him and
his colleagues at the Pic du Midi observatory to our present knowledge
of the fine structure in the solar atmosphere.

The sun is the only star, where we are able to observe the inhomogeneities of the physical parameters directly and there is no doubt that
other stars have similar, stronger or weaker, fine structure. This implies that the results of solar physics could be the key to a correct
and more detialed analysis of stellar atmospheres in general.

In this connection I only have to remind you of the relationship between granulation and convection. In the discussion about how chromosphere and corona are heated the probability of small scale magnetic
fields or fine structure playing an important role is more than zero.

Another aspect is important for astrophysics in general: the abundance
of the elements in stellar atmospheres. To determine this we have to
measure the intensities in different spectral lines. But the relation
between the measured intensity and the temperature is not linear which
means that one can not equate a temperature deduced from a spatially
averaged intensity and the mean temperature deduced from resolved elements with different temperatures. To determine the abundances we might
have to take into account the temperature fluctuation. Hermsen (1982)
has done this and found that for the sun the abundance of iron is un-

certain to the same order as oscillator strength, whereas of oxygen the
difference could be a factor of two when using homogenous models or
taking into account either temperature fluctuations or inhomogenous models. This is only one example of the importance of the konwledge of
solar fine structure.

What is the solar fine structure, or rather: which are the typical scales
when talking about fine structure? There is no consensus but I think
the scale length is of the order of arc seconds or less. Doubtlessly
a normally developed sunspot is not in the fine structure regime and
this applies to the oscillation too. But penumbral filaments as well
as umbral dots are typical fine structures.

On the other hand, there is resolution limit, set by the atmosphere.
Kneer (1978) has discussed the horizontal radiative exchange and the
possible loss of information before the photons reach the instrument.

All measurements of the solar surface in the range of about one second
of arc are influenced by the limited resolution of the instrument and,
in the case of ground based observations, the seeing.

I think it useful to consider this a little more quantitatively. In
general the resolution of an instrument is given by the Raleigh criterion. There is a "rule of thumb" for the visible region: an instrument
with a 10-cm aperture has a resolution of approximately 1 second of arc:
20 cm = 0".5; 1 m = 0".1. When comparing these values to the modulation
transfer function of the instruments we have to consider that they yield
about 10% only of the original contrast. If we accept a reduction of
the original contrast by a factor of two, we need a 25-cm instrument to
resolve 1 second of arc: 50 cm for 0".5; and a 1-m telescope for 0".25.
This has to be kept in mind when discussing qualitative data of solar
fine structure.

The second important source of distortion of the observation is the dis-
turbance by the terrestrial atmosphere: the seeing. Let me demonstrate
this with an extremely simple model. The intensity distribution of the
granular pattern be approximated by a two dimensional cosine function

$$\Delta I(x,y) = \Delta I_o \cdot \cos \frac{2\pi x}{L} \cdot \cos \frac{2\pi y}{L} \ .$$

L being the wavelength of the structure, the point spread function be
a Gaussian

$$P(x,y) = \frac{1}{2\pi S^2} \cdot \exp -\left[\frac{x^2+y^2}{2S^2}\right],$$

S being the dispersion parameter (2.4·S is the full half width). Assuming this we can easily solve the convolution integral. The result is

$$\Delta i = \Delta I \cdot \exp -\left[4\pi^2\left(\frac{S}{L}\right)^2\right],$$

Δi is the apparent, ΔI the true intensity difference. In table I the effect of atmospheric disturbances shows something remarkable:

Tab. 1: The ratio $\Delta i/1I$ as a function of S

L\S	0".0	0".1	0".2	0".3	0".4	0".5	0".6
2".5	1.000	0.939	0.777	0.566	0.364	0.206	0.103
2".0	1.000	0.906	0.674	0.411	0.206	0.085	0.029
1".5	1.000	0.839	0.496	0.206	0.060	0.012	0.002

a reduction of the amplitude by a factor of two corresponds to S = 0".3 (L = 2".5) or S = 0".2 (L = 1".5). But these values of seeing are obtained only under extremely favourable conditions.

When using a spectrograph (to measure velocities, etc.), an additional distortion comes into being: the finite slit width. Using high resolution spectrographs and focal length of more than 10 m the optimised slit width is of the order of 50-100 um, which means that with a solar image of about 30 cm diameter the slit width corresponds to 0".5. Frequently one enlarges the width in order to reduce the exposure or integration time. This magnitude is not small in comparison to the diameter of a granule close to the limb and probably larger than a filigree element or an umbral dot.

Keeping in mind all these difficulties it is not difficult to understand why there is no common picture of the fine structure. In my opinion the divergent results originate in the uncertainties of observation and hence the modulation transfer function.

GRANULATION

The small scale intensity fluctuation on the solar disc, the granulation, has been known for more than a hundred years. A few months ago the monograph "The Solar Granulation" by Loughhead, Bray and Durrant was published. For many details I refer you to this book and also to the gossary of Bruzek and Durrant (1977) and to the review article of Wittmann (1979) in the proceedings of the Freiburg colloquium. Unfor-

tunately the proceedings of the Sac Peak meeting in 1983 have not been
published to this day.

The mean distance between the granules or a characteristic cell size
has a wide scattering but a good mean value is 1."8-2."0 or 1400 km. This
has been challenged twice only: 1."4 by Namba and Diemel(1969); and 1."6 by
Hejna (1977).

Photometric measurements of single granules have been carried out by
Karpinsky (1981), the intensity profiles have been approximated by a
simple analytic function. The maximum horizontal intensity variation
corresponds to a horizontal temperature gradient of 5-10 K/km, compared
to $dT/dh(\tau =1) = 20$ K/km for the vertical gradient in homogeneous solar
models.

The evolution and life-time of the granules has been discussed in detail by Mehltretter (1978) and also by Altrock and Musman (1981). Using
the data obtained with the balloon-borne telescope "Spectrostratoscope",
the correlation coefficient of the intensity fluctuation as a function
of time can be expressed by an exponential function with a decay time
of 5.9 minutes, which is in agreement with a life-time of about 15 minutes.

Contrast: The intensity ratio between the bright granules and the intergranular lanes as a function of λ has been summarized by Bray (1982).
The observed general trend with λ is apparently in contradiction with
the expected wave length dependence when this ratio is a function of
the temperature difference only. But uncertanties are large and the observations have been performed by different observers using different
instruments at different times.

Alissandrakis et al. (1982) compared observed contrast measurements obtained during long time intervals, the result is extremely interesting
for the whole physics of convection and the solar cycle. Figure 1 shows
a possible cycle-dependence: at the maximum the contrast is larger than
at the minimum. This could be a second indication of a cycle dependent
structure of the convection zone, the first being Albregtsen and Malby
(1981) and Albregtsen et al. (1984) who have shown that the mean sunspot intensity depends on the phase of the solar cycle.

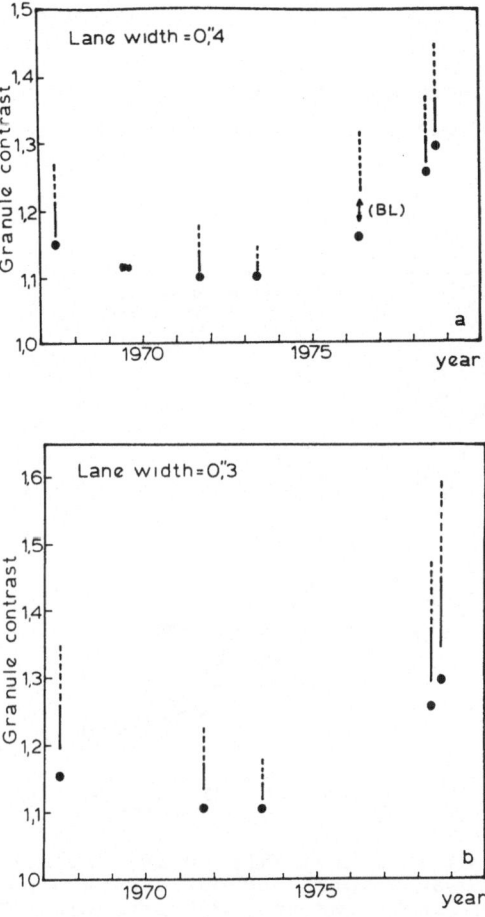

Fig.1. Time variation of granule-intergranule contrast at 5200 Å.
Filled circles represent observed (i.e. uncorrected) values,
solid bars the range of corrected values for the intersection
of three intergranular lanes, dashed bars the range of corrected
values for the middle of a lane. The corrected and uncorrected
measurements of Bray and Loughhead (1977) at 5500 Å are marked:
BL. (a) and (b) refer to assumed lane widths of 0.4 and 0.3,
respectively. Alissandrakis et al. (1982).

Intensity fluctuation: Determining the contrast or the intensity ratio
problems arise when defining the dark intergranular lane. But on the
other hand this is the simplest way of deriving the temperature difference of approxmately 300 K.

To measure the full intensity distribution well-known statistical methods
are available and here the most convenient parameters are power spectra
and rms-values. By transforming the data into Fourier space it is (math-

ematically) easy to correct the data in case the MTF is known.

Schmidt et al. (1981) have computed different MTF's by assuming one real power spectrum for the granulation and comparing observations. Fig. 2 shows the result

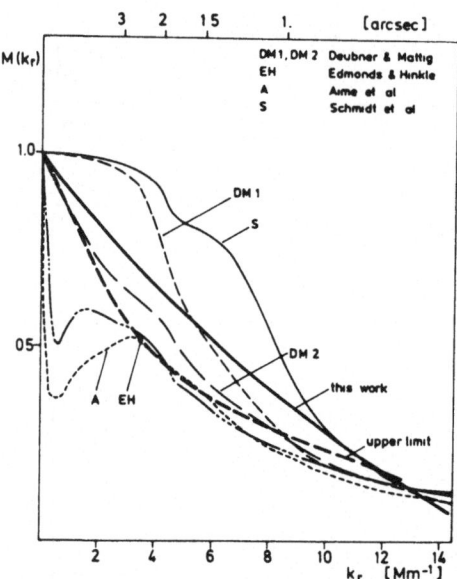

Fig. 2. Comparison of experimental MTFs derived from the various published true power spectra and our observed power spectrum PS1.
The full thick line shows the most probable MTF obrained in this
work. The broken thick line shows the MTF which we used to check
the upper limit of I_{rms}. Schmidt et al. (1981).

Because of Bray, Loughhead and Durrant's (1984) detailed discussion I will only review some recent data.

Due to the necessity of knowing the MTF in order to correct the data most recent observations have been carried out from balloon-borne tele-scopes, during partial eclipses or alternatively using a theoretical atmospheric MTF. In the first case no atmospheric disturbances are as-sumed to exist and in the second the moon's limb is used to determine the combined MTF of telescope and atmosphere. In a series of papers Ricort et al. (1979, 1981, 1982) and independently von der Lühe (1984) have developed a method of deriving an atmospheric MTF under the assump-tion of Korff's theory on atmospheric turbulence. The first method was also tested during a partial eclipse.

The result of careful new reductions of observations with the "Spectro-stratoscope" by Durrant et al. (1983) is seen in fig. 3. It depicts the

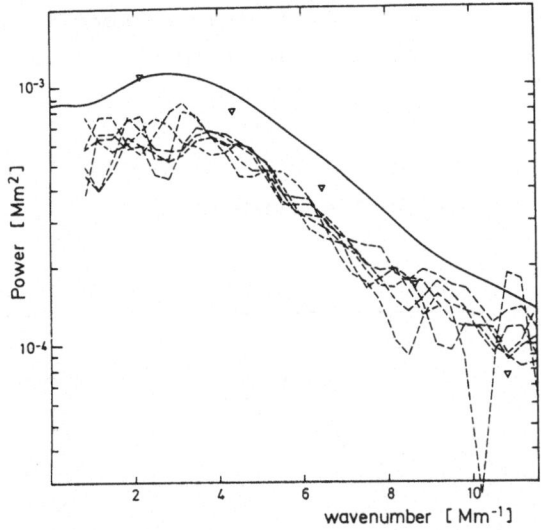

Fig.3. Two dimensional power per unit wavenumber suface element of the relative intensity fluctuations at the centre of the disk. The dotted lines show the six examples described here, the heavy line that of Ricort and Aime (1979), the triangles that of Deubner and Mattig (1975). Durrant et al. (1983).

two-dimensional power spectrum of the centre of the disc; the rms-value, the integration over the power spectrum is 11.3%. The upper curve is the result of Ricort et al. (1979): 17% rms. The difference is rather large but do pay attention to the fact that the smaller values have a limit of integration and that the largest value assumes a theory of atmospheric turbulence. I believe that, regrettably, these values àre within the limit of uncertainty. The centre-to-limb variation of the rms-intensity contrast is remarkably small.

In order to study the height dependence of granular structure Kneer et al. (1980) and Durrant and Nesis (1981) have analysed the intensity fluctuation in the wing of the Mg b 5172 line. There is a remarkable decrease of rms-value near the line centre and a strong decrease of correlation or coherence. This results in a model of T(h) with a steep gradient in the deeper layers and a constant (T) for higher levels. This is in agreement with the Nelson-Musman model (1977). Models of this kind are also well suited to describe the centre-to-limb variation. The existence of a strong T-gradient now seems well established.

C-shape: The direct methods of determination of temperature fluctuations as above mentioned are supplemented by another method, namely the
interpretation of carefully measured completely spatially averaged line
profiles; the slogan is C-shape. The line is asymmetric due to temperature differences between bright and dark elements as well as velocity
shifts.

In a recently published review article Dravins (1982) has discussed
the observations and the theoretical aspects.

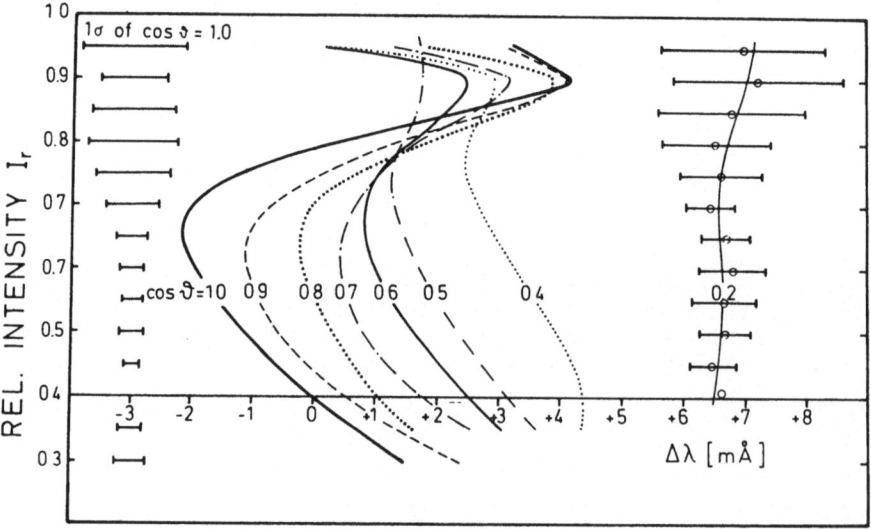

Fig. 4. The center-to-limb variation of the C-shape and line shift of
λ 5576. The horizontal bars indicate the rms variation of the
bisectors at cos ϑ = 1.0 and 0.2. Brandt and Schröter (1982).

Brandt and Schröter (1982) have measured the C-shape as a function of
cos ϑ and near the limb it is not detectable. This rendered evident
how to explain the limb shift as has been done by Schröter (1957)
25 years ago.

Kaisig and Durrant (1982) and Kaisig and Schröter (1983) have used a
solar model with perturbations of temperature and velocities. The
agreement with observation is fairly satisfactory as they had to assume a steep T-gradient as well.

At this stage new concepts enter the discussion: perturbations and correlations of T and V. The group in Naples (Caccin and Marmolino (1980),
Marmolino and Severino (1981), Buonaure and Caccin (1982)) have developed new methods for deriving the velocity fluotuation in the small
scale regime of the atmosphere.

But also experimental improvements have been made, mainly by eliminating
the instability of spectrographs. The Florence group Cavallini et al.
(1980, 1982) use a Fabry-Perot interferometer. The Freiburg group use
iodine absorption cells to determine exact wavelengths.

New aspects were introduced by Keil and Yackovich (1981) who attempted
to measure the bisectors of the element itself in highly resolved spectra. Here a possibility of studying the granules and the intergranular
space directly opens up. The important result is a T-scale height of
100 km.
In all of the above mentioned results the C-shape, that is the bisector of the absorption line was the charactreristic criterion for further
explanation. Recently Pierce (1984) proposed to determine the differences between line profiles and Gaussinas. A detailed theoretical analysis
of the utility of this proposal is necessary.

Velocities: In connection with the line asymmetries we had to take into
account small scale velocities. Again I will refrain from mentioning
more than just the most recent results. Beckers (1981) has contributed
"Dynamics of the solar photosphere" in the NASA and NCRS monograph
"The Sun as a star". As in the case of intensity fluctuations the problem of correction of instrument/atmosphere arises.

An additional problem has to be considered: the measured Doppler shifts
of different origin, viz. the granulation; the oscillation and the
supergranulation (velocities of possible smaller scales, the filigree,
are ignored). Here methods have to be found to separate the different
components.

The influence of the various quantities is seen very clearly in fig.5,
showing the correlation of intensity and velocity. If the granulation
is of convective origin we expect a clear correlation. The original
data yields poor correlation but filtering the data, in this case by
means of running means the correlation coefficient increases remarkably.

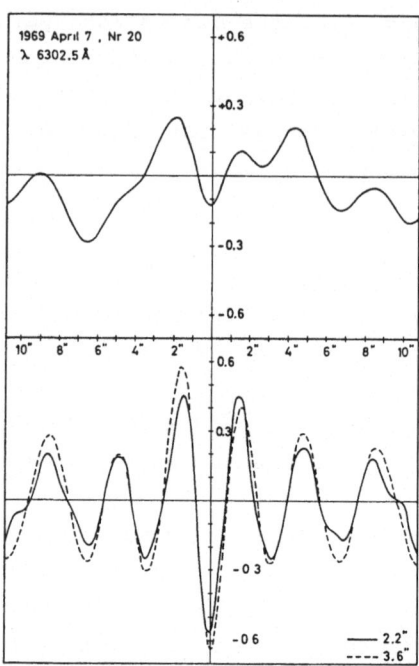

Fig.5. Crosscorrelation function of brightness and velocity.
Above: original values; below: residue from running mean with
(2n+1) = 2."2 and 3."6. Mattig et al. (1969).

The discussion of height dependance of velocity fluctuation introduces
yet another problem, namely radiative transfer. In order to determine
δ v(h) it is necesarry to derive a velocity weighting function and to
use a velocity model. Only by comparing this with observation the results become reliable (Canfield (1976)).

The most surprising result is the discrepancy between Canfield (1976)
and Durrant et al. (1979). Unfortunately Canfield had no empirical
knowledge of the MTF, whereas Durrant et al. analysed spectra obtained
during a partial eclipse. This has remained unresolved until this day.
Keil and Canfield (1978) and Keil (1980a,b,c,) confirm the steep velocity gradient, whereas Mattig (1980) and Bässgen and Deubner (1982) favour
a rather flat gradient. I believe this problem will not be solved until
observations with good knowledge of all distortions are performed.
The only two sets of velocity fluctuations as a function of position on
the solar disc by Keil and Canfield (1978) and Mattig et al. (1981) are
in good agreement. In both cases the largest fluctuations are away from
the solar centre. In the uncorrected data there is at first an increase

and closer to the limb a decrease. But please take into account the strong increase of corrections approaching the solar limb. From this centre-to-limb variation we have to conclude that the horizontal small scale velocity fluctuation is a factor 1.5-2 larger that the vertical. This was confirmed in an analysis of Kondrashove (1983) where unresolved spectra of 437 iron lines were used to calculate macroturbulent motions.

Models: In the last few years our knowledge of special details on the granulation has increased but we are still far away from a unique picture and a clear physical understanding.

Nelson and Musman (1977) and Beckers and Nelson (1978) have derived models with vertical and horizontal velocities but only the most recent Dravins et al. (1981) and Nordlund (1982) are beginning to arrive at a general picture of the granulation. Dravins presented a theoretical atmosphere model incorporating radiation coupled time dependent hydrodynamics of solar convection. His method is equally useful for interpreting stellar line profiles. Nordlund has simulated time dependent behaviour of granulation using basic hydrodynamical and radiative transfer equations.

But, gathering all the information on granulation I have the impression that much higher resolved observations are needed. We have to consider the structure and evolution of single granules (an analysis of Namba et al. (1983) is a step in the right direction). The statistical description can only be the first step towards an understanding of the phenemenon of granulation.

Granulation in the vicinity of sunspots: 30 Years ago Macris (1953) and later Schröter (1962) have shown, that the granular structure in the surroundings of large sunspots changes counsiderably. Macris (1979) has confirmed thus and has demonstrated a strong relation between the diameter of the granules and the umbral magnetic field strength. A remarkably larger velocity fluctuation in active regions was found by Nesis and Mattig (1974, 1976) and Kaisig and Schröter (1983). Brandt and Schröter (1984) reported a significant change of the C-shape which yields a difference in velocity-height scale compared to the undisturbed photosphere.

Mesogranulation: A new kind of pattern has been detected recently by

November et al. (1981), the so-called "mesogranulation". This fairly
stationary pattern of cellular flow has a characteristic scale length
of 5-10 Mm (7-14") and a spatial rms-velocity amplitude of about 60 ms^{-1}
only. The life-time of mesogranules appear to be at least 2 hours.

SUB-GRANULAR STRUCTURES

In the last ten years high resolution observations have revealed new
fine structures in the solar atmosphere. Filtergrams taken in photospheric lines or in the wings of strong lines, as Hα or CaII K and H,
show that the bright photospheric network is visible across the entire
solar disc, in both active and quiet regions. The network has been identified with white-light faculae by Mehltretter (1974) and with the solar
filigree, discovered by Dunn and Zirker (1973), one solar cycle ago.
The filigrees are embedded in the granular pattern.

These structures, these small bright points have a size remarkably less
than one arc sec. Coming back to the introductary remarks about the disturbances of fine structures, this means that it is extremely difficult
to observe these fine structures. Large telescopes and good seeing conditions are needed to identify and to analyse these features.

All quantitative results have considerable uncertainties. Koutchmy (1978)
came to the conclusion that the mean filigree intensity is two times
the averaged granular intensity which means that the filigree-element is
1000 K hotter than the surroundings. But because of the difficulties in
the restoration, fig. 6, he pointed out that he does not believe that
he has obtained the true value of the intensity.

Recently Müller and Keil (1983) have repeated such measurements and
found a value in the range of 1.3 to 1.5 of the mean photospheric value
(for the characteristic intensity of the facular points - or filigree.)
For the typical size in the quiet photosphere they found 0.2 ≙ 150 km.
To insure a proper identification of facular points they observed calcium pictures with a pass-band of 15 Å too. In contradiction to this,
Spruit and Zwaan (1981) find a much wider size distribution but in active regions and by using Mg b$_1$ filtergrams.

Independent of the intensities and the widths Muller (1983) pointed out
that:

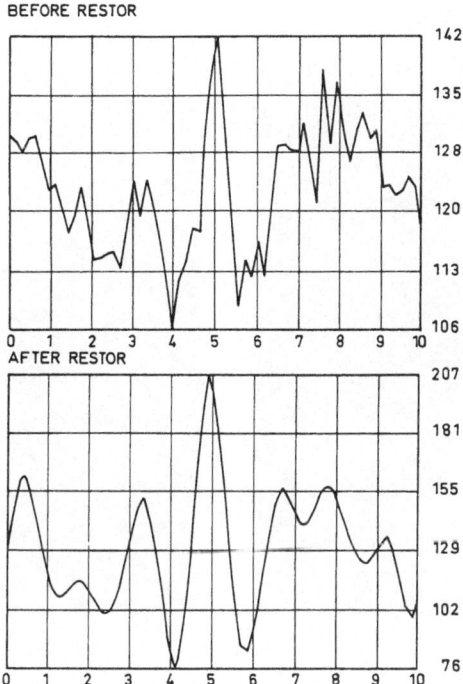

BEFORE RESTOR

AFTER RESTOR

Fig.6. Intensity of a filigree structure. Upper part shows the original
       data and the lower half the result of the deconvolution. Ordina-
       te are linear with respect to arbitrary units. Koutchmy (1978).

"(a) the facular points appear at the supergranular boundaries rarely
     inside the cells;
 (b) they appear in spaces in the junction of several granules, never
     inside a granule or the space between two granules only;
 (c) the mean life-time is 18 min.;
 (d) they remain in intergranular lanes during their whole life".

In addition to this, P. Wilson (1981) concludes again that white light
faculae, the Hα filigree and the calcium bright points are different
manifestations of the same photospheric features by comparing observations in Ca K, Mg b_1, Hα + 1.5 Å and Hα continuum.

Kitai and Muller (1984) have studied the relation between chromospheric
and photospheric fine structures in active regions. They confirm the
result by Wilson (1981) that the bright points in the Hα wing, the filigree are cospatial with facular points.
In addition to Wilson, Kitai and Muller suppose that faint moustaches

(a)   λ3933   (b)   λ5700

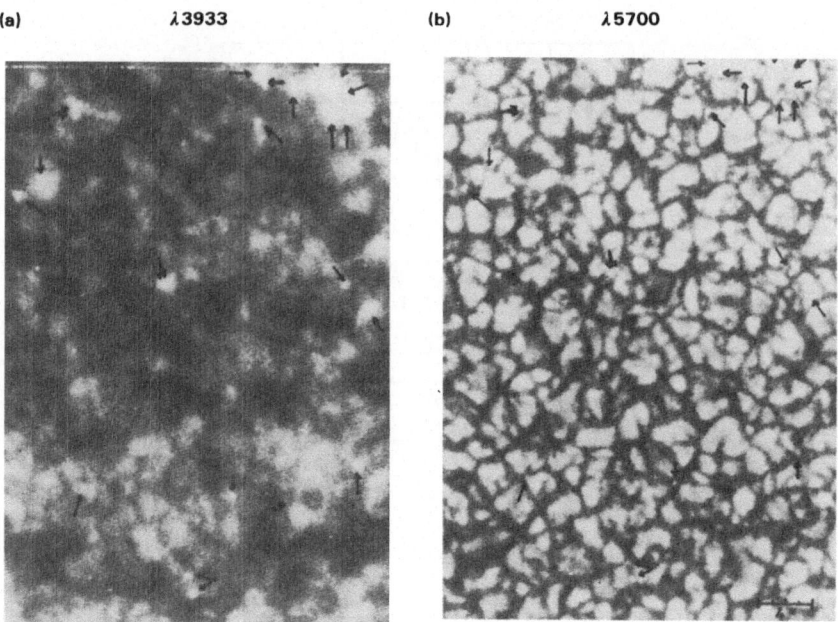

Fig.7. A subset of the observed regions is shown in both calcium and
in whight light. The location of some facular points are indicated with arrows.

and filigree have a common physical origine. Using the new 60 cm domeless Solar Telescope Hida, Kurokawa (1982) has studied the features of
Ellermann bombs or moustaches. The apparent length is 1.1 arc sec while
80% have a diameter of less than 0.6 arc sec. The life-time is about
12 min but the first maximum in the life history was attained in about
2 min. A mean upward velocity will be around 8 km s^{-1} which is in good
agreement with a spectral analysis of line profiles by Kitai (1983).

It has been well established that chromospheric plage and the active or
bright network as shown on Ca II K_{232} spectrograms, are cospatial with
magnetic fields. The observations of small scale photospheric magnetic
fields are reviewed by Harway (1977). Here I would like to notice that
the diameters of the flux tubes with 1" to 2", observed in spectra by
Simon and Zirker (1974), are remarkably larger than the crinkle or filigree elements.

In addition to this Muller (1981) pointed out that strong magnetic
fields concentrated in small tubes are associated with photospherics

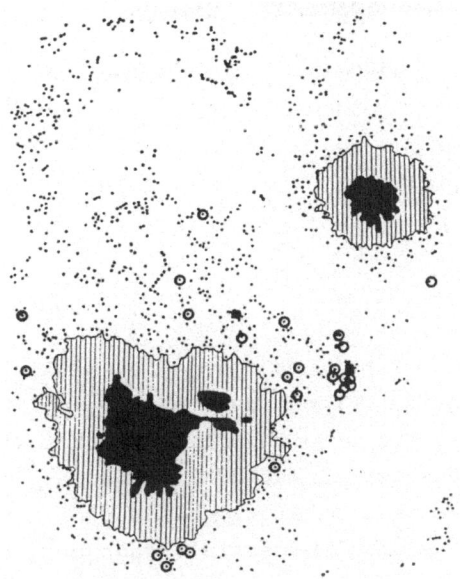

Fig.8. Spatial distribution of the bright points in the Hα - 0.75 Å
filtergram. Small dots indicate the location of the bright
points. The bright points associated with facular granules are
indicated by encirclements. Kitai and Muller (1984).

bright points.
Daras-Papamargarites and Koutchmy (1983) have studied quiet sun magnetic network elements using spatially highly resolved spectra and confirmed that strong magnetic fields are found at the location of the network element. The corresponding magnetic field structure has a size of
1-3 arc sec.

By using magnetograms in the red iron line λ 8688 Å Sivaraman and
Livington (1982) have shown that weak magnetic elements also lie at the
roots of the K_{2v} points. They pointed out that it is now clear that
there is a one to one correspondence between bright points and the magnetic structure in the interior of the network.

The three-dimensional structure of the magnetic field, especially the
distribution of the field with height has been discussed by Jones and
Giovanelli (1982) and Giovanelli and Jones (1982). A field configuration
which forms horizontal magnetic canopies over (lower level) field-free
regions was discussed extensively.

I would like to mention this one paper only: Tarbell and Title (1982)
have pointed out that observations with a resolution better than 0".5
show offsets between intensity and magnetic maps of faculae predicted
by magnetic fluc tube models. Beyond this a variability in the relation
between downdrafts speed and magnetic flux - and a relationship between
granulation and flux tubes of different sizes are illustrated.

SPACE OBSERVATIONS

During two rocket flights Bonnet et al. (1980, 1982) have obtained filtergrams in Ly α and in the UV-continuum at 160 nm. The resolution is
better than one arc sec. For Ly α this is the first one arc sec filtergram. The two filtergrams have a marked difference: broader structures
in Ly α are obvious. Another rocket flight in the same 160 nm region
by Cook et al. (1983) was also successful.Bruckner (1980) has reported
on earlier high resolution observations from space on different occasions for example at the joint discussion during the IAU meeting 1979.
Concerning the XUV structure of active regions I refer you to Dere (1982)
who has discussed all the structures.

A few remarks on the statistical analysis by Foing and Bonnet (1984a,
1984b): In the temperature minimum region (160 nm) three characteristic
scales are found: the small scale; and intermediate scale and the plage
area scale. The intermediate scale of 8000 km could be related to the
meso-granulation or associated with the oscillation.

The dimensions of the small scale cell grains are found to be of the
order of 600 km or 0".9, the mean distances being 1400 km. The network
and plage elements are brighter and broader, in agreement with other
ground based results. The temperature excess of the grains is in the
range of 80-360 K. In addition to this Cook et al. (1983) have found
that in the cell centers the bright points are highly variable on a
1 min time scale.

On the basis of OSO 8 observations Dumont et al. (1982) and Mourdian
et al. (1982) have studied the structure and physics of solar faculae,
especially in the transition region between the chromosphere and the
corona, by analysing the profiles of SiII, SiIV, CIV and OVI lines.
The high resolution EUV structure above a sunspot was discussed by
Nicolas et al. (1982). The derived typical umbral model for the transi-

tion region.

Earlier observations are discussed during the IAU-meeting in 1979. A
high resolution view of the Chromosphere and Corona was given by
Brückner (1980). The results from "Skylab Solar Workshop III on Active
Regions" are published by Orrall (1981).

SUNSPOTS

Penumbra: The characteristic scale of the penumbral fine structure is
noticeably smaller than the mean distances of the granules. The progress
of collection of more accurate observationed results as well as the
progress of the theoretical understanding of sunspots in general has
been discussed in the summer of 1981 at the Sac Peak Observatory and
is published in the proceedings of this meeting. Two contributions by
Moore (1981) and Muller (1981) are of special interests in connection
with penumbral fine structure. In addition to this I want to report on
a determination of the brightness distribution by Großmann-Doerth and
Schmidt (1981). The intensity distribution is almost symmetrical and
single peaked which implies that this is in disagreement with a two
component model of the penumbra. The rms intensity fluctuation is 13%
and the widths of the features is not smaller than 250 km or 0."35. This
is in agreement with another determination of the width of penumbral
filaments by Bonet et al. (1982) who used observations performed during
a partial eclipse. Their mean value is 0."37. Both values are also in
agreement with an earlier measurement by Muller (1973).

From the discussions on 0."2 resolution photographs obtained with the
65 cm gregorian telescope at Big Bear Moore (1981) came to the conclusion that many, and possibly most dark fibrals, lie above and partially
mash lower bright structures. This is in contradiction to previously
results and means that the dark component of the penumbra is not physically analogous to the intergranular lanes in the photosphere.

Measurements by Tonjes and Wöhl (1982) confirmed the old results by
Muller (1973) that penumbral bright grains move towards the umbra but
the maximum proper motion is not near the umbral border, as Muller
pointed out. The life-time of about 3 hours in the middle of the penumbra is confirmed by these measurements.

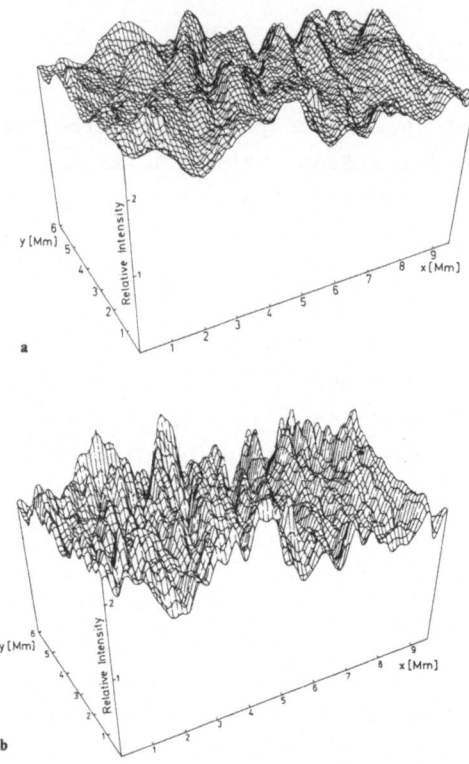

Fig.9. Intensity contours of a penumbral area before a and after b image
restoration. Großmann-Doerth and Schmidt (1981).

New efforts to obtain more spectroscopic information of the penumbral
fine structure have been carried out by Stellmacher and Wiehr (1980,
1981). From nearly resolved spectra only they came to the conclusion
that assuming the bright features at rest, the outflow in dark regions
is nearly 3 km s^{-1}. That is the Evershed-effect. Looking to the magnetic fine structure they found no systematic fluctuations of the magnetic field strength of the order of 1000 gs. Highly resolved spectroscopic observations are needed to effectively attack the problem of the
Evershed effect.

Umbral dots: In a comprehensive paper Bumba and Suda (1980) discribed
the inhomogeneity in a sunspot umbra "as a dark network of cellular
elements, the center of each being filled in by a bright grain. This
structure does not differ from the network of the photospheric granules,
only the apparent widths of the dark intergranular space grows and its

darkness deepens with increasing intensity of the magnetic field".

Disputing this Loughhead, Bray and Tappere (1979) came to the conclusion that the two patterns, photosphere and umbra, appear progressively
more dissimilar as the resolution is improved.
I will not discuss this here in any detail but I think it more appropriate to talk about umbral dots or grains or finestructure.
By using the name umbral granulation we are tempted to identify these
features with a convective phenomenon and I think this is still an open
question.

A characteristic scale length of 1''2-1''4, no changing of the general
structure with the evolution of the spot and no change with the solar
cycle are the main results of Bumba and Suda (1980). The life-time is
in the range of 10-30 min. Loughhead, Bray and Tappere (1979) find a
typical separation in the range of 0''9-1''3, with an apparent diameter
of 0''4-0''5.

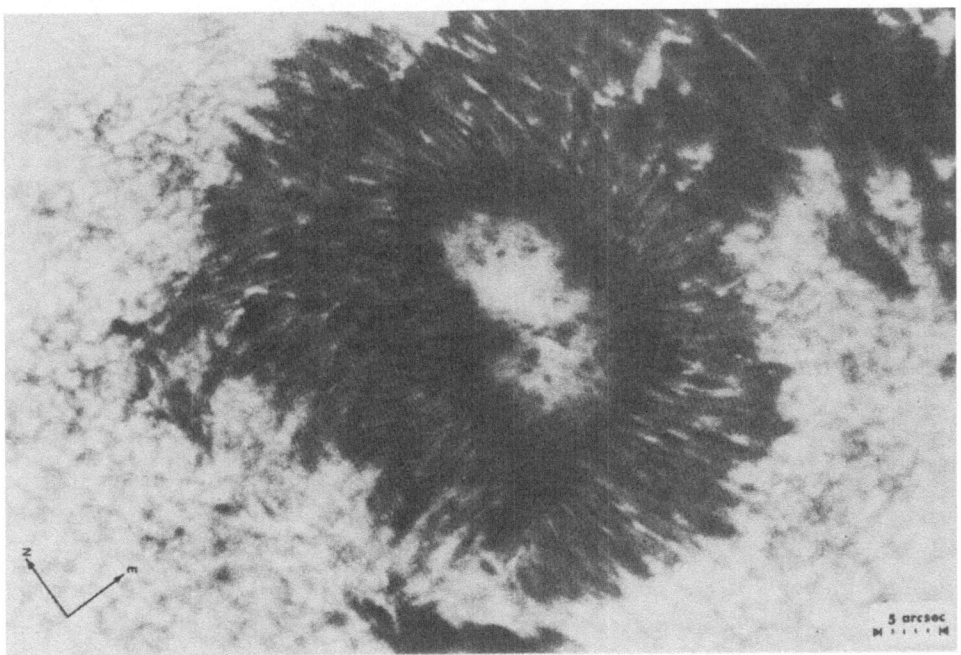

Fig.10. Negative (sunspot-core) - positive(penumbra and photosphere)
        compose the sunspot studied by Adjabshirzadek and Koutchmy
        (1980). Compare the size of umbral dots and the size of photo-
        spheric granules.

Adjabshirzadeh and Koutchmy (1980) derived a larger apparent half-width, but the life-time of 26 min is in agreement with Bumba and Suda's result as well as with the value of the pioneeing work by Beckers and Schröter (1968).

Abdussamatov (1980) noticed that the true corrected mean brightness of the bright features is smaller than the nondisturbed photosphere. This was confirmed by Parfinenko (1981): for the mean contrast of bright points points as referred to the photosphere a value of 0.69 is given after correction for diffraction. Only for bright points on the boundary of the penumbra the contrast is hihger and reaches 1.14. This is in contradiction to a photometric analysis by Koutchmy and Adjabshirzadeh (1981). Using pictures in two wavelengths they conclude, that the intensity of the dots is the same as the intensity of the photosphere. The corrected mean diameter was calculated to 190 km or 0.25. In center-to-limb variation studied by Adjabshirzadeh and Koutchmy (1982) the bright points show no foreshortening like granular structure.

Let us return to the morphological structure. Soltau (1982) pointed out that there is a close relation between bright penumbrae filaments and umbral dots.

In an unpublished paper Großmann-Doerth, Schmidt and Schröter (1984) distinguished between umbral dots in a more or less undisturbed central umbra and dots near the penumbrae border. In any case the temperature is less than the photospheric value, for the central umbral dots only in the range of 4400-5000 K.

Unfortunately new spectroscopic information is extremely rare. Adjabshir-zadeh and Koutchmy (1983) studied a sunspot umbra and a bright umbral dot. A two component model was presented. The temperature distribution is nearly in agreement with photospheric models, the dark component does not differ essentially from older umbral models which is illustrated in the λ-dependence of the umbral intensity.

One remark in connection with the magnetic field. Sattarov (1982) has shown that in a cluster of umbral dots the Fe-line λ 6303 Å shows a double splitting only, whereas in the other region three components are visible. This means that the field is longitudinal in the dot's cluster.

Aknowledgement: I am greatful to Mr. A. Hessenbruch for his kind help at preparing the manuscript.

161

References:

Abdussamatov, H.I.: 1980, Soln. Dannye No. 11, 99.
Adjabshirzadeh, A., Koutchmy, S.: 1980, Astron. Astrophys. 89, 88.
Adjabshirzadeh, A., Koutchmy, S.: 1982, Solar Phys. 75, 71.
Adjabshirzadeh, A., Koutchmy, S.: 1983, Astron. Astrophys. 122, 1.
Albregtsen, F., Joras, P.B., Maltby, P.: 1984, Solar Phys. 90, 17.
Albregtsen, F., Maltby, P.: 1981, Solar Phys. 71, 269.
Alissandrakis, C.E., Macris, C.J., Zachariadis, Th.G.: 1982, Solar Phys. 76, 129.
Altrock, R.C., Musman, S.: 1981, Bull. Amer. Astr. Soc. 13, 879.
Bässgen, M., Deubner, F.L.: 1982, Astron. Astrophys. 111, L1.
Beckers, J.M.: 1981, in "The Sun as a Star", SNRS and NASA monograph, page 11.
Beckers, J.M., Nelson, G.D.: 1978, Solar Phys. 58, 243.
Beckers, J.M., Schröter, E.H.: 1968, Solar Phys. 4, 303.
Bonet, J.A. Ponz, J.D., Vasquez, M.: 1982, Solar Phys. 77, 69.
Bonnet, R.M., Bruner, E.C., Acton, L.W., Brown, W.A., Décaudin, M.: 1980, Astrophys. J. 237, L47.
Bonnet, R.M., Bruner, M., Acton, L.W., Brown, W.A., Décaudin, M., Foing, B.: 1982, Astron. Astrophys. 111, 125.
Brandt, P.N., Schröter, E.H.: 1982, Solar Phys. 79, 3.
Brandt, P.N., Schröter, E.H.: 1984, in Sacramento Peak workshop, in press, private communication.
Bray, R.J.: 1982, Solar Phys. 77, 299.
Bray, R.J., Loughhead, R.E., Durrant, C.J.: 1984, The Solar Granulation, Cambridge University Press.
Brückner, G.E.: 1980, Highlights of Astron. 5, 557.
Bruzek, A., Durrant, C.J.: 1977, Illustrated Grossary for Solar and Solar-terrestrial Physics, Reidel, Dordrecht.
Bumba, V., Suda, J.: 1980, Bull. Astron. Inst. Czechosl. 31, 101.
Buonaura, B., Caccin, B.: 1982, Astron. Astrophys. 111, 113.
Caccin, B., Marmolino, C.: 1980, Astron. Astrophys. 83, 73.
Canfield, R.C.: 1976, Solar Phys. 56, 239.
Cavallini, F., Ceppatelli, G., Righini, A., Barletti, R.: 1980, Astron. Astrophys. 85, 255.
Cavallini, F., Ceppatelli, G., Righini, A.: 1982, Astron. Astrophys. 109, 233.
Cook, J.W., Brückner, G.E., Bartoe, J.-D., F.: 1983, Astrophys. J. 270, L 89.
Daras-Papamargaritis, H., Koutchmy, S.: 1983, Astron. Astrophys. 125, 280.
Dere, K.P.: 1982, Solar Phys. 75, 189.
Deubner, F.L., Mattig, W.: 1975, Astron. Astrophys. 45, 167.
Dravins, D.: 1982, Ann. Rev. Astron. Astrophys. 20, 61.
Dravins, D., Lindegren, L., Nordlund, A.: 1981, Astron. Astrophys. 96, 345.
Dumont, S., Mouradiau, Z., Pecker, J.-C.: 1982, Solar Phys. 78, 71.
Dunn, R.B., Zirker, J.B.: 1973, Solar Phys. 33, 281.
Durrant, C.J., Mattig, W., Nesis, A., Schmidt, W.: 1983, Astron. Astrophys. 123, 319.
Durrant, D.J., Mattig, W., Nesis, A., Reiss, G., Schmidt, W.: 1979, Solar Phys. 61, 251.
Durrant, C.J., Nesis, A.: 1981, Astron. Astrophys. 95, 221.
Foing, B., Bonnet, R.M.: 1984 a, Astrophys. J. 279, 848.
Foing, B., Bonnet, R.M.: 1984 b, Astron. Astrophys. 136, 133.
Giovanelli, R.G., Jones, H.P.: 1982, Solar Phys. 79, 267.
Grossmann-Doerth, U., Schmidt, W.: 1981, Astron. Astrophys. 95, 366.
Grossmann-Doerth, U., Schmidt, W., Schröter, E.H.: 1984, private communication.
Harvey, J.: 1977, Highlights of Astron. 4, part II, 223.
Hejna, L.: 1977, Bull. Astr. Inst. Czech. 28, 126.

Hermsen, W.: 1982, Astron. Astrophys. 111, 233.
Jones, H.P., Giovanelli, R.G.: 1982, Solar Phys. 79, 247.
Kaisig, M., Durrant, C.J.: 1982, Astron. Astrophys. 116, 332.
Kaisig, M., Schröter, E.H.: 1983, Astron. Astrophys. 117, 305.
Karpinsky, V.N.: 1981, Soln. Dannye No. 1, 88.
Keil, S.L.: 1980 a, Astrophys. J. 237, 1024.
Keil, S.L.: 1980 b, Astrophys. J. 237, 1035.
Keil, S.L.: 1980 c, Astron. Astrophys. 82, 144.
Keil, S.L., Canfield, C.: 1978, Astron. Astrophys. 70, 169.
Keil, S.L., Yackovich, F.H.: 1981, Solar Phys. 69, 213.
Kitai, R.: 1983, Solar Phys. 87, 135.
Kitai, R., Muller, R.: 1984, Solar Phys. 90, 303.
Kneer, F.: 1978, in proceedings of the JOSO workshop "Future Solar Op-
tical Observations", page 204, Florence.
Kneer, F.J., Mattig, W., Nesis, A., Werner, W.: 1980, Solar Phys. 68,
31.
Kondrashova, N.N.: 1983, Soln. Dannye No. 1, 89.
Koutchmy, S.: 1978, Astron. Astrophys. 61, 397.
Koutchmy, S., Adjabshirzadeh, A.: 1981, Astron. Astrophys. 99, 111.
Kurokawa, H., Kawaguchi, I., Funakoshi, Y.: 1982, Solar Phys. 79, 77.
Loughhead, R.E., Bray, R.J., Tappere, E.J.: 1979, Astron. Astrophys.
79, 128.
v.d. Lühe, O.: 1984, J. Opt. Soc. Am. A, 1, 510.
Macris, C.: 1953, Ann. d'Astrophys. 16, 19.
Macris, C.J.: 1979, Astron. Astrophys. 78, 186.
Marmolino, C., Severino, G.: 1981, Astron. Astrophys. 100, 191.
Mattig, W.: 1980,Astron. Astrophys. 83, 129.
Mattig, W., Mehltretter, J.P., Nesis, A.: 1969, Solar Phys. 10, 254.
Mattig, W., Mehltretter, J.P., Nesis, A.: 1981, Astron. Astrophys. 96,
96.
Mattig, W., Nesis, A.: 1974, Solar Phys. 38, 337.
Mattig, W., Nesis, A.: 1976, Solar Phys. 50, 255.
Mehltretter, J.P.: 1974, Solar Phys. 38, 43.
Mehltretter, J.P.: 1978, Astron. Astrophys. 62, 311.
Moore, R.L.: 1981, Astrophys. J. 249, 390.
Moore, R.L., 1981, in The Physics of Sunspots, Proceedings Sac. Peak
Obs., page 259.
Mouradian, Z., Dumont, S., Pecker, J.-C., Chipman, E., Artzner, G.E.,
Vial, J.C.: 1982, Solar Phys. 78, 83.
Muller, R.: 1973, Solar Phys. 29, 55.
Muller, R.: 1981, in Proceedings of the Japan-France Seminar on Solar
Physics, page 142.
Muller, R.: 1981, in The Physics of Sunspots, Proceedings Sac. Peak
Obs., page 340.
Muller, R.: 1983, Solar Phys. 85, 113.
Muller, R., Keil, S.L.: 1983, Solar Phys. 87, 243.
Namba, O., Diemel, W.E.: 1969, Solar Phys. 7, 167.
Namba, O., Hafkenscheid, G.A.M., Koyama, S.: 1983, Astron. Astrophys.
117, 277.
Nelson, G.D., Musman, S.: 1977, Astrophys. J. 214, 912.
Nicolas, K.R., Kjeldseth-Moe, O., Bartoe, J.-D.F., Brückner, G.E.: 1982,
Solar Phys. 81, 253.
Nordlund, A.: 1982, Astron. Astrophys. 107, 1.
November, L.J., Toomre, J., Gebbie, K.B.: 1981, Astrophys. J. 245, L 123.
Orrall, F.O.: 1981, Solar Active Regions, Colorado Associated University
Press.
Parfinenko, L.D.: 1981, Soln. Dannye No. 2, 96.
Pierce, A.K.: Solar Phys. 90, 195.
Ricort, G., Aime, C., Deubner, F., Mattig, W.: 1981, Astron. Astrophys.
97, 114.
Ricort, G., Aime, C.: 1979, Astron. Astrophys. 76, 324.
Ricort, G., Borgnino, J., Aime, C.: 1982, Solar Phys. 75, 377.

Sattaroo, I.: 1982, in W. Fricke a. G. Telebi (eds.), Sun and Planetary Systems, p. 129.
Schmidt, W., Knölker, M., Schröter, E.H.: 1981, Solar Phys. 73, 217.
Schröter, E.H.: 1957, Z. Astrophys. 41, 141.
Schröter, E.H.: 1962, Z. Astrophys. 56, 183.
Simon, G.W., Zirker, J.B.: 1974, Solar Phys. 35, 331.
Sivaraman, K.R., Livingston, W.C.: 1982, Solar Phys. 80, 227.
Soltau, D.: 1982, Astron. Astrophys. 107, 211.
Spruit, H.C., Zwaan, C.: 1981, Solar Phys. 70, 207.
Stellmacher, G., Wiehr, E.: 1980, Astron. Astrophys. 82, 157.
Stellmacher, G., Wiehr, E.: 1981, Astron. Astrophys. 103, 211.
Tarbell, T.D., Title, A.M.: 1982, Bull. Amer. Astron. Soc. 14, 924.
Tönjes, K., Wöhl, H.: 1982, Solar Phys. 75, 63.
Wilson, P.R.: 1981, Solar Phys. 69, 9.
Wittmann, A.: 1979, in proceedings of the Freiburg colloquium "Small Scale Motions on the Sun", page 29, Freiburg.

Discussion

Righini: You have shown the effect of MTF on the contrast of fine structures, however the instrumental stray light must be taken into account, especially in instruments in which many reflections are involved.

Mattig: That is correct. The MTF should include the stray light, in particular the instrumental part.

E. Wiehr: Filigree easily cause "abnormal granulation" already at very slightly reduced seeing conditions. Might the reduced rms contrast of such "abnormal granulation" explain: a) the dependence on cycle via variation of filigree population? b) the dependence on spot magnetic field via density of filigree as function of field strength c) might "abnormal granulation" possibly yield different granule sizes?

Mattig: As far as I know there is no observational data available on either of your questions.

R. Muller: A comment on the Ron Moore photograph of sunspot penumbra: The quarter of a sunspot shown suffers from a strong atmospheric astigmatism which artificially smoothes the structure along the filament.

FINE STRUCTURE AND EVOLUTION OF SOLAR GRANULATION

D. Dialetis, C. Macris, Th. Prokakis, E. Sarris.
Astronomical Institute, National Observatory of Athens.
Athens, 11810, P.O. BOX 20048, GREECE.

Abstract

We studied some aspects of the temporal and morphological evolution of
the fine structure of the solar granulation.

1.   Introduction.

This paper is a contribution to the study of photospheric granulation,
based on a large number of excellent photographs of the solar photosphere taken by S.Koutchmy at the Sacramento Peak Observatory with the 75
cm tower telescope on September 25, 1975 near the minimum of the solar
cycle. The average time resolution between successive series of about
18 pictures is 15 sec. The mean spatial resolution is better than 0.3".
From each series of 18 pictures we have chosen the best one for our study.
We measured the mean life time, and studied the morphological evolution
of granules, the different types of birth and decay the relation between
area and brightness, and the time evolution of the area for different
types of granules.
Time-dependent theoretical models of the granulation have only recently
begun to be investigated, an exact knowledge of the manner in which granules involve
and how it varies e.g. with granule size could throw useful light on
the physics of the non-linear interactions. For these purposes we used
photometrical and photographical methods.

2.   The mean life time of granulation.

Many investigators (Macris 1953, Bray and Loughhead 1958 Rösch and

Hugon 1959, Bahng ans Schwarzschild 1961, Macris and Prokakis 1963)
have determined, by using different methods, the mean lifetime of granules.
The main difficulty in the determination of the lifetime of a granule
is to determine the terms "birth " and "death". The adopted criteria
are based on the study of the morphological evolution of granules.

The criteria are the following :
1. A granule is born when it is formed into the intergranular region
from the merging of two or more smaller granules, or from the division
of a biger one.
2. A granule dies when it is diluted into the intergranular region
when it is divided to two or more smaller granules or when it merges
with other granules.

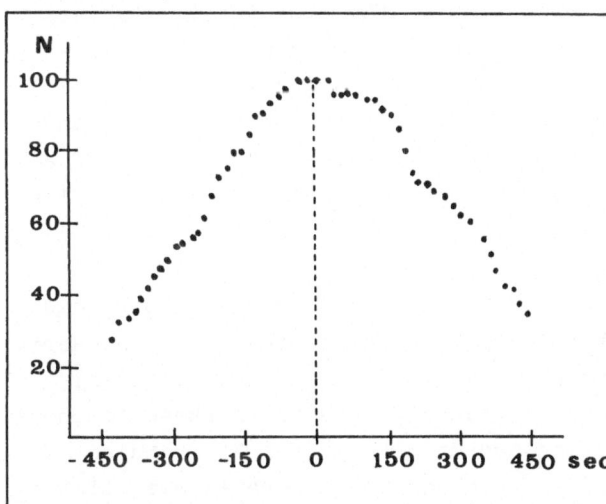

Fig. 1. Distribution of the
number of existing granules,
from a sample of 100 granules,
as a function of time. N is
the number of granules existing after or before the time
zero.

For the determination of the mean lifetime we have study the temporal
evolution of 100 granules during a period of 18 min. By using the method
proposed by Macris (1962), we found a mean lifetime of 11 min (fig. 1).
This lifetime is a little greater than that given by previous authors.
That was expected because of the excellent quality of pictures as it was
easier to distinguish the granules in the successive pictures for linger time.

b.   *Evolution of Structure.*

The evolution of structure of solar granules has been studied by several
uthors (i.e. Rösch and Hugon, 1959). A qualitative and quantitative

study is presented here based on our photographic material.

From the study of the morphological evolution of 100 granules for a period of 18 min we have defined the statistical distribution of different types of birth and death of a granule.

For the case of birth, we have found 25 appearances from the intergranular region, 10 appearances of a new granule from the merging of two or more smaller granules, and 41 appearances of a new small granule from the fission of a big granule.

For the case of death, we have found 13 disappearances into the intergranular region, 14 disappearances from the merging with another granule, and 47 disappearances from the fission.

We have also examined the morphological and temporal aspect of these evolutions.

We give, in fig. 2, the temporal evolution of the area for six typical cases of granules. Of course the measurement of a relative area is a complex photometric problem involving not only the determination of intensity, but also various corrections for the instrumental profile and for the atmosphere. As we can see for all the cases, the fission of a granule to two or more smaller ones occurs when the granule reaches its maximum area. That may indicate that at this moment an instability developes.

The granules show a diversity in their brightness. Bray and Loughhead (1967) gave the first statistics for the brightness-area relation. Here, we come back to this subject. Our good quality photographic material was used for taking isophots of the granules. Based on these isophots, we studied the relation between size and brightness for 70 granules. The results are given in fig. 3 where it is clear that there is weak linear relation between these two quantities.

Loughhead and Bray (1967) mention that there is almost complete absence of large, faint granules, and that in case of small granules the bright ones are very rare and less numerous than faint ones. Our present study verifies these conclusions.

The main differences between Loughhhead and Bray's results and our results is that we found many small, faint granules (40% of our total number). It appears likely that this discrepancy is partly due to a difference in the methods of performing the measurements and partly to the high spatial resolution of our material, so that we can distinguish many small granules in our study.

We can conclude the following :

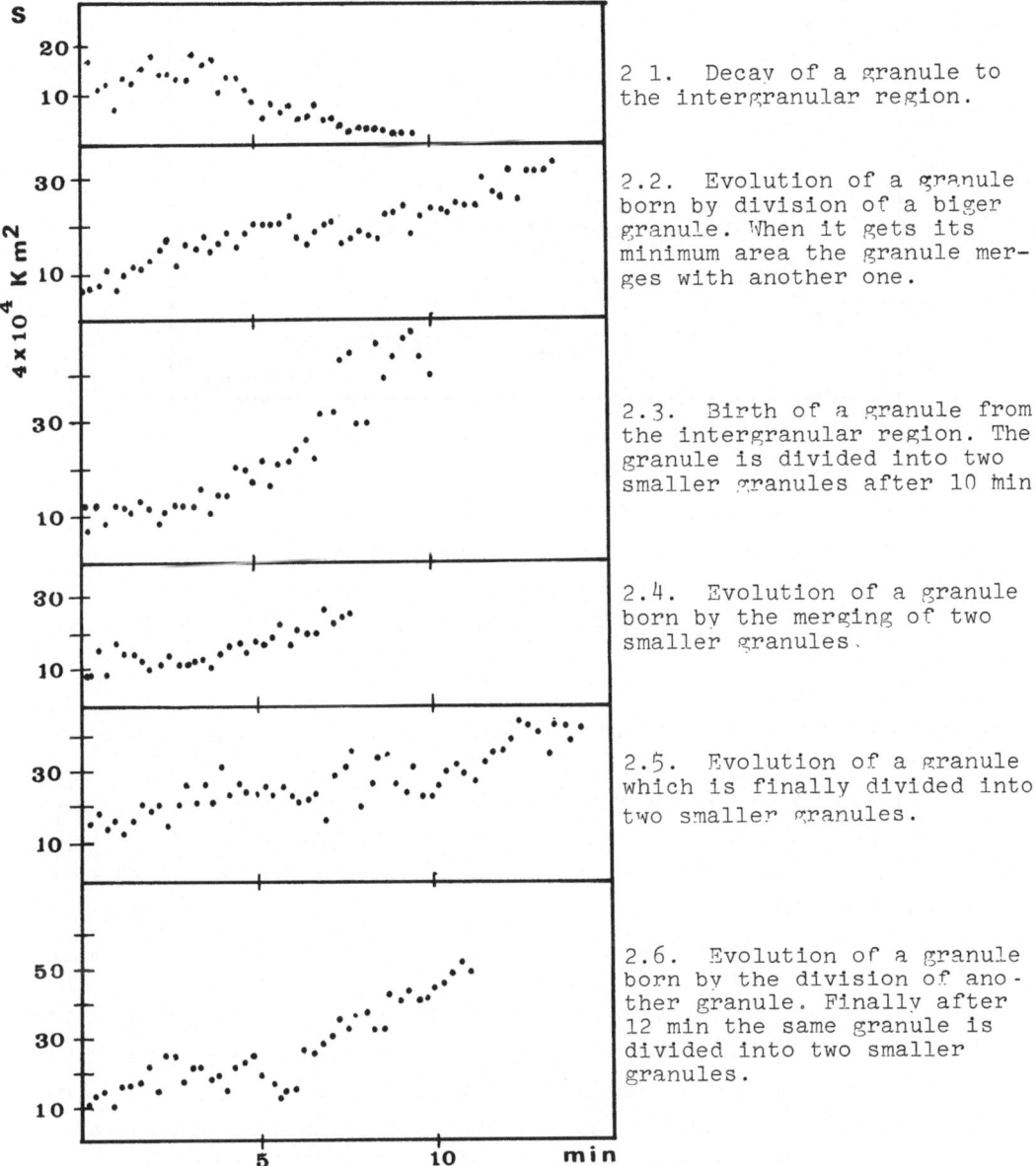

2 1. Decay of a granule to the intergranular region.

2.2. Evolution of a granule born by division of a biger granule. When it gets its minimum area the granule merges with another one.

2.3. Birth of a granule from the intergranular region. The granule is divided into two smaller granules after 10 min.

2.4. Evolution of a granule born by the merging of two smaller granules.

2.5. Evolution of a granule which is finally divided into two smaller granules.

2.6. Evolution of a granule born by the division of ano- ther granule. Finally after 12 min the same granule is divided into two smaller granules.

Fig. 2. Evolution of the area of different types of granules as a function time.

a. Only a small percentage of granules are born from the intergranular region or are diluted to it. For most granules, birth and death is only a stage of a continuous process where big granules are divided to smaller, or smaller granules merge to a new bigger one.

b. There is a weak relation between area and intensity of a granule.

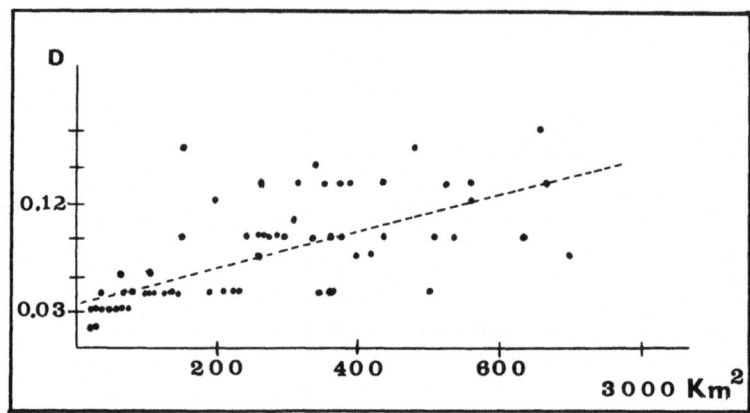

Figure 3. Brightness of granules as a function of area.

The larger granules are the brighter.

c. An instability may occur when a granule attains its maximum area.

d. There are several different types of granule evolution.

e. The mean life measured with the adopted criteria is about 11 min.

References.

Bahng J.D.R., and Schwarzschild M. :1961 " Lifetime of solar granules"
    Astrophys. J., 134, 312.

Bray R. J., and Loughhead R. E. : 1958. " Observations of changes in the
    photospheric granules ", Aust. J. Phys. 11, 507.

Bray R. J. and Loughhead R.E. :1967, "The Solar Granulation ´ Ed.
    Chapman and Hall Ltd.

Macris C.J. :1953, "Recherches sur la granulation photospherique ",
    Ann. Astrophys. 16, 19.

Macris C. J. :1962, Mem. della Soc. Astr. Italiana vol XXXIII. No 1 85.

Macris C. J. and Prokakis T. J. :1963, "New results on the lifetime of the
    solar granules ", Mem. Nat. Obs. Athens, Series 1, No 10.

Rösch J., and Hugon M. :1959, ´Sur l' evolution dans le temps de la
    granulation Photospherique ", C.R. Acad. 249 625.

Discussion.

Z. Mouradian : The lifetime distribution can be decomposed into two
curves this means two different classes of granules. Is it correct ˙.

D. Dialetis : It is true that the lifetime curve seems to be composed by
two different curves. But it is not easy to conclude that we have two different types
of granules. This question needs a new statistical study of our population of granules.

TEMPERATURE GRADIENTS IN THE SOLAR GRANULATION

Günther Elste
Astronomy Department
The University of Michigan
Ann Arbor, Michigan 48109, U. S. A.

Introduction. In earlier investigations Elste and Teske (1978, 1979) attempted to obtain information about temperatures, velocities and longitudinal magnetic fields in the solar photosphere. Hyperfine structure broadened manganese lines were chosen to avoid the influence of non-thermal velocities, i.e. microturbulence, on equivalent widths, Table 1. The nearby iron line is a normal Zeeman triplet and not too strong, so it originates still in the photosphere.

Observations at Kitt Peak with apertures 3″5 x 2″5 and 1" x 1" revealed a decrease of the negative correlation between the continuum intensity and the manganese line equivalent width with increasing resolution. The coherence analysis showed that this becomes understandable if finer details are picked up, in which the quantities correlate differently. This was supported by the results of Edmonds, Michard and Servajean (1965), EMS hereafter, who studied the high excitation carbon line at 5052 Å with respect to the continuum. Attempts to observe Mn 5395 and C 5380 simultaneously with the same setup at Kitt Peak failed. Therefore a new observing program was undertaken using two CCD's in the focal plane of the Echelle grating spectrograph of the Vacuum Tower at Sacramento Peak Observatory.

Observing Program. One CCE recorded the profile of C 5380 and its short wavelength neighborhood for a reliable continuum. The other CCD recorded Mn 5395 in two polarization directions to be used as an indicator for the presence of longitudinal magnetic fields. Scanning perpendicular to the direction of the slit a total area of 30" x 32" of the photosphere was covered with a spatial resolution of 0″3 x 0″4. Scanning was sufficiently fast to enable repeating a scan of the same area every 150 seconds. Combining two successive images will therefore permit a correction for the 5-minute oscillations.

TABLE 1.   Spectral Lines   Standard Deviations

		χ_{os} eV	W_λ mA	log τ formation	3″5 x 2″5	1" x 1"	0″3 x 0″4	EMS photographic spectral
Fe I	5393	3.2	153	−1.7				
Mn I	5420	2.1	78	−1.1	0.020	0.042		
Mn I	5395	0.0	74	−1.5	0.027	0.049	0.124	
C I	5380	7.7	26	0.0			0.133	0.042 (C 5052.2)
Continuum					0.010	0.024	0.044	0.016
Vel.	km/s				0.16	0.24	0.37	0.24 (Cr 5051.9)

Results. Preliminary results from one of our better frames are presented
here. The standard deviations, Table 1, show the expected increase with
resolution. Our values are about a factor three greater than those from
photographic spectra by EMS. The correlations are shown in Table 2.
Their behavior with increasing resolution is quite striking, in particular the change in sign for the correlation between continuum and manganese line strength. Continuum intensity and carbon line equivalent
width are strong positively correlated in agreement with EMS for C 5052
and continuum. Both carbon lines originate in the same deep layers as
the continuum and become stronger with increasing temperature. The manganese line, which weakens with increasing temperature, also shows a
positive correlation with the continuum intensity or the carbon line
equivalent width. For the interpretation we must realize that the ground
level manganese line is formed about 200 km higher than the carbon line.
Consequently there must be regions, where the deep layers are hotter,
while the temperature in higher layers is lower than average, and vice
versa. In other words, temperature gradients tend to be steeper in
regions with higher temperatures in deep layers and flatter in regions
with lower temperatures in deep layers.

TABLE 2

Aperture	$3\overset{''}{.}5$ x $2\overset{''}{.}5$	1.0 x 1.0	0.3 x 0.4		EMS photographic spectra
Correlating					
Cont. and V(Fe)	-0.25	-0.37	-0.49	(Mn)	-0.28 (Cr 5051.9)
Cont. and W(Mn)	-0.44	-0.09	+0.58		
W(Mn) and V(Fe)	-0.19	-0.16	-0.32	(Mn)	
Cont. and W(C)			+0.92		+0.81 (C 5052.2)
W(Mn) and W(C)			+0.60		

In order to show the spatial arrangement of the varying temperature gradients, the points are divided into three groups according to their continuum intensity. The divisions are set by the average plus and minus
50% of the standard deviation, resulting in equal numbers in each group.
Figure 1 shows the location of brighter continuum intensities as heavily
shaded, while locations of fainter continua remain unshaded. The intermediate brightness range is lightly shaded. In a similar way Figure 2
shows areas where temperature gradients are steeper and flatter, based
on the criterium that the equivalent widths of both lines, carbon and
manganese, are above or below average plus or minus 50% of the standard
deviation respectively. Comparing the two figures shows quite nicely
the variation of the temperature gradients with respect to the variation
of the temperature in deep layers.

The analysis of the data after correction for the influence of the 5-minute oscillations and hopefully after an image improvement procedure will be presented at a later time. In particular the study of the influence of a magnetic field concentration on the temperature gradients is expected to be of interest.

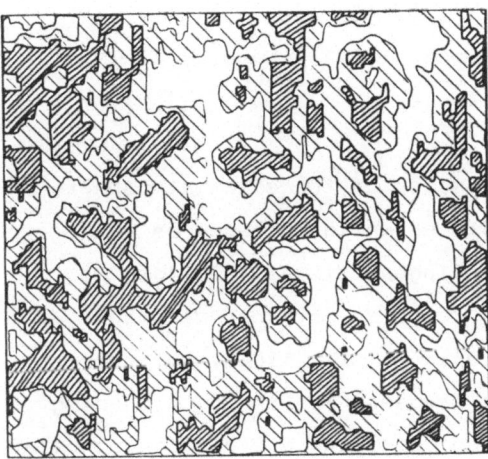

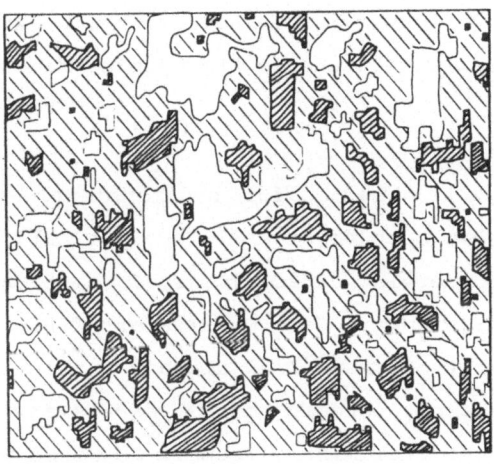

Figure 1. Continuum intensity above or below average plus or minus σ/2.

Figure 2. Equivalent widths of both lines above or below average plus or minus σ/2.

References

Edmonds, F.N.Jr., Michard, R. and Servajean, R. 1965,
    Ann. Astrophys. 28, 534.
Elste, G.H. and Teske, R.G. 1978, Solar Phys. 59, 275.
Teske, R.G. and Elste, G.H. 1979, Solar Phys. 62, 241.

Question from L.M.B.C. Campos to G. Elste:
Can you use the correlation data values to estimate a correlation scale? Could the correlation scale be interpreted as a difference between the depth of formation (of the two lines whose correlation was measured)?

Answer from G. Elste: If two quantities correlate well we may assume that they refer to the physics at the same depth, and the weaker the correlation is, the less coupled their physics appears, the further apart the quantities are probably formed in the atmosphere. But for obtaining a measure in geometrical depth we do need model calculations with spectrum synthesis and certainly several iterations.

LINE PROFILES AND LONGITUDINAL VELOCITY FIELD IN SEEING LIMITED SMALL-SCALE ATMOSPHERIC STRUCTURES

(1) F. Falchi, (1) R. Falciani, (2) L.A. Smaldone

(1) Osservatorio Astrofisico di Arcetri, I-50125 Firenze, Italy

(2) Osservatorio Astronomico di Capodimonte, I-80131 Napoli, Italy

A method for bidimensional spectroscopy of the solar atmosphere
has been developed (Caccin et al., 1983), essentially based on the
observing capabilities of Universal Birefingent Filter (UBF) of the
Sacramento Peak Observatory (SPO). Some results, obtained with this
method, have been already published (Caccin et al., 1984), but they
mainly concern active regions and flare studies. We like to present here
the preliminary results obtained when the observing and diagnostic capabilities of this method are applied to small spatial, seeing limited
structures in the solar atmosphere.

More details concerning the observing program and the capabilities
of all our methods can be found in Caccin et al.(1984).

The complete set of an H_α restored and calibrated filtergram series
is shown in Fig. 1.

We can determine the line profiles for every pixel in the FOV
frame; the units are relative to the profiles of the mean quiet atmosphere with a rms error of about 10%. We can also derive the longitudinal velocity field pattern for the four observed lines, by using a
linear approximation and by interpreting the measured line asymmetry as
due only to line-of-sight Doppler shifts.

We applied our method to the small spatial structures, as the
bright points visible in the H_α wings (see Fig. 1.). We divided them
in two separate classes: IBP, the isolated bright points an example of
which is indicated in Fig. 1, and EB, the bright points outside the
penumbra of the large spot on W side of Fig. 1 (EB means Ellerman bombs
according to the most accepted definition for these structures).

For each selected point we determined the line profiles and the velocity field as a mean over a minor of 3x3 pixels. Then, the effective spatial resolution was of 1.5x1.5 arcsec^2.

The relative variations of the H$_\alpha$ and H$_\beta$ line profiles among the different bright points are of the same order of those measured at

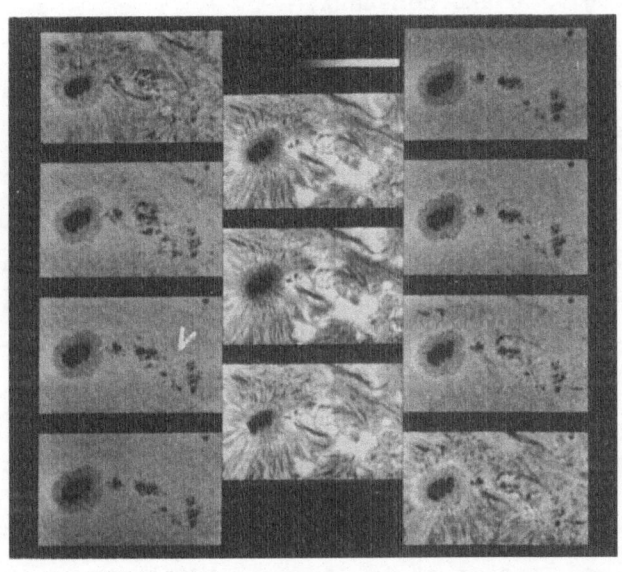

Fig. 1. The complete filtergram series, at 21:07 UT-13-6-1980 from -2.0 A (lower-left) to +2.0 A (upper-right) through H$_\alpha$ -0.0 (middle). The linear scale is given in arcsec and in the grey look-up table the mean quiet atmosphere corresponds to 1.0. North of the bottom, West on the left. The arrow indicates an isolated bright point.

different times. For each class we considered independent the change due to different positions or different times of observations.

The shape of all the H$_\alpha$ line profiles for IBP's is shown in Fig. 2. Similar behaviour is also observed for H$_\beta$ line profiles, but with a lower "sensitivity". If we morphologically compare the mean general trend of H$_\alpha$ and H$_\beta$ for IBP and EB points with the mean quiet line profiles, we can draw the following qualitative conclusions:

IBP's - the contrast $C = \dfrac{I(IBP)-I(quiet)}{I(quiet)}$ is always > 0 in the line wings, but C(blue wing) is much greater (and more variable than C(red-wing);

- C in the line center is always > 0 and roughly constant;

- the scatter strip of the various profiles at the points of inflection is narrower on the blue side with respect to that on the red
side;

- the line-of-sight H_α velocity (measured at ± 0.6 A from the
line core) is always downwards, with typical values of 1.5÷2.5Km/sec.
EB's   - C(blue wing) > 0 and constant;

- C in the line center is ≈ 0;

- the scatter strip at the points of inflection is narrow on
both sides of the line profiles;

- no clear line-of-sight velocity pattern is evident, but certainly H_α velocities can be considered zero within our error limits.

Quantitative parameters, describing the above mentioned properties
can be obtained by computing the mean value $<I(\lambda)>$ of different profiles
and the corresponding standard deviation $\delta(\lambda)$ for the two bright point
classes.

Table 1.- Mean contrasts along the H_α and H_β line profiles evaluated
         for isolated bright points (IBP) and for Ellerman bombs (EB).

H_α + $\Delta\lambda$	IBP	EB	H_β + $\Delta\lambda$	IBP	EB
- 2.0	0.01±0.04	0.05±0.05	- 1.0	0.02±0.04	0.04±0.04
- 1.5	0.03±0.06	0.08±0.06	- 0.8	0.05±0.05	0.07±0.03
- 0.9	0.08±0.08	0.08±0.07	- 0.6	0.09±0.06	0.11±0.05
- 0.6	0.08±0.09	0.05±0.09	- 0.4	0.14±0.06	0.14±0.08
- 0.3	0.19±0.14	0.01±0.09	- 0.2	0.17±0.06	0.10±0.07
0.0	0.24±0.18	0.03±0.07	0.0	0.14±0.09	0.07±0.06
+ 0.3	0.23±0.16	0.08±0.08	+ 0.2	0.02±0.11	0.06±0.09
+ 0.6	0.12±0.16	0.16±0.11	+ 0.4	0.01±0.11	0.09±0.05
+ 0.9	0.00±0.11	0.09±0.08	+ 0.6	0.01±0.08	0.08±0.04
+ 1.5	0.01±0.06	0.06±0.04	+ 0.8	0.02±0.04	0.06±0.03
+ 2.0	0.02±0.03	0.05±0.03	+ 1.0	0.01±0.03	0.05±0.03

These H_α parameters are compared with the mean quiet H_α profile in Figure 3 for the IBP's. The corresponding C (λ)'s can be consequently computed and are quoted in Table 1. for H_β and EB's points too.

We can derive a very first and qualitative explanation of our
results. Our IBP's belong substantially to active regions and are probably the same structures called facular knots by Spruit and Zwaan
(1981) and facular granules by Mehltretter (1974). According to the
results of Wilson (1981), Spruit and Zwaan (1981) and Kitel and Muller
(1984), these structures represent spectral signatures of the presence

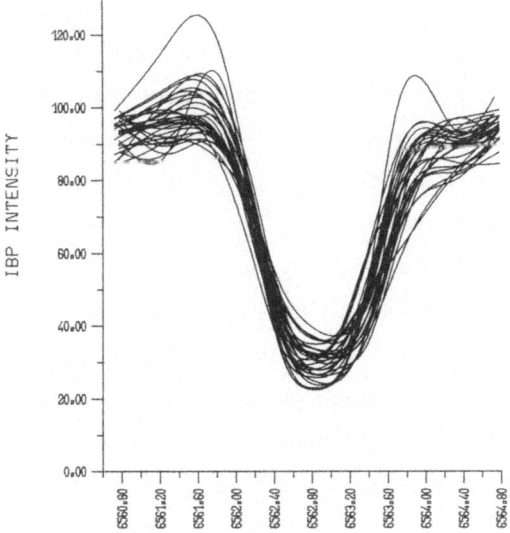

Fig. 2. The profiles for all the analyzed isolated bright points (IBP's) (see Fig. 1).

Fig. 3. Comparison of the H_α profile (continuous) averaged over all the IBP's with the quiet H_α profile (thick).

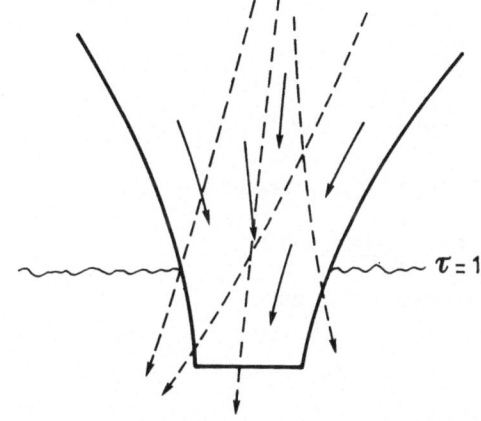

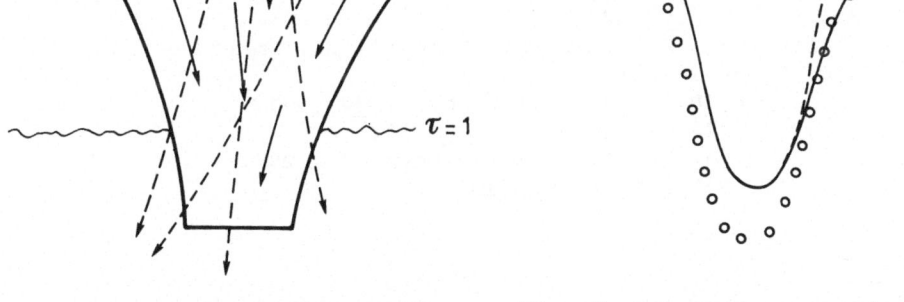

Fig. 4. Sketch of a magnetostatic flux tube in the solar quiet atmosphere; dashed arrows indicate different line-of-sights and continuous arrows represent downward motions.

Fig. 5. Schematic comparison of the quiet H_α profile (circles) with that (dashed) given by the "wall effect" (see Fig. 4) and with the final asymmetric profile (continuous), caused by downward motions.

of elementary magnetic flux tubes. A very schematic sketch for a magnetostatic flux tube is given in Fig. 4, together with the associated downward motions. Various line-of-sights are mixed altogether (see Fig. 4)
within a pixel in our images. This fact originates a "wall effect"
(Caccin and Severino, 1979), because the mean optical path within the
tube allows to "see" deeper photospheric layers, i.e. layers in which
the source function is greater than in the mean quiet photosphere. In
Fig. 5 the circle profile qualitatively represents the line profile due
to the "wall effect"; it is symmetric and has a positive contrast relative to the mean quiet profile. Since many line-of-sights intercept
downward moving elements within the tube, the observed profile can be
supposed as due to a convolution of the red side dotted profile with a
suitable velocity distribution function. The final profilè (continuum
in Fig. 5) fits the above stated morphological properties derived for
the IBP's.

As far as the EB's line profiles are concerned, their main feature
is a positive wing contrast and this might be simply explained by a
density increase at lower photospheric levels.

References
Caccin, B., and Severino, G.: 1979, Astrophys. J., 232, 397.
Caccin, B., Falciani, R., Roberti, N., Sambuco, A.M., Smaldone, L.A.:
     1983, Solar Phys. 89, 323.
Caccin, B., Falchi, A., Falciani, R., Roberti, G., Smaldone, L.A.:
     1984, Adv. Space Res. (in press).
Kitai, R. and Muller, R.: 1984, Solar Phys., 90, 303.
Mehltretter, J.P.: 1974, Solar Phys., 38, 43.
Spruit, H.C. and Zwaan, C.: 1981, Solar Phys. 70, 207.
Wilson, P.R.: 1981, Solar Phys. 90, 303.

A. RIGHINI:

Are you sure that the red asymmetry that you observe on the H_α might not be due to a minor blue shift of the center of the upper part of the line due to the presence of a magnetic field which channels the convection excluding therefore an inward mass motion?

R. FALCIANI:

We calibrate our mean reference profiles on two areas (15 x 25 arc-sec^2 each) of quiet atmosphere, assumed then to have no line-of-sight velocity. Consequently, we have to interpret the measured red asymmetries on IBP's as due to downward motions relative to the mean quiet atmosphere.

C. ALISSANDRAKIS:

Did you observe any difference in the pattern or in the velocity of the Evershed effect between H_α and H_β? Our observations at the core and the wings of H_α suggest a decrease of the value of the velocity with height.

R. FALCIANI:

The general H_α Evershed pattern around the big spot (on the W side of our frames) morphologically agrees with the results of Maltby (Solar Phys. 43, 91, 1975). Representative H_α line-of-sight velocities are of the order of 2.5 Km/sec towards the disk center and 0.5 Km/sec in the opposite direction in the mean Evershed channels. Similar patterns, but with lower values, are observed in H_β. Our dopplergram determinations refer to the maximum sensitivity part of the line profiles (i.e., to the inflection points). Care has to be used to compare line-of-sight velocity values derived from different portions of the line profile.

DETERMINATION OF MAGNETIC FIELDS IN UNRESOLVED FEATURES

M. Semel
Observatoire de Meudon
92195 - Meudon Principal Cedex

ABSTRACT

Observations of solar faculae and networks reveal the unresolved nature of the magnetic field configurations, with present achieved spatial resolutions. While the apparent field is weak (from a few gauss to several hundreds) the "true" field is likely to
be much stronger (up to kilogauss fields).
There are various methods to determine the field strength in the unresolved flux tube.
They vary according to the choice of spectral lines, the weight given to the observables (intensity, profiles, polarisation ...), model assumptions and physical considerations.
The purpose of this paper is to give a review of these methods and assessment of their
capacities and limits.

INTRODUCTION

Observation of solar magnetic fields obtained with high spatial resolution and away
from sunspots exhibit a distribution from a few gauss up to kilogauss fields. See for
example observations obtained at Sac Peak (Simon and Zirker, 1974) and magnetic map
(Fig. 1) obtained by S. Koutchmy (Daras-Papamagaritis and Koutchmy, 1983).
The question to be discussed here is whether photospheric magnetic fields are unresolved and thus quite strong even when apparently they are weak.
It is still too difficult to give a complete synthesis of all methods used. Observational conditions and the particular solar features observed might have been different in each experiment. When spatial resolution is poor, it is not easy to evaluate
the "contribution" or the weight of each of the details seen in a high resolution observation (Fig. 1). Even if a "typical" flux tube may satisfy the observation it is
hazardous to claim that all flux tubes still unresolved in Fig. 1 are the same. The
main effort in this review is to describe each method in the simplest way and also
to mention its main limitation due to model dependence
There is some confusion on the way to handle radiative transfer. On the one hand we
may use a powerful numerical program to solve the equations. But since the true atmosphere is generally not known, we may doubt about the significance of the result
and loose insight on the way the method used depends on the uncertainty in the model

assumed.

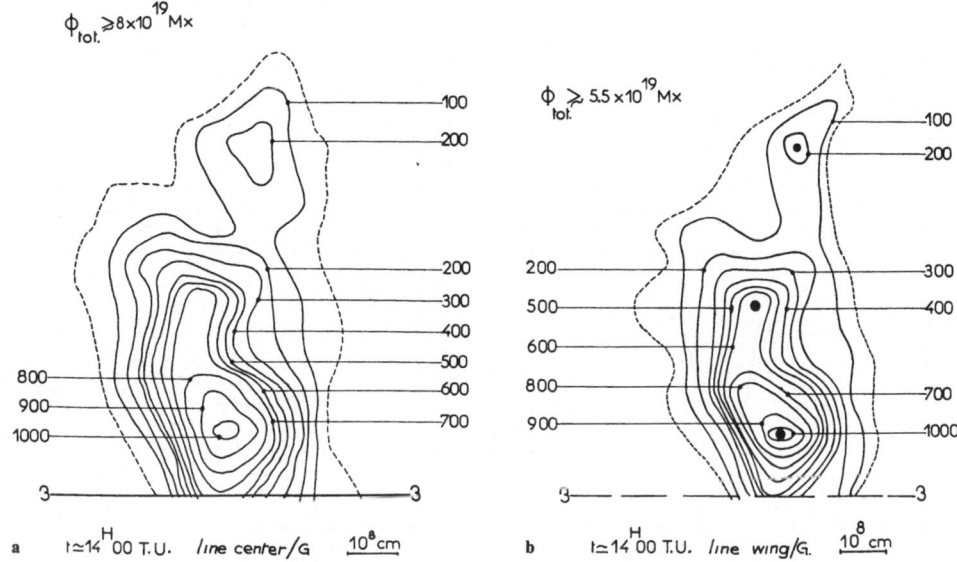

Figure 1 : Photospheric Networks

a Maps of the intensity of the magnetic
field deduced from measurements in the
line center of Fe λ = 6302.5A.

b As in a but in the line wings of
Fe λ = 6302.5A.

On the other hand, we know that model dependency may be reduced in the measurements
of magnetic fields. For example, in the case of homogereous magnetic field, we may use
Milné-Eddington model to fit the observed Stokes parameters. The surprising result is
that this inversion program turns out to be a reliable method to determine magnetic
fields, if the parameter of Milné-Eddington model are kept free in the processus of
iteration. Thus these parameters and the magnetic field should be determined simultaneously. While Milné-Eddington model is a poor approximation of the true atmosphere,
it is still a powerful mathematical tool provided its parameters are not predetermi-
ned.

The assumption of purely longitudinal field will bring about simple analytic expressions for the intensity and circular polarization profiles. Thus the description of
the various methods will become quite clear with little loss of generality. Indeed
observers often considered explicitly or implicitly purely longitudinal fields.

Depth dependence of magnetic and velocity fields cannot be considered so simply. Such
gradients may lead to effects not to be reviewed here (Semel et al., 1980)

Figure 2 is quite instructive. Distinct σ components are visible in I_m profile, (fig.
2b) but no more in the observed profile I_t (2a). The peak positions in V_t the observed circular polarization (fig. 2d) correspond to those of the σ, the separation of
which is a measure of the field strength. But the same peaks separation of the V

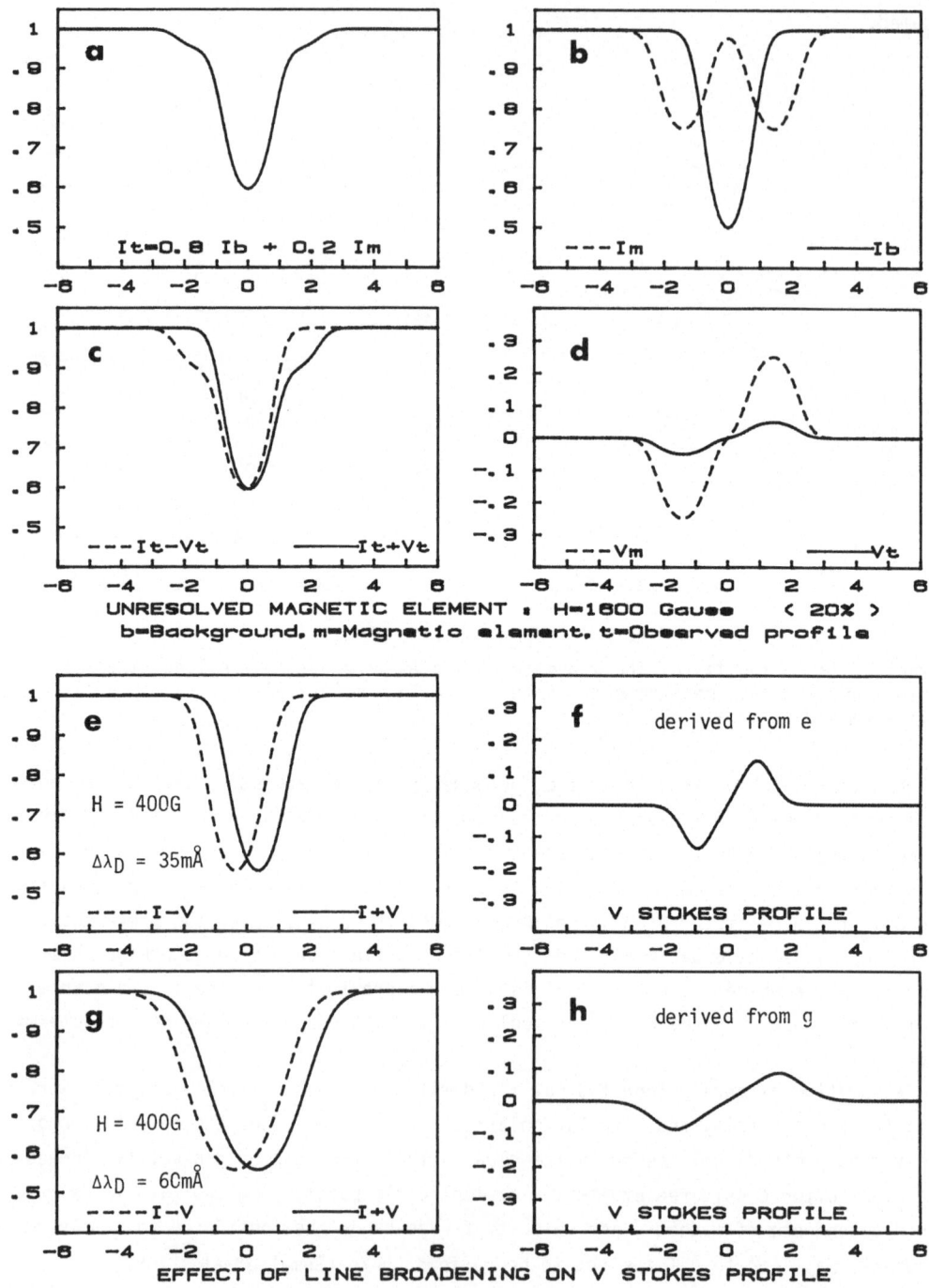

Figure 2

profile may be due to weak field as well depending on line width (compare fig. 2d with 2h and 2f). Doppler broadening must be considered in this context. In any case when a small V_t is not proportional to $\frac{dI}{d\lambda}$ it indicates unresolved magnetic field.

I. THE SIMPLEST TWO COMPONENTS APPROACH

Radiative transfer is reduced to its simplest expression in the following way :

1) Magnetic field will be assumed to be longitudinal.

2) All spectral lines considered as Zeeman triplet, but with different Landé factors g.

The unresolved structure will be reduced to two components only.

- A magnetic element with relative surface S, magnetic field H. $f_m(\lambda)$ denotes the line intensity profile if it were not affected by Zeeman effect (g = 0).

- A background element with relative surface 1 - S. $f_B(\lambda)$ will be the line intensity profile in absence of magnetic field. The possibility of a weak magnetic field H_B will be considered.

For the magnetic element the intensity and circular polarization profile will be given by :

$$I_m = \frac{1}{2} \left[f_m (\lambda - g\Delta\lambda_H) + f_m (\lambda + g\Delta\lambda_H) \right] \tag{1}$$

$$V_m = \frac{1}{2} \left[f_m (\lambda + g\Delta\lambda_H) - f_m (\lambda - g\Delta\lambda_H) \right] \tag{2}$$

Where $\Delta\lambda_H$ is the Zeeman displacement for a normal triplet given by

$$\Delta\lambda_H = C \cdot H \tag{3}$$

where C is a known constant.

Noting by I_B the intensity in the background, the observed intensity I is then given by :

$$I = (1 - S) I_B + S \cdot I_m \tag{4}$$

In absence of magnetic field in the background ($H_B = 0$), the observed circular polarization is given by :

$$V = S \cdot V_m \tag{5}$$

If a weak field H_B is assumed to be in the background

$$\Delta\lambda_B = CH_B \tag{3a}$$

the circular polarization from the background element will be given by :

$$V_B = \frac{1}{2} \left[f_B (\lambda + g \Delta\lambda_B) - f_B (\lambda - g\Delta\lambda_B) \right] \tag{2a}$$

eq. 5   should be replaced by :

$$V = S \cdot V_m + (1 - S) V_B \tag{5a}$$

II. BABCOCK TYPE MAGNETOGRAPH

1) Magnetograph response

By development in series of the profile $f_m(\lambda)$ equation 2 for V_m becomes

$$V_m = f'_m(\lambda)g\Delta\lambda_H + \frac{1}{3!} f_m^{(3)}(\lambda) (g\Delta\lambda_H)^3 + \ldots \ldots \tag{6}$$

The signal measured by the magnetograph $V = S.V_m$ is reduced to a magnetic signal H_M by

$$H_M = \frac{V}{g.c.f'_c(\lambda)} \tag{7}$$

where $f'_c(\lambda)$ is the calibration of the magnetograph by a profile $f_c(\lambda)$ either calculated or taken equal to the line profile in the quiet photosphere.

Inserting equation 6 in equation 7 we obtain :

$$H_M = S.H. \frac{f'_m(\lambda)}{f'_c(\lambda)} \left[1 - \theta g^2 H^2 + \ldots \ldots \right] \tag{8}$$

$$\text{where} \quad \theta = -\frac{1}{3!} \frac{f_m^{(3)}}{f'_m} \cdot c^2 \tag{9}$$

For weak field (or for small g factor) and if $f_c = f_m$ the magnetograph signal H_M is equal to the magnetic flux = S.H.

An inadequate choice of $f_c(\lambda)$ would be the origin of a first order error.

Simultaneous observations with several spectral lines have shown significant disagreements due to that kind of error.

The second and higher terms in equation 8 describe the magnetograph saturation when Zeeman displacement is not small.

2) Line ratio technique

Consider two identical lines that differ only by their g factors. For example :

Fe λ = 5250 Å g_1 = 3

Fe λ = 5247 Å g_2 = 2

H_1 and H_2 are the magnetic fields measured by these two lines.
The ratio r is given by using equation 8.

$$r = \frac{H_1}{H_2} = \frac{1 - \theta \, g_1^2 \, H^2 + \ldots..}{1 - \theta \, g_2^2 \, H^2 + \ldots..} \tag{10}$$

S, H, f_m, and f_c being the same for the two lines

Keeping only the first two terms in the series development :

$$H^2 \simeq \frac{1 - r}{\theta \, (g_1^2 - g_2^2 \, r)} \tag{11}$$

and in the present case

$$H^2 \simeq \frac{1 - r}{\theta(\lambda) \left[5 + 4\,(1-r)\right]} \tag{12}$$

when $r \to 1$ $H^2 \to 0$. Otherwise we have an indication for strong field. To determine
the field from equation 12 it is still necessary to determine θ. Thus the result is
still model dependent (the explicit choise of $f_m(\lambda)$). However this dependence may be
reduced by measuring r at different wavelengths in the line. Successful choice of $f_m(\lambda)$
should yield a determination of H^2 independant on wavelength.

Now the problem of field distribution becomes important. In the case of two components
model it means a weak field in the background   The contributions from strong field
element to the two lines are given by H_1 and H_2. δH the contribution from the background identical for the two lines (no saturation effect for weak field). The total
field measured by the two lines would be $H_1 + \delta H$ and $H_2 + \delta H$.

Put $\alpha = \dfrac{\delta H}{H_2}$ and then :

$$r_\alpha = \frac{H_1 + \alpha H_2}{H_2 \, (1 + \alpha)} \qquad\qquad r = \frac{H_1}{H_2} \tag{13}$$

$$1 - r = (1 - r_\alpha) \, (1 + \alpha) \tag{14}$$

equation 12 becomes

$$H_\alpha^2 \simeq \frac{1 - r}{\theta \left[5 + 4 (1 - r) \right]} = \frac{(1 - r_\alpha)(1 + \alpha)}{\theta \left[5 + 4 (1 - r_\alpha)(1 + \alpha) \right]} \tag{12a}$$

when $1 - r_\alpha$ is small

$$H_\alpha^2 \simeq H_o^2 (1 + \alpha) \tag{12b}$$

α is a free parameter and the solution for the strong field is α dependent. Choosing α positive will increase the field, a choice of negative α will lead to a weaker field.

The approximated analytical approach has been given here for the purpose of demonstration only. It may be valid when the field is not too strong and the derivation of equation 11 from equation 10 is possible.

3) Observation and numerical computations of r.

Frazier and Stenflo (1978) measured r at four wavelengths in these lines :

$\pm \Delta\lambda =$	22,5	47,5	72,5	97,5	mÅ
r =	0.76	0.85	0.98	1.08	

They found a field strength H = 960 gauss.
When a smooth distribution of magnetic field has been assumed, the maximum field strength became much higher H_{max} = 2000 gauss.
This corresponds to $\alpha > 0$ in the analytical example.
It is easy to conceive the case with negative α.
To illustrate it I assumed a two component model :
   - strong field element with H = 675 gauss about 30% less than singled value solution mentioned above (960 gauss).
   - a background with weak field with a flux = β . S.H
A Milné-Eddington has been employed with the following parameters with the usual terminology. Background has been chosen equal to photosphere.

	Doppler broadening $\Delta\lambda_D$ mÅ	Damping Constant	Line to continuum Absorption coefficient ratio η_0	Magnetic field gauss
background = photosphere	31.9	0.2	2.2	$\beta.H$
strong field element	50	0.2	0.5	$H = 675$

- S has been fixed to 0.5 e.g. same surface for the two elements.
- The parameters for the strong field element has been fixed after some iteration and are very close to those found by Harvey et al. (1972). See § 3.
- Slit width and instrumental profile have been considered by integration over 31 mÅ.
- No velocities have been assumed.

Results of computation and comparison with observation is given in the following table

	$\pm \Delta\lambda$ mÅ	23	48	73	98
	observed	0.76	0.85	0.98	1.08
r	calculated $\beta = 0$	0.921	0.934	0.984	1.067
	calculated $\beta = -0.2$ ($\alpha < 0$)	0.761	0.842	0.977	1.079

Note that the relative moderate field 675 gauss may satisfy the observation very well when negative flux ($\alpha < 0$) is added.
Thus quite a wide range of magnetic fields may satisfy the same observed ratios r.

4) Doppler shifts and intensities.
Frazier and Stenflo also measured velocities and brightness in the two lines : the surprising result was that no significant difference between the lines could be detected. Equation 1 for example predicts difference in intensity between the lines if strong magnetic field were present. Even, more sophisticated computations confirm this prediction.
At present, the best candidate to explain this contradiction would be the assumption that magnetic field varies strongly with depth. If that is true, then also the present application of the line ratio technique to the polarization measurement should be reviewed. Till now no computations have been successful to satisfy all observables.

5) Line ratio technique with three lines has been applied by Wiehr (1978). Another experiment by the same author was to measure magnetic fields at three different wavelengths in the same line simultaneously. The results obtained by Wiehr indicate that the magnetic field was not unique but had a range from 1200 to 1800 gauss. The range was found to be the same in plage in active regions and in networks.

Beside the advantage of using three lines instead of two, there is an interest in using lines "insensitive" to temperature. Indeed the lines used by Wiehr are much better in this respect than lines 5250 and 5247. In our analytical example that means that $f_m(\lambda)$ differs only very little from $f_B(\lambda)$. But the lines used by Wiehr are not really identical. We may suspect that $f_m(\lambda)$ for each line does change slightly due to temperature and velocity fluctuations, and that these changes are not the same for those lines.

III. FITTING OF OBSERVED I AND V BY USE OF MILNE-EDDINGTON MODEL

The solution of radiative transfer equations for homogeneous field and in the case of Milné-Eddington has been given by Unno (1956).

These equations have been used to determine the magnetic field in unresolved structures.

1) Beckers and Schröters (1968) observed magnetic knots in vicinity of sunspots. The apparent field was typically 250-400 gauss. The Milné-Eddington parameters for the line used, 6173 Å, were fixed from the photospheric profile.

A good fit of observed circular polarization profiles was obtained by assuming stray light of the order of 67%. The typical range for the magnetic field was 600 to 1400 gauss. When stray light is assumed the magnetic field was found to be almost vertical.

The determination of the magnetic field in this case is model dependent through the assumption that Milné-Eddington parameters are the same in the photosphere and in the magnetic knots.

In terms of the model described in § I it means that $f_m = f_B = f_{photosphere}$. Also the determination of the parameters of Milné-Eddington in the photosphere may not be unique. For example, in this model the Planck function B is assumed to be linear with τ optical depth in the continuum.

$$B = B_0(1 + \beta_0 \tau)$$

From this, the center to limb effect for the normalized line profile is given by the factor :

$$A(\theta) = \frac{\beta_0 \, \cos \theta}{1 + \beta_0 \, \cos \theta}$$

$$f_p (\lambda, \theta) = A (\theta) f_p (\lambda, o)$$

Where θ is the heliographic angle and $f_p (\lambda, \theta)$ and $f_p (\lambda, o)$ are the normalized line profiles at θ and at disc center respectively.

Now in general the center to limb effect for photospheric lines is small.

Observers therefore tend to put $\beta_o \to \infty$ and $A = 1$.

However the true interpretation should be that Milné-Eddington model cannot describe center to limb effect. And in general A should be one of the parameter to be determined and may be quite different in photosphere and in facula.

2) Harvey et al. (1972) observed I and V profiles for the line 5250 Å and also magnetograph measurement with the line 5233 Å. The last measurements were considered to be the true average field.

A two-components model has been assumed, the non-magnetic component to be identical to the photosphere.

Milné-Eddington model has been used through Unno equations. β_o has been assumed very large (A = 1), but all other parameters were free.

It resulted that magnetic fields range from 200 to 700 Gauss.

Doppler broadening in the magnetic element was found to be larger by about $\frac{2}{3}$ than in the photosphere. The interpretation advanced was that this is due to dispersion of magnetic fields of the order of 1000 gauss. So that when the observed field is 500 gauss there is still a range of fields of ± 500 gauss.

This interpretation excludes other causes of broadening and perhaps rely too much on the physical meaning of Milné-Eddington's parameters.

3) Comparison and discussion of the two experiments.

The results in 1) and 2) are different.

In paper 1 (Beckers and Schröters, henceforth B.S.), the magnetic field range is

600 < H < 1400 gauss

In paper 2 (Harvey et al., henceforth H.L.S.) the range of the average field in the magnetic element :

200 < H < 700 gauss

Kilogauss fields are then obtained through the assumption of non homogeneous field and interpretation of line broadening by "Zeeman microturbulence".

The difference between these results could be attributed partially to different observational conditions and selection effects, e.g. B.S. could distinguish magnetic knots in the high resolution photographic spectra.

One fundamental difference in the two approaches deserves particular attention. In both cases the authors were led to analyze the data as if the field were practically longitudinal. So implicitly they solved equations 1 to 5 described in § I in terms of which the discussion is simplified.

In both cases f_B was chosen equal to the quiet photosphere profile, while B.S chose $f_m = f_B$, H.L.S let f_m to be free, this is a crucial difference. By inspection of the V profile it is easy to see that it may be approximated by different combinations of $f_m(\lambda)$ and $\Delta\lambda_H$. If the profile of $f_m(\lambda)$ is chosen wide, that is large FWHM, then $\Delta\lambda_H$ will be found small and vice versa. In other words there is a kind of competition between $\Delta\lambda_D$ and $\Delta\lambda_H$. Useless to say that the best fit will be obtained when $f_m(\lambda)$ is let free, which is the case in paper 2 (H.L.S.). Indeed they find that $f_m(\lambda)$ is much wider than the photosphere profile, that is $\Delta\lambda_D$ is increased substantially, and as a consequence the magnetic field was found much weaker.

While in H.L.S.'s paper, H results from a better fit, the estimate of the field dispersion is not certain. These authors are right to assume a distribution of magnetic fields that increases line width. However the quantitative interpretation of the increased broadening in terms of "Zeeman microturbulence" is hard to accept :

   a) The quantitative physical meaning of Milne-Eddington parameters is limited, even when a good fit to observed profile is obtained.

   b) True Doppler broadenings may be quite different in flux tubes and in the quiet sun.

   c) Radiative transfer is not consistent. Either magnetic field is assumed to vary strongly along the line of sight and the radiative transfer used is wrong, or if magnetic field is assumed constant along the line of sight and varies laterally then Stokes parameters have to be obtained first and then summed over the area observed. The amount of dispersion in magnetic fields might be quite different than what has been determined so simply by H.L.S.

IV. FOURIER TRANSFORM OF THE V PROFILE. Tarbell and Title (1977)

As usual Fourier transform is a very elegant technique. The F.T. of the V is defined by :

$$F_V(x) = \int V \ell^{ix\lambda} d\lambda \qquad (15)$$

applied to equation 5 :
$$F_V(x) = - S.i.F_m(x) \sin (g\Delta\lambda_H \, x) \qquad (16)$$

If $f_m(\lambda)$ is a Gauss, Lorentz or Voigt function, its Fourier transform $F_m(x)$ will have no zeros.
So that the equation :

$$F_V(x_o) = 0 \tag{17}$$

is reduced to

$$Sin\ (g\Delta\lambda_H x_o) = 0 \tag{18}$$

and the first interesting zero is given by

$$g\Delta\lambda_H \cdot x_o = \Pi \tag{19}$$

or $\quad g\Delta\lambda_H = \dfrac{\Pi}{x_o} \tag{20}$

No assumption of line profile is necessary to determine the magnetic field by equations 17 and 20.

Title and collaborators have controlled that the method may be applied to much more complicated cases, e.g. general inclination of magnetic fields. Because of signal to noise ratio the method was sensitive only to fields stronger than 1000 gauss. The authors found typical fields of the order of 1600 gauss. The same authors compared these results with those obtained by fitting I and V through Milné-Eddington model (see § III). This last method gave similar results only when Doppler broadening was assumed to be within 10% of the photospheric value !

Now assume that weak magnetic field may reside in the background and substitute equation 5a for equation 5 to obtain $F_V(x)$ through equation 15.

$$F_V(x) = -i\ \left[SF_m(x)\ sin\ (g\Delta\lambda_H x) + (1-S)\ F_B(x)\ sin\ (g\Delta\lambda_B x)\right] \tag{16a}$$

where $\quad F_B(x)$ is the F.T. of $f_B(x)$.

The equation $F_V(x_o) = 0$ is no more simple.

This time the solution depends on all the parameters S, $\Delta\lambda_H$, $\Delta\lambda_B$, $f_m(\lambda)$ and $f_B(\lambda)$!!!

To illustrate this, assume

$$S = \frac{1}{2} \quad , \qquad \Delta\lambda_B = -\frac{\Delta\lambda_H}{2} \quad , \quad f_m(\lambda) = f_B(\lambda)$$

equation 17 then becomes

$$cos\ (g\ \frac{\Delta\lambda_H}{2}\ x_o) = \frac{1}{2}$$

and $g\Delta\lambda_H = \dfrac{2}{3}\ \dfrac{\Pi}{x_o} \quad$ to compare with equation 20.

V. FOURIER TRANSFORM OF I PROFILE; ROBINSON METHOD.

The F.T. of the intensity profile I is given by

$$F_I(x) = \int I \ell^{ix\lambda} d\lambda \tag{21}$$

To give the most simple demonstration of the Robinson method I will be taken from equation 4.

$$F_I(x) = (1-s) F_B(x) + S F_m(x) \cos (g\Delta\lambda_H x) \tag{22}$$

where $F_B(x)$ and $F_m(x)$ are the F.T. of $f_B(\lambda)$ and $f_m(\lambda)$ respectively.
Now assume another line identical in every respect to the former line except its insensitivity to Zeeman effect (g = o). The F.T. of the intensity profile is denoted by $F_r(\lambda)$.

$$F_r(x) = (1-S) F_B(x) + S F_m(x) \tag{23}$$

combining equations 22 and 23

$$S \left[\cos (g\Delta\lambda_H x) -1\right] = \left(\frac{F_I}{F_r} - 1\right)\left(\frac{F_B}{F_m} (1-S) + S\right) \tag{24}$$

if we assume $f_m(\lambda) = f_B(\lambda)$

$$S \left[\cos (g\Delta\lambda_H x) -1\right] = \frac{F_I}{F_r} - 1 \tag{24a}$$

the right hand side is determined from observation and equation 24a may be solved easily to find $\Delta\lambda_H$ and S.
The method has been developed by Robinson to a general distribution of magnetic fields and also for the use of a reference line which is not completely insensitive to Zeeman effect.
The method is sensitive to strong fields >1000 gauss. At present this method is the only one to measure directly magnetic fields related to stellar activity.
The method has been applied to faculae (Robinson, 1980). The field found was of 1600 gauss occupying 10% of the surface.
The first serious limitation of the method comes from the assumption $f_m(\lambda) = f_B(\lambda)$ that allows to use equation 24a, the solution of equation 24 is model dependent through the ratio

$$\frac{F_B}{F_m} \quad \text{(unknown in general).}$$

A second problem is to find a couple of identical lines. In practice the two lines are "slightly" different and a "correction" is necessary. The method becomes reliable only when the effect due to magnetic field is much stronger than the correction. A third limitation comes from the use of Sears formulae when the magnetic field is not longitudinal.

VI. THE INFRARED LINE. Harvey and Hall (1975)

For the same velocity distribution, magnetic field and g factor, the ratio of Zeeman splitting to Doppler broadening is proportional to wavelength only. At 1.6 μ Zeeman splitting will be relatively three times larger than in the visible. One may expect that σ components becomes completely separated and their wavelengths given by the peaks of the V profile.

Harvey and Hall (1975) observed Zeeman effect with a line at 1.6 μ. They found H = 1600 gauss and the relative surface occupied by the field was about 10% (if f_m is assumed to be equal f_B). Also, they found a downward motion in the flux tube with a velocity of about 2 Km/sec. For photospheric lines observed in the visible, downward motions of about 0.5 Km/sec. correlated with magnetic fields in faculae and network were practically the rule, and mentioned by almost all observers. Some doubts about this red shift arose from the questions whether it comes from the flux tube only or from the surrounding as well. Also the nature of the red shift is not completely clear since it is generally measured relatively to the quiet photosphere. The lines in the quiet photosphere are not symmetric and possess some blue shifts due to the convection. The apparent observed red shift needs some correction before it may translate to "absolute" velocities. But in any case the red shift in the infrared line seems to be so strong that the only conclusion could be that velocity in flux tube increases with depth. The infrared lines at 1.6 μ being formed deeper than the lines in the visible.

More observations have been taken by Giovanelli (1977) and all the data was corrected and compiled to give a picture of flux tube with increasing downward motion with depth (Giovanelli et Slaughter, 1977).

The result of these observations is that it is difficult to conceive strong and thin flux tube with the observed regime of velocities.

We may imagine stationary situation at least as a first approximation. For a thin flux tube the following equations apply (Unno and Ribes, 1979).

$$P_i + \frac{B^2}{8\pi} = P_e \qquad (25) \qquad \frac{dP}{dz} = -\rho g - \rho v \frac{dv}{dz} \qquad (26)$$

$$B = A \frac{P_i}{T_i} v \qquad (27) \qquad P = \frac{R}{\mu} \rho T \qquad (28)$$

Where P_i and P_e are gaz pressure inside and outside, B - magnetic field, v - velocity, and T - temperature.

The first equation represents pressure balance between inside and outside this magnetic tube. The second equation stands for the equation of motion parallel to magnetic field (stationary). Third equation results from magnetic flux conservation and frozen matter condition in magnetic field (stationary). A is a constant. The last one is the perfect gaz state equation.

We define scale height H_x for a variable x by

$$\frac{1}{H_x} = - \frac{\frac{dx}{dz}}{x} \tag{29}$$

where z is the height variable. Then it is easy to derive

$$\cdot \beta' = \frac{P_e}{\frac{B^2}{8\pi}} = \frac{H_{P_e}}{H_{P_e} - H_{P_i}} \left(1 - 2 \frac{H_{P_i}}{H_B}\right) \tag{30}$$

where H_{P_e}, H_{P_i} and H_B are the scale height for P_e, P_i and B respectively.

Since v << sound velocity the scale height for pressure H_p will be proportional to the temperature.

$$\beta' \approx \frac{T_e}{T_i - T_e} \left(1 + 2 \frac{H_{P_i}}{H_v} - 2 \frac{H_{P_i}}{H_T}\right) \tag{31}$$

Where T_e and T_i are the temperature outside and inside the tube and H_v and H_T scale heights for velocity and temperature respectively.

First conclusion from equation 31 is that for the condition $\beta' > 0$ the temperature inside must be higher than outside $T_i > T_e$. This is accepted by many observers.

We may now give typical estimates for the quantities in equation 31.

$$H_T \sim 1200 \text{ Km} \qquad H_{P_i} \sim 130 \text{ Km} \qquad \frac{H_{P_i}}{H_T} \ll 1$$

It is difficult to assume an excess of temperature more than 1000 degrees

$$T_i - T_e \sim 1000° \text{ K}$$

From this observation H_v is positive and roughly $\frac{H_p}{H_v} \sim \frac{1}{2}$

$$\beta' \approx 12 \tag{32}$$

β' is very high and as a consequence magnetic field cannot be very strong in a thin magnetic tube. Scale height for magnetic field is small.

$$H_B = \frac{2}{3} H_{P_e} \tag{33}$$

That is, magnetic fields decrease very rapidly with height. In view of these difficulties, Giovanelli (1977) suggested a departure from "frozen" matter conditions.
It seems we have to abandon something : either the image of "thin" magnetic flux
tube that is drop equation 25 (valid in absence of magnetic tension), or part of the
data obtained with the infrared line : strong field ? or velocity ?

VII. CENTER OF GRAVITY METHOD (Rees et al. 1979 ; Semel, 1981).

1) Description of the method.

By definition the centers of gravity of the profiles (I ± V) are given by

$$\lambda_{G\pm} = \frac{\int (I_c - (I \pm V)) \lambda \, d\lambda}{\int (I_c - (I \pm V)) \, d\lambda} \tag{34}$$

where I_c is the continuum intensity.

When equations 1 to 5 in § I apply, it is easy to show that the apparent magnetic
field H_A is given by

$$H_A = \frac{\lambda_{G-} - \lambda_{G+}}{2C \, g} = \frac{S \cdot I_{mc} W_m H}{I_c W} \tag{35}$$

$$W = \frac{S I_{mc} W_m + (I - S) I_{Bc} W_B}{I_c} \tag{36}$$

where I_{mc} and I_{Bc} are the continuum intensities in the magnetic and background elements respectively. W_m, W_B and W are the equivalent widths of the line in the magnetic, background and observed elements respectively.

-When the magnetic field is not purely longitudinal equation 35 is still a good approximation, where H stands for the longitudinal component of the field.

-Rees et al. (1979) extended equation 35 to 3 dimensional configurations of the magnetic field.

Observational test of this method in sunspots was satisfactory. However, in faculae
simultaneous observations with several lines have shown serious discrepancies indicating unresolved structure : the only possible interpretation (Semel, 1981).

If H_i and H_j are the observed fields with lines i and j respectively by equation 35:

$$\frac{H_i}{H_j} = \frac{W_{mi}}{W_i} \Big/ \frac{W_{mj}}{W_j} \tag{37}$$

where W_{mi}, W_i, W_{mj}, W_j are the equivalent widths in the magnetic and background elements and for the lines i and j respectively.

Thus the observed ratio $\dfrac{H_i}{H_j}$

tells us about the relative weakening of the corresponding lines in the magnetic element.

A correlation was found between the observed magnetic fields and equivalent widths : the stronger the line, the stronger the field observed.

We may conclude that, in flux tubes, weak lines weaken relatively more than strong lines. Curve of growth is likely to change appreciably in the magnetic element : lines tend to be less saturated.

2) Tentative to determine H (Bezanger, in prep.)

From eqaution 35 we can obtain easily

$$H = \frac{H_i \dfrac{W_i}{W_{Bi}} - H_j \dfrac{W_i}{W_{Bj}}}{\dfrac{W_i}{W_{Bi}} - \dfrac{W_j}{W_{Bj}}} \qquad (38)$$

where W_{Bi} and W_{Bj} the equivalent widths in the background for the two lines respectively are the only unknown on the right hand side.

Note that equation 38 is more general than equation 35 : it may be applied even if $H_B \neq 0$.

There might be different ways to estimate W_{Bi} and W_{Bj}, for example assuming that they are the same as in the quiet photosphere. Here, it was assumed that the background is the same in different points in the faculae and that W_{Bi} can be determined by linear regression.

Consider the relation :

$$W_{i, k} = W_{Bi} + \alpha_i H_{i, k} \qquad (39)$$

where different k means different point in the same faculae.

It was found that the correlation (39) was good enough for the ensemble of points in the same faculae. This led to the determination of W_{Bi}.

When several lines are observed simultaneously, it is easy to see that plotting

$$H_i \frac{W_i}{W_{Bi}} \qquad \text{against} \qquad \frac{W_i}{W_{Bi}}$$

should yield a straight line. The slop of the line determines H.

Observations of four lines in several faculae have been performed by means of the HAO Stokesmeter at Sac Peak Observatory. The result was a distribution of fields from 400 to 940 gauss. Best correlation coefficients were obtained in two cases where the magnetic fields were 500 and 600 gauss.

No model of radiative transfer is necessary in this method. However an assumption concerning the background is necessary and is the source of main limitation in the present two components model.

VIII. FIT OF OBSERVED LINE PROFILES THROUGH COMPUTATION OF A FACULAR MODEL

By computing line formation in LTE, Rees (1971, 1974) has fitted 14 observed lines. Chapman (1977) considered 10 lines and included depth variation of magnetic fields. For best fit it was necessary to consider
   - unresolved structure,
   - kilogauss fields.
Thus, this approach joins the other experiments discussed obove. It is different from other methods in considering in detail radiative transfer and a physical model for faculae. Complete description cannot be given here for such an elaborated calcul-ation. All observables cannot be completely satisfied, for examole, because of NLTE effects, Chapman disregarded four lines considered too strong. But few aspects di-rectly related to effects of magnetic fields will be discussed here.

   - Wilson depression has been determined by Chapman requiring pressure balance according to equation 25. The level of continuum-optical depth = 1 is 160 km lower in the flux tube than in the surrounding. That means a temperature difference of more than 3000° between inside and outside the tube at this level !! From this must result a very strong heat-flow into the cool flux tube. This problem has been dis-cussed by Spruit (1977).

   - The Zeeman effect on intensity line profiles is basically given by equation 1 § I. It must be included to give the best fit for the observed line profiles and lead to the determination of magnetic fields in the present context. While several effects are in competition, it is not easy to evaluate this method even when the calculated line profiles match fairly well the observed ones. But it is possible to discuss the case when the Zeeman effect may be dissociated from others.
In § II the line ratio technique has been described by using two lines identical ex-cept their g factors. When Frazier and Stenflo observed such lines e.g. 5250 (g = 3) and 5247 (g = 2) they found that their intensity profiles were practically the same in faculae as well as in quiet sun. So that line ratio technique applied to intensi-ty profiles of these line indicated weak field. The authors then suggested that un-detected difference between intensities of these lines is due to depth dependence of the magnetic field and cannot be considered when Milné-Eddington model is used.

Now Chapman considered depth dependance of magnetic field with a realistic line formation theory, thus the problem can be rediscussed now. The calculated profiles are quite close to the observed ones and we may think that the problem is settled at first approximation. However the difference between the calculated profiles of the two lines is often not as small as the observed one. We may wonder what would be the resultant magnetic fields if in the processus of fitting a particular weight were given to the difference in intensity of the two lines ?

IX. CONCLUSION

How strong is the magnetic field in an unresolved structure ? The answers given are by no means unique and the divergence is quite significant. The first reason for this is simply that the magnetic field is not unique : the distribution of the field may be wide, perhaps from 600 to 1600 gauss. The magnetic field may be time dependent (Wilson, 1983) and changes with height. Photospheric and chromospheric fields are quite different.

The second reason for the divergence in field determinations may come from too simple assumptions. In general the model of two components assumed is even more degenerate than the one given in §I. For example, assumptions that Doppler broadening is the same everywhere or that the non magnetic component has the same atmosphere as in the quiet sun etc. That may explain why on the one hand the determination of magnetic field is not unique and on the other sometimes it is not possible to satisfy all observations.

Our judgement should not be too severe in spite of the discrepancies. Let us remember that often the apparent field is quite weak, say 100 gauss, and we are looking for a method to infer from it the "true" field about one order of magnitude stronger. This is by no means a "small correction", the task is really difficult.

A given solar feature with apparent weak field may be analyzed by two different methods. Both may certify that the magnetic structure is unresolved and that the "true" field is "several" times stronger than the apparently observed one. Quantitatively, in this particular case "several" may mean seven for the first method and fourteen for the second. We may disregard these fields strengths as overdetermined obtained through far-reaching extrapolation, and still consider both methods as powerfull and usefull.

Naturally solar physicists cannot be satisfied with strong and undetermined magnetic fields, somewhere between 700 and 1400 gauss. The next step is obviously to improve spatial resolution with instruments like spectrotourelle in Pic-du-Midi, the future VTT and THEMIS in Canary islands. Again methods like those discussed here must be used to test whether magnetic structure has been resolved. Eventually it may be still necessary to find the "true" field but this time with much smaller "correct-tion" than those needed now. In the next decade LEST and especially SOT in space will attain highest spatial resolution.

BIBLIOGRAPHY

Beckers, J.M. and Schröter, E.H., 1968, Solar Physics, 4, 142.
Bezanger, C. (in preparation).
Chapman, G.A., 1977, Ap. J. supplement series, 33, 35.
Daras-Papamargaritis, H. and Koutchmy, S., 1983, A & A, 125, 280.
Frazier, E.N. and Stenflo, J.O., 1978, A & A, 70, 789.
Giovanelli, R.G., 1977, Solar Phys., 52, 315.
Giovanelli, R.G. and Slaughter, C., 1977, Solar Phys., 57, 255.
Harvey, J. and Hall, D., 1975, Bull. of Am. Astron. Soc., 7, 459.
Harvey, J.W., Livingston, W.L., and Slaughter, C., 1972, Line Formation in the
    Presence of Magnetic Field, HAO.
Rees, D.E., 1971, Ph. D thesis, University of Sydney.
Rees, D.E., 1974, IAU Symposium 56, Chromospheric Fine Structure, ed. R.G. Athay.
Rees, D.E. and Semel, M., 1979, A & A, 74, 1.
Robinson, R.D., 1980, Ap. J., 239, 961.
Semel, M., 1981, A & A, 97, 75.
Semel, M., Ribes, E. and Rees, D., 1981, Proceedings of the Japan-France Seminar on
    Solar Physics, ed. F. Moriyama and J.-C. Hénoux.
Simon, G.W., and Zirker, J.B., 1974, Solar Physics, 35, 331.
Spruit, H.C., 1977, thesis, Utrecht, Netherlands.
Tarbel, T., Title, A., 1977, Solar Physics, 52, 13.
Unno, W., 1956, Publ. Astr. Soc. Japan, 8, 108.
Unno, W. and Ribes, E., 1979, A & A, 73, 314.
Wiehr, E., 1978, A & A, 69, 279.
Wilson, P.R. and Simon, G.W., 1983, Ap. J., 275, 805.

COMMENTS

From : A. Righini
I want to make a remark about inward motion in flux tubes. Recently we have observed the line bisector in some photospheric iron lines in active regions. The conclusion is that the line approaches the gravitational red shifted laboratory line suggesting that the red shift, often interpreted as an inward motion, is in reality a minor convective blue shift of the line.

From : C. Zwaan
Together with Jack Harvey we measured I and V profiles in the IR using the KPNO FT Spectrometer. We found for the majority of active and quiet regions typical field strength between 1200 and 1800 gauss, and - for all regions except emerging flux regions - no significant downflow.

ACKNOWLEDGEMENTS

I am gratefull to my colleagues Z. Mouradian, J. Rayrole and Cléo Pace for helpfull discussions.

EVERSHED EFFECT AND MAGNETIC FIELD IN PENUMBRAL FINESTRUCTURES

E. Wiehr, Universitäts-Sternwarte, 3400 Göttingen, W.-Germany
G. Stellmacher, Institut d'Astrophysique, 75014 Paris, France

Individual penumbral finestructures occur in highly resolved white light photographs as a 20 - 30% intensity fluctuation with typical widths of 200 - 300 km (Muller 1973). In contrast, best resolved penumbral spectra (see e.g. Mattig and Mehltretter 1968; Beckers and Schröter 1969; Stellmacher and Wiehr 1980, 1981) typically show continuum streaks with 5 - 8% intensity modulation and widths of 1000 km. Consequently, the investigation of penumbral spectra yields only restricted information about true penumbral finestructures. From averaged intensities of locally bright and dark penumbral finestructures published by Muller (1973) Kjeldseth-Moe and Maltby (1974) deduce two corresponding temperature stratifications. However, Stellmacher and Wiehr (1980) were unable to fit observed line asymmetries with superposed line profiles calculated with these models.

From an exceptionally highly resolved penumbra spectrum taken at the evacuated Locarno telescope on July 4, 1983 at $\theta = 30^{\circ}$, Wiehr et al. (1984) confirm previous findings by Stellmacher and Wiehr (1980) that the line core shifts are predominant in dark, the line asymmetries, however, in bright spectrum streaks (Fig. 1). In addition, Wiehr et al. find for the $Fe^{+}5264.8$ line a *continuous dependence* of these line parameters on the continuum brightness in the sense: *the darker the continuum, the smaller the line asymmetry and the larger the line core shift* (Fig. 2). This result fits the suggestion by Stellmacher and Wiehr (1980) that the line asymmetries are a result of the limited spatial resolution achieved in the spectra, which yields a superposition of differently bright structures with their corresponding different line core shifts. The only parameter which then varies with continuum brightness would be the line displacement, thus suggesting an increase of the Evershed flow with decreasing temperature of the penumbral finestructures.

Wiehr et al. (1984) also investigate the line widths determined from the steeper (i.e. 'symmetric') part of the profiles - thus disregarding the profile asymmetries. They find for $Fe^{+}5264.8$ (g = 0.1 $\hat{=}$ 2mÅ penumbral splitting) an increase of these line widths with decreasing continuum brightness (see Figs.1 and 2). Considering that the thermal line widths are slightly smaller in the cooler dark structures, this increase might be interpreted as an enhancement of the *microturbulence* with decreasing continuum brightness, thus supporting the idea by Beckers and Schröter (1969) of an enhanced microturbulence generated by the Evershed flow. In contrast, Stellmacher and Wiehr (1981, end of §6) give an interpretation in terms of *macroturbulence*: If the relation between velocity and *true* structure brightness is non-linear (still compatible with the *spectrum* brightness dependence in Fig. 2a!), a superposition of unresolved structures might well yield broader profiles for darker spectrum streaks.

The systematic variation of the line width with structure brightness has an important consequence for the discussion of magnetic field fluctuations in penumbral finestructures: From a highly resolved penumbral spectrum of Fe 6302.5 (g = 2.5) Stellmacher and Wiehr (1981) find no systematic fluctuations of the line broadening by Zeeman effect. Their numerical simulation of line profiles at limited spatial resolution (~ 2" in the spectra) show that field strength fluctuations up to 1000 Gs would be compatible with that result. These calculations consider thermal but not turbulent line broadening: The above discussed turbulence increase in darker structures may compensate an enhanced Zeeman broadening in brighter structures. On the other hand, a possible enhancement of the magnetic field in the dark structures could only be considerably below that 1000 Gs excess, otherwise the additional turbulence broadening found would yield a measureable line broadening in the spectra.

References

Beckers, J.M., Schröter, E.H.: 1969, Solar Physics 10, 384
Kjeldseth-Moe, O., Maltby, P.: 1974, Solar Physics 36, 101
Küveler, G., Wiehr, E.: 1984, Astron. Astrophys. (in press)
Mattig, W., Mehltretter, J.P.: 1968, I.A.U. Symp. No. 35, 187
Michard, R.: 1953, Ann. d'Astrophys. 16, 217
Muller, R.: 1973, Solar Physics 32, 409
St.John, C.E.: 1913, Astrophys. J. 38, 341
Stellmacher, G., Wiehr, E.: 1980, Astron. Astrophys. 82, 157
Stellmacher, G., Wiehr, E.: 1981, Astron. Astrophys. 103, 211
Wiehr, E., Koch, A., Knölker, M., Küveler, G., Stellmacher, G.: 1984,
     Astron. Astrophys. (in press)

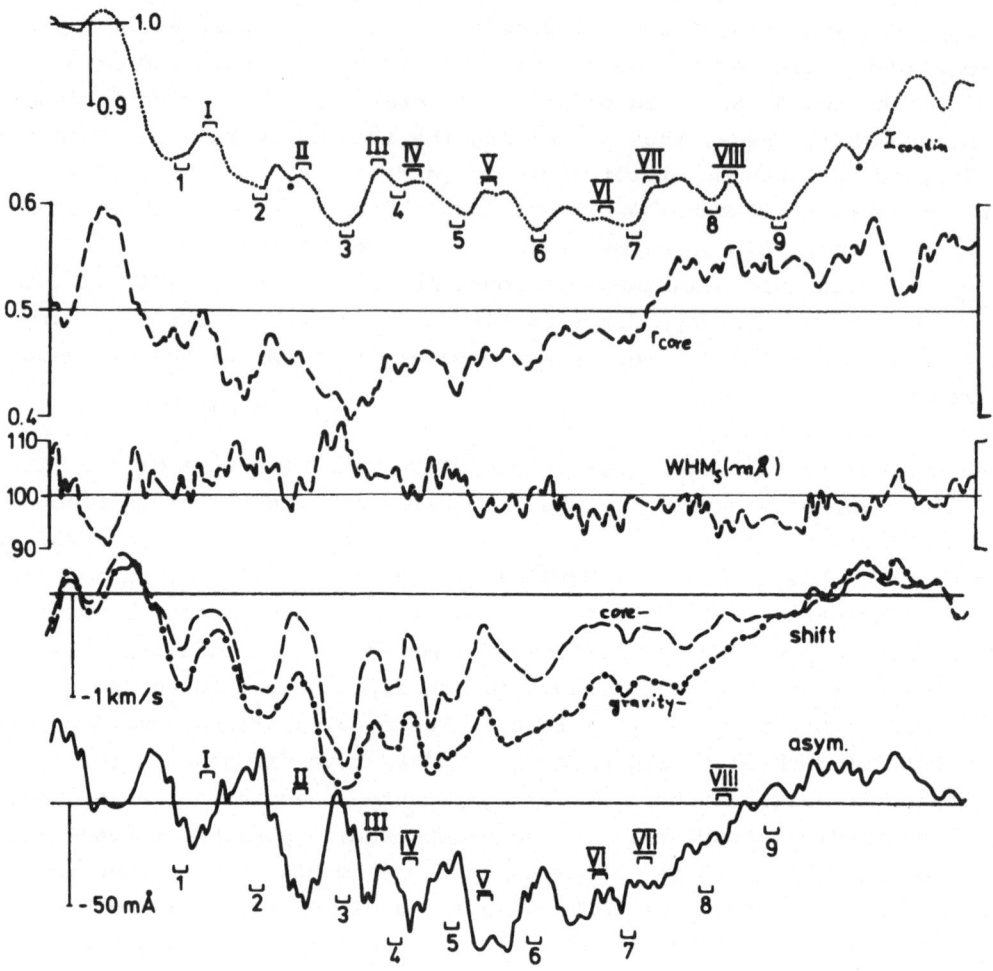

Fig. 1: Variation of Fe^+ 5264.8 Å over the centerside penumbra from a spot at $\theta = 30°$ observed on July 4, 1983 at the evacuated Locarno telescope.

I_{cont} = continuum intensity in units of the neighbouring photosphere;

r_{core} = central line depth;

WHM_s = width at half maximum of the steeper (i.e. 'symmetric') line flank;

shift = displacement of core, respectively, gravity center of the line;

asym = bisector inclination extrapolated from continuum to zero level.

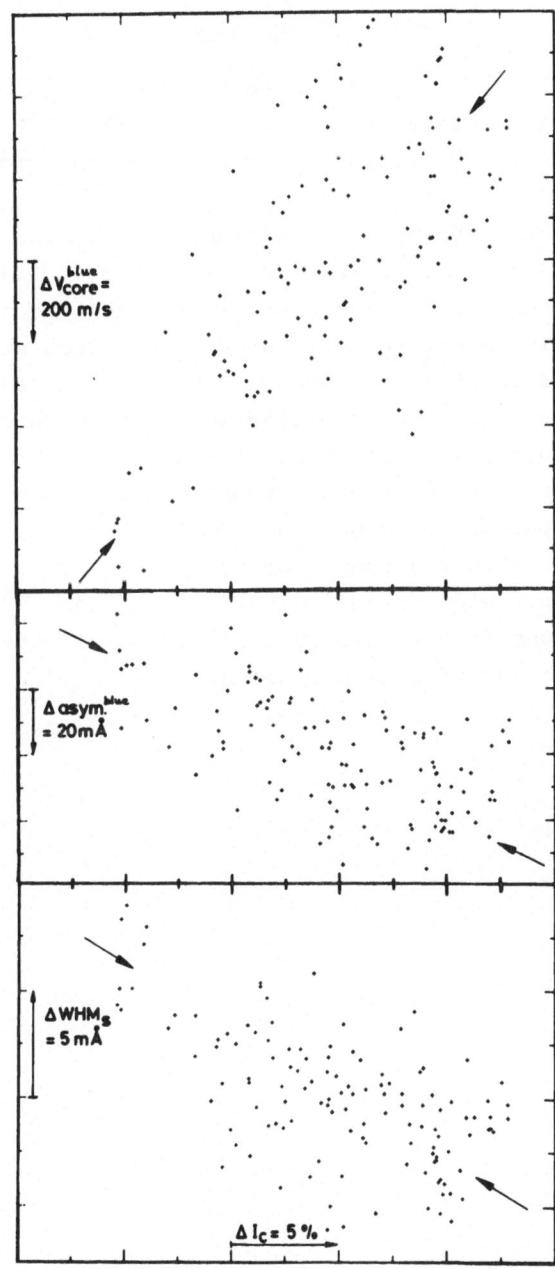

Fig. 2: Dependence of some parameters from Fig. 1 vs. continuum
intensity after elimination of the overall trend (quadratic
mean).

Discussion

Mattig : Are you sure, that the change or not change of halfwidths of the lines is really related to the magnetic field ? Dark and bright structures could have different other effects than magnetic field strength.

Wiehr : I have shown in my second slide a systematic increase of the width of a non-magnetic line (Fe$^+$ 5264,8; g = 0.1 1.3 mÅ/SOOO Gs) with decreasing continuum brightness. This might indicate enhanced microturbulence in darker penumbra structures. Such effect, of course, has to be subtracted from the magnetic broadening. Additional information might be obtained when investigating lines with different Zeeman broadening but equal excitation and strengths.

Alissandrakis : Do you observe an abrupt change in the magnetic field at the penumbra-photospheric boundary ?

Wiehr : We do not find contradiction to Beckers and Schröter who report a decrease of the field strength within 1"-2" from B^{max}/2 to zero. But such investigations require highly resolved polarization profiles rather than our simple study of fine broadening.

BRIGHT POINTS IN Hα WINGS AND CONNECTED MASS FLOWS IN THE SOLAR CHROMOSPHERE

P. Mein, J.-M. Malherbe, Z. Mouradian, N. Mein and B. Schmieder
Observatoire de Paris, Section de Meudon
92195 Meudon, Principal Cedex, France
and
R. Muller
Observatoire du Pic-du-Midi et de Toulouse
65200 Bagnières-de-Bigorre, France

ABSTRACT

Many mass flows in the chromosphere are physically connected with bright photospheric points. Bidimensional spectroscopy with high spatial resolution is very well suited for the study of both kinds of structures. We give some preliminary results derived from recent observations of the MSDP at the Pic-du-Midi.

I - INTRODUCTION

It is well known that photosphere and chromosphere are strongly coupled in all ther-
modynamical processes. Dark surges visible in the Hα core are rooted in bright faculae. Little surge-like features can be associated with "moustaches" or Ellerman
bombs. Perhaps some connection does exist between spicules and bright points of filigrees. Any investigation of such structures requires simultaneous observations at
different places of the sun (bidimensional field) and different wavelengths (spectroscopy). Moreover, high spatial resolution is necessary in most of the cases. These
required capabilities are joined together in the MSDP observations, which consist in
several simultaneous images of the same field, in a set of wavelengths corresponding
to a whole line profile. The "Multichannel Subtractive Double Pass" spectrographs
have been described in previous papers (Mein, P., 1977; 1980; 1984).

II - MSDP AT THE TURRET DOME OF THE PIC DU MIDI : HIGH RESOLUTION IN BIDIMENSIONAL SPECTROSCOPY

A time sequence was obtained on October 19, 1983, at the Pic-du-Midi, with good seeing conditions. The field (4' x 30") included a spot located 4°S, 26°W. The spectral
resolution was 0.26 Å (local bandpass 0.09 Å). The exposure time was short enough
(0,2 s) to ensure high quality pictures.
The figure (1) shows the result of data-processing. The field is restricted to

2' x 23", and the pixel size is 0.22". The four maps are respectively :

1. Intensity fluctuations in the Hα wing (∿ ± 1Å). We can notice the spot and surrounding bright points.

2. Intensity fluctuations in Hα core (± 0.32 Å). The fibrils are visible, as well as many little absorbing features.

3. Bright regions in the wings, and dark regions in the core, have been visualized simultaneously in order to show the geometrical connections.

4. Dopplershifts in bright points (at ± 0.65 Å) and dark features (± 0.32 A). Upward velocities are represented by bright areas.

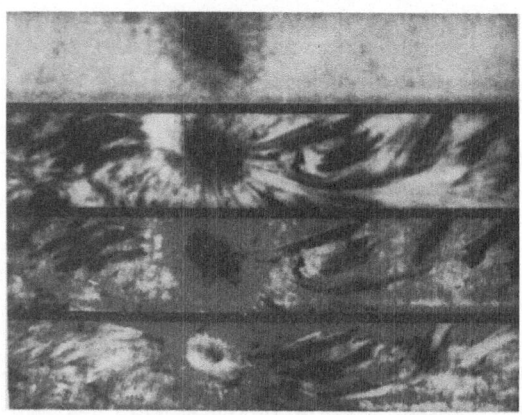

_ Figure 1.

III - SELECTED STRUCTURES ; DOPPLERSHIFTS AND LINE PROFILES

We have selected two structures consisting in a bright point and the connected absorbing feature. The first one A is located between the spot and the disk-centre. The second one B is on the other side (fig. 2). In both cases :

- The dark feature points towards the bright point.
- The dark feature does not extend beyond the bright point.
- The radial velocity is approximately continuous between dark feature and bright point.

The sizes of the selected structures are very small : 10" x 1" (A) and 5" x 1" (B). The figure (2) shows the intensity curves (broken line) and dopplershifts (solid line) along each structure. Blueshifts are noted positively.

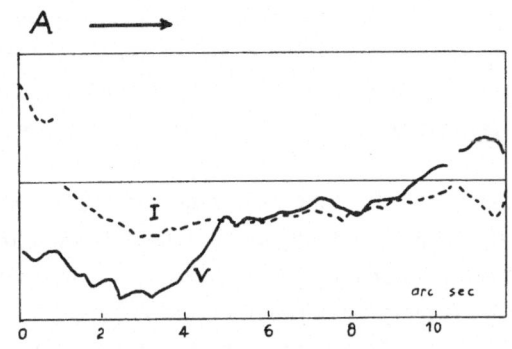

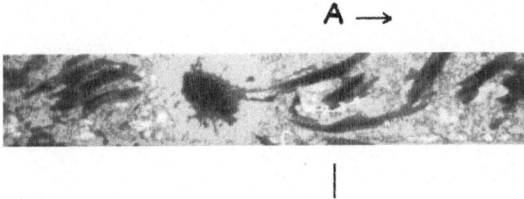

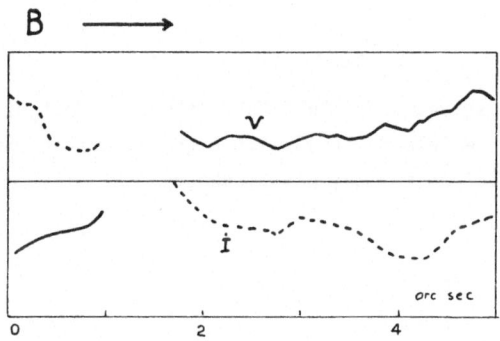

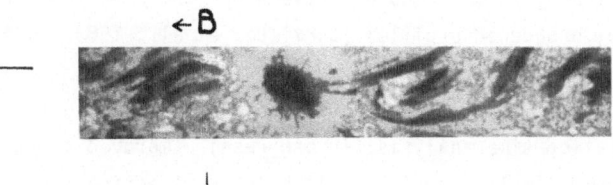

— Figure 2.

The dynamical behaviour is similar in A and B : downward motions in the bright point and the beginning of the dark feature, bueshifts at the end of the dark feature. The well-known reverse Evershed effect (Dialetis et al., 1984) can be recognized by the relative length of the redshifted part in the absorbing feature, which is greater in the case of A.

Since we did not analyze so far the time evolution of the structure, we can only suggest a few models which could account for the instantaneous observations (fig. 3): an Evershed flow in curved stationary magnetic lines (a), opposite flows in straight lines (b), a free fall inside rising magnetic lines (c), ... In fig. 3, blueshifts are represented by solid lines. Arrows indicate the material velocity.

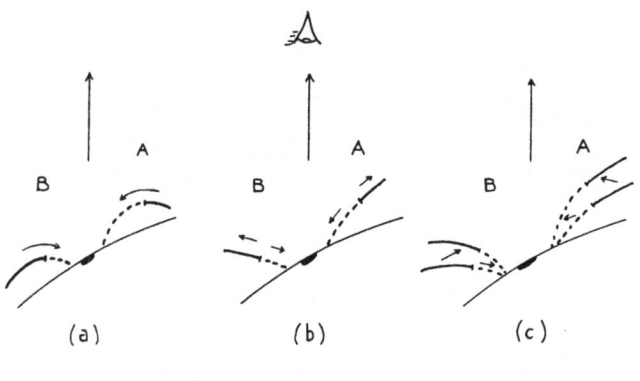

—Figure 3.

Lines profiles can be derived from the 9 MSDP channels. Most of them are asymmetric and can be explained by a "cloud model" with high velocities. Some of them are broadened, and could be explained in many ways : shears, helicity, microturbulence ...

IV - CONCLUSION

Many of the observational aspects of the structures A and B are similar to surge characteristics : continuity in the doppler curve along the structure, redshifted foot and blueshifted top, broadened profiles (Schmieder et al., 1983; 1984, Mein, P. et al., 1983)

Do similar behaviour imply similar acceleration mechanisms ? The answer cannot be given without a detailed time analysis (in progress). Improved solutions of transfer and MHD problems, used simultaneously, could help in rejecting unrealistic dynamical models.

V - DISCUSSION

O. Engvold. Your velocity images showed very pronounced blueshifts in the sunspot umbra. Is this mainly an instrumental effect due to low light level or does it represent a true mass flow ?

P. Mein. Dopplershifts in the umbra are computed in the line core, where the light level is not too low compared with the surrounding chromosphere. Such blueshifts should be checked by further measurements, and also by analysis of time evolution.

C.E. Alissandrakis. The first structure that you showed resembles arch filament systems. Did you check the polarity of the magnetic field at the location of the bright points ?

P. Mein. Unfortunately, we have not any simultaneous magnetograms of this active region. But it is very likely that magnetic field gradients play an important role in acceleration mechanisms.

BIBLIOGRAPHY

Dialetis, D., Mein, P., Alissandrakis, C.E. : 1984, Astron. Astrophys. (submitted).

Mein, P. : 1977, Solar Phys. 54, 45.

Mein, P. : 1980, Proceedings of the Japan-France Seminar on Solar Physics, p. 285.

Mein, P., Schmieder, B., Vial, J.-C., Tandberg-Hanssen, E. : 1983, VIIth European Regional Astronomy Meeting (Florence).

Mein, P. : 1984, Workshop "Chromospheric Diagnostics and Modelling" (Sacramento Peak) to be published.

Schmieder, B., Vial, J.-C., Mein, P., Tandberg-Hanssen, E. : 1983, Astron. Astrophys 127, 337.

Schmieder, B., Mein, P., Martres, M.-J., Tandberg-Hanssen, E. : 1984, Solar Phys., to be published.

QUANTITATIVE FILTERGRAM IMAGERY OF THE SOLAR ACOUSTIC OSCILLATIONS

Audouin Dollfus - Observatoire de Paris-Meudon, France
Didier Dillard, Ecole Polytechnique, France.
Jean-Marc Monguillet, Ecole Polytechnique, France.

ABSTRACT

The solar birefringent filter FPSS is used at the focus of a 28cm diameter solar refractor, at Meudon Observatory, to produce images in a large field of view, with a spectral resolution of 4.6×10^4. The instrument is calibrated for absolute Doppler velocity measurements in the wings of the strong line NaD1, and is presently used for quantitative imagery of the radial velocity motions in the photosphere with a resolving power of near 0.8 arc sec. The coherent five minutes acoustic oscillations of large spherical harmonics degree ℓ (of a few hundreds) are directly imaged. The power spectra of the velocity structures are derived over the fields; Fourier transform spectra are computed. Several maxima are observed which characterize at least 3 superimposed coherent oscillations systems of typical elements diameters 14, 4.3 and 2.1 Mm and probably a system with 22 Mm. Vertical velocities are derived. Spatial filtering by optical and by computerized techniques produces separate images for each of these coherent oscillating modes.

1 - THE SOLAR INSTRUMENT FPSS

The solar birefringent filter FPSS (Filtre Polarisant Solaire Sélectif) of Observatoire de Meudon is presently used at the focus of a refractor of 28cm diameter, directly pointed at the Sun, and produces an on-axis and polarization-free image of the Sun, enlarged by a Barlow lenses system to a diameter of 84mm in a wide-field mode, or 135mm for high resolution purpose. The 30mm diameter entrance circular field of view isolates a portion of the Sun of respectively 10.9 arc min or 6.8 arc min diameter, which is re-imaged at the same size at the exit focal plane of the instrument, on a Kodak 2415 emulsion. The FPSS instrument will be described elsewhere (Dollfus et al, 1985). Information is already found in Dollfus (1984).

For the work which follows, the instrument was used to record radial velocities by Doppler shifts in the wings of the NaD1 5896 Å line. The FWHM of the line, of 0.38 Å, is 2.75 times larger than the width 0.138 Å of the band transmitted by FPSS. We record image sequences at several distances $\Delta\lambda$ from the line axis, in order to explore the radial velocity field at different depths into the photospheric layers. The smallest working distance is ±0.05 Å and corresponds to a height above the $\tau 5000=1$ reference level of around 500 km, which is near the minimum of temperature. The images which are analysed here are taken with $\Delta\lambda=0.08$ Å and their height of formation is around 400 km.

The intensities are directly converted into radial velocities; the transmission band is tuned to scan along the wing of the line; the signal averaged over a large portion of the field feeds a photomultiplier and a recorder. The response is almost linear for a velocity range of ±5 km sec^{-1}. A longitudinal motion of 100 m sec^{-1} produces a contrast of 1.3%. The threshold sensitivity is around 50 m sec^{-1}.

The images are analysed quantitatively by digitalization on a PDS microdensitometer with 12 intensity bits. For the large field images, the scanning aperture is a square of 25μm size (0.6 arc sec. at the solar surface, or 400 km). For the high resolution images, the size aperture of 20 μm corresponds to 0.3 arc sec. or 200 km.

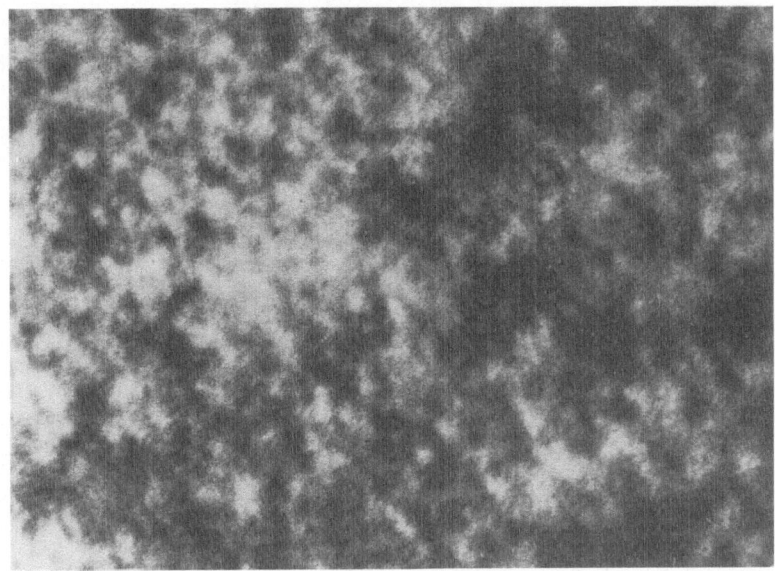

Figure 1 : *Acoustic oscillations of high degree in the photosphère. Velocities are dark for upward, bright for downward. Filter FPSS at D_1 -0.09 Å Field 124 x 90 Mm² - July 13, 1983 at 11:56 UT.*

2 - DIRECT IMAGES OF THE 5 MIN. ACOUSTIC OSCILLATIONS

Figure 1 shows the portion of an image taken with the high resolution 84mm diameter solar size. The area is centered at 0.21 $R_\odot$ from disk center, at latitude 15°N and at central meridian. The frame size is 124x90 Mm². The filter is tuned at -0.09 Å from the axis of NaD1 5896 Å. The angular resolution reachs 0.75 arc sec. The bright elements are tracing the downward motions, the dark patches are the upward moving areas. Lifetime analysis shows that these structures are oscillatory phenomena. They are produced by the 5 minutes acoustic oscillations of large spherical harmonics degree ℓ in the photosphere (Deubner, 1981).

The somewhat irregular mottled patterns which are disclosed are similar but sharper than those described by other observers (Noyes, 1967, Sheeley and Bhatnagar, 1971, Bray et al., 1974) using spectroheliograms of filtergrams taken in the same and other lines. Other images taken with the wide-field mode have been published by Dollfus et al (1984) and by Dollfus (1984).

3 - SIZE OF THE OSCILLATING ELEMENTS

Fourier transforms of the images have been computed, using a FFT algorithm :

$$F(u,v) = \frac{1}{N} \sum_{j=1}^{N} \sum_{k=1}^{N} f(j,k)\, e^{-\frac{2\pi i}{N}\left[u(j-1) + v(k-1)\right]}$$

u and v are the spatial frequencies. The features appear in the four corners of a square, they are reframed at center.

For the wide field images, a zone of 400 X 400 pixels was explored by steps of 2. The Fourier image, averaged in polar coordinates, produces the radial distribution of amplitudes shown in fig. 2(a) in which the horizontal axis is the size of the oscillatory elements, or half the spatial frequency. A maximum characterizes a typical size of 4.3 ± 0.2 Mm for the oscillating elements, hereafter designated Mode I. Another maximum is observed at 13 ± 1 Mm and corresponds to a distinct Mode II. A Mode III is suspected at 22 ± 3 Mm although the field of 160 X 160 Mm² involves only few of these elements.

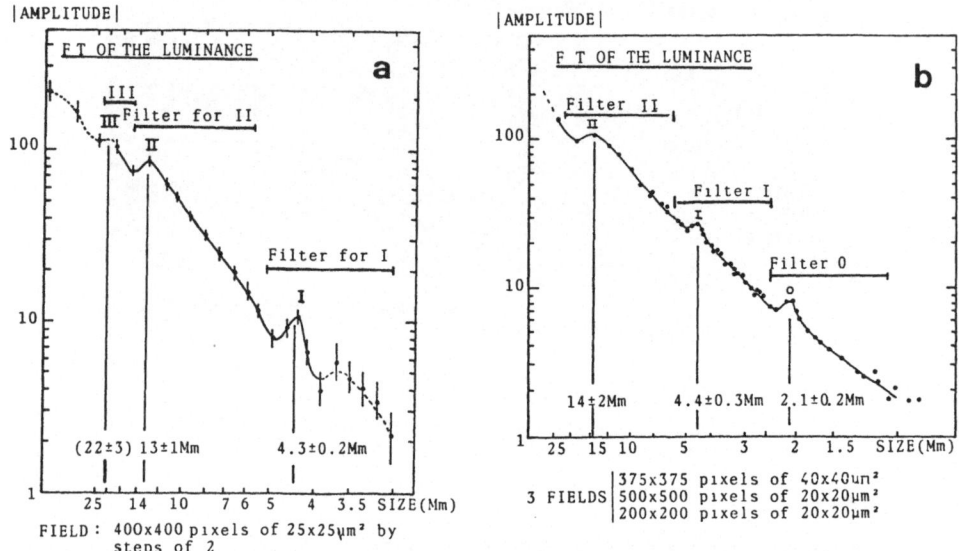

Figure 2 : *Amplitude of the Fourier Transform*

a : *wide field image*   b : *High resolution image*

For the higher resolution image of fig. 1, the FT curve was derived from three areas , as indicated in figure 2(b). The Modes I and II are confirmed, with typical sizes of respectively 4.4 ± 0.3 and 14 ± 2 Mm. A new Mode is exhibited hereafter designated Mode O, with a typical oscillating element size of 2.1 ± 0.2 Mm.

4 – VERTICAL VELOCITIES

Velocity histograms were derived. For the wide field images, in order to minimize the large scale luminance variations, four areas covering each a solar surface of around 570 Mm² are used. These measurements are normalized and added to produce the histogram of fig. 3(a). The histogram of a function is the derivative of the inverse function. The vertical oscillations produce a sinusoïdal distribution of contrast over the field which, along a direction x, is $\Delta i/i = a \sin x$. The derivative of arc sin x is $y = (1 - x^2)^{-1/2}$ and produces two symmetric asymptotic rises at ± a from the zero contrast which is the zero velocity reference. In the fig. 3(a), several symmetric peaks are identified. They are interpreted as the presence of asymptotic raises smoothed by gaussian spreadings. Several superimposed oscillations mode are again indicated; their maximum vertical velocities, derived from the distances between the symmetric peaks, are respectively 220 ± 40 m/sec., 600 ± 60 m/sec. and 1080 ± 80 m/sec. Direct analysis of the images identifies these values respectively with Mode I, Mode II and the less clearly established Mode III.

The same approach was developed with the high resolution frame of fig. 1 for which five areas were selected, 150 X 150 pixels each, covering a surface of 900 Mm² per area. The resulting histogram, in fig. 3(b), isolates again peaks, for Mode II with 500 ± 40 m/sec, for Mode I with 250 ± 25 m/sec and a temptative value of 100 ± 20 m/sec for Mode O which is a lower limit because the contrasts are already reduced near the cutt-off of the modulation transfer function.

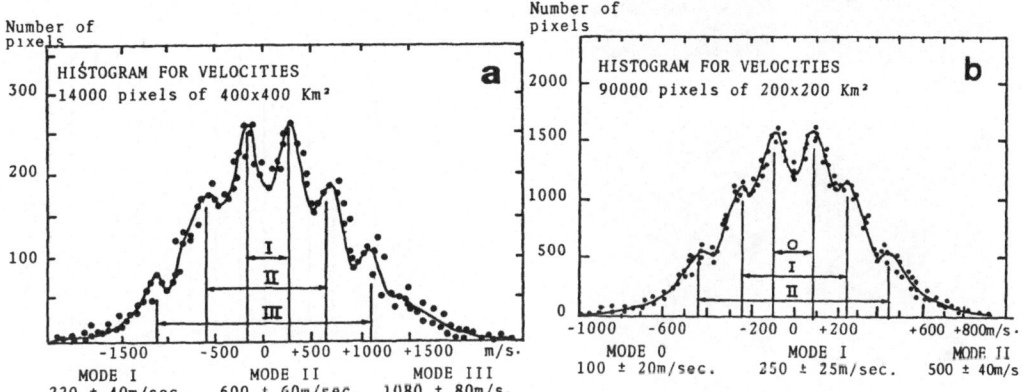

Figure 3 : Histogram for velocities

a : Wide field image   b : High resolution image

5 - PERIODS

The oscillatory character of the structures was analysed after digitalization of the first 40 images of a film, taken on July 15, 1983, starting 07h20m UT for a total duration of 30 min. covering 6 complete cycles of photospheric oscillations, at the rate of an image every 15 sec.. All the images have been re-framed in x and y by feature correlations; they were corrected from vignetting and for large-scale luminance variations, by subtracting a fitted quadratic polynomal field. Then, a same area of 196 X 220 pixels covering a surface of 39 X 44 Mm² was selected on each image.

A temporal Fourier transforms of the image was applied as follows :

$$I(i,j) = \left[[\sum APO(k) \times I_k(i,j) \times \sin(2\pi \times 15 \times (k-1)/T)]^2 + \sum APO(k) \times I_k(i,j) \times \cos(2\pi \times 15 \times (k-1)/T)]^2 \right]^{1/2}$$

in which APO (k) is an apodization function independent of the period. Only the pixels I(i,j) having the proper oscillation period will add coherently their effects to produce a sizeable intensity in the image. Six of these images are reproduced in the fig. 4, for periods increasing by steps from 120 to 420 seconds. There is no area of the field with a period at 120 sec. The maximum is reached at around 300 sec. The histogram (figure 5) shows a maximum at T = 300 sec. (the background is due to permanent defects in the images, which add coherently). Detailed analysis of the scans assign periods of 300 ± 30 sec. for Mode I and 330 ± 60 sec. for Mode II. The periodic character of Mode O was recognized.

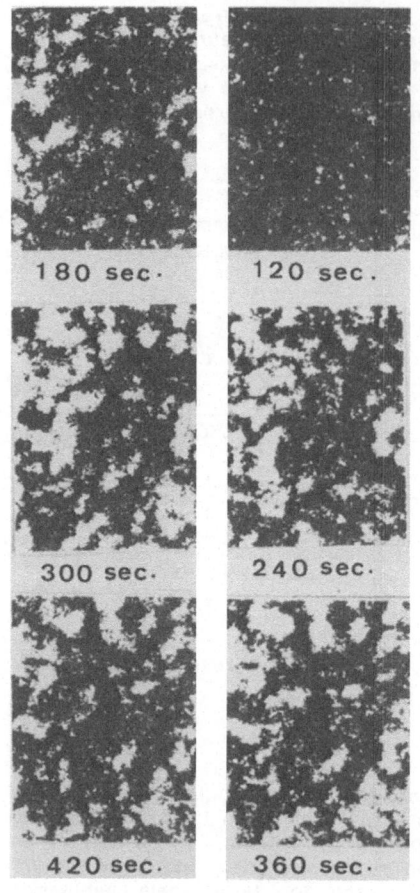

180 sec. 120 sec.

300 sec. 240 sec.

420 sec. 360 sec.

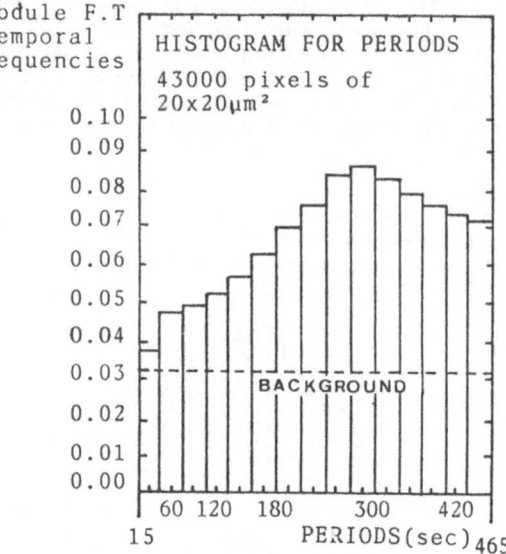

Module F.T
Temporal
Frequencies

HISTOGRAM FOR PERIODS

43000 pixels of
20x20μm²

Figure 5 : *Histogram from the Fourier Transform images.*

Figure 4 : *Temporal Fourier Transform sequence of images.*

6 - FILTERED IMAGES

The distinct oscillation modes were images separately by spatial filtering. Modes I and II were isolated by the optical Abbe method . Modes I, II and III were also filtered and imaged with the computer after digitalization (Dollfus et al, 1984, Dollfus, 1984). We produce here on figure 6 filtered images for the Modes I and O. The transmission bands used are shown in the figure 2(b).

CONCLUSION

The coherent five minutes acoustic oscillations of large spherical harmonics degree ℓ were directly imaged over the photosphere with the instrument FPSS. They appear to split into several distinct typical coherent systems. The numerical data which were derived from quantitative imageries are grouped in the table.

ACKNOWLEDGEMENTS

The authors are particularly indebted to the technical and scientific contribution by J. Advielle, G. Banos, D. Crussaire, L. Damé, M. Jordy, G. Macaisne, V. Michau, T. Morice and P. Titeux.

PHOTOSPHERIC OSCILLATIONS

MODE	SIZE(Mm)	VELOCITY(m/sec)	PERIOD(sec
O	2. 1±0. 2	⩾ 100±20	–
I	4. 5±0. 3	250±25	300±30
II	13 ±3	450±30	330±60
(III)	(22±3)	1080±60	–

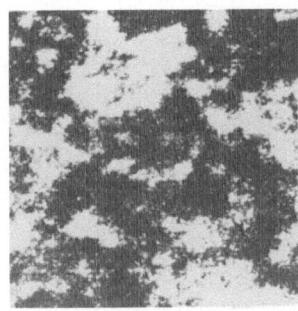 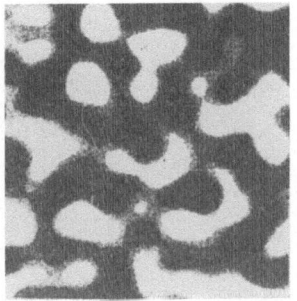 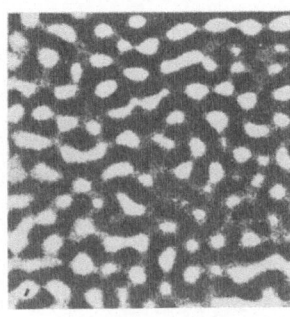

10 Mm

FIELD	MODE	MODE O
200 x 200 Pixels	Filter 2.2 - 6.6 Mm	Filter 1.2 -2.2 Mm

Figure 6 : *Spatial filtering of the oscillations. Modes I and O.*

REFERENCES

- Bray R. J. Loughead R. E. Tapperte E. J. (1974), High resolution photography of the solar chromosphere-XV : Preliminary observations in Fel 6569. 2. Solar Physics 39, 323–326.

- Deubner F. L. (1981), Pulsation and oscillations, in "the Sun as a Star", S. Gordon, editor, NASA SP-450, 65–84.

- Dollfus A. (1984), Stokes parameters modulator for birefringent filters. NASA Workshop Measurements of Solar Magnetic Fields, Huntsville, Alabama, Proceedings.

- Dollfus A., Banos G., Crussaire D., Jordy M., Michau V, Morice T., Titeux P. (1984), Imagerie quantitative des oscillations photosphériques. Compte-Rendus Acad. Sci. Paris 298, serie II, n° 17, 735–740.

- Dollfus A., Colson F., Crussaire D., Launay F., (1985), The birefringent filter FPSS for solar quantitative imagery (to be published).

- Noyes R. W., (1967), Solar velocity fields, in "Cosmical gas dynamics", R. N. Thomas editor, IAU Symp. 28, 297–326.

- Sheeley N. R., Bhatnagar A. (1971), Measurements of the oscillatory and slowly varying components of the solar velocity field. Solar Physics 18, 379–384.

3. THEORETICAL INTERPRETATION OF THE SMALL SCALE SOLAR FEATURES

THEORETICAL INTERPRETATION OF SMALL-SCALE SOLAR FEATURES

N.O. Weiss
Department of Applied Mathematics and Theoretical Physics
University of Cambridge
Silver Street
Cambridge CB3 9EW, England

Summary

The structure of photospheric magnetic fields is dominated by the interaction between granular convection and isolated flux tubes. Our current understanding of compressible convection, and of both kinematic and dynamical aspects of magnetoconvection is summarized. These theories are related to the formation and location of intense magnetic fields within the photospheric network. The overall structure of sunspots is reconsidered and related to umbral and penumbral features.

1. Introduction

In the first part of this meeting we learnt about observations: now comes the turn of theory. This is the proper ordering, since theory usually struggles to keep up with observational results: as Parker puts it, "Nature is cleverer than we are". The aim of this review is to analyse the interaction between small-scale convection and magnetic fields, to ask questions and to present a few results. I shall be particularly concerned with the relationship between small flux tubes and the granulation, and with fine structure in sunspots. In addition, I shall try to indicate what we would be able to learn by increasing the resolution of our telescopes.

2. Granulation

All our notions about convection in stellar atmospheres really depend on the behaviour that is observed in the solar photosphere. This has already been described by Mattig here and in the recent book by Bray, Loughhead and Durrant (1984). Theory cannot add much to the observations of photospheric granulation, though more sophisticated numerical simulations have become available. Two-dimensional studies of nonlinear compressible convection (Hurlburt, Toomre and Massaguer, 1984) and three-dimensional anelastic computations by Nordlund (1984) show that compressibility leads to strong horizontal flows and vigorous downdrafts. Moreover, overshooting convection excites internal gravity waves in stably stratified regions (Toomre, Hurlburt and Massaguer, 1984) as discussed by Nesis.

Solar convection seems to show a hierarchy of scales, from giant cells (which have never been observed) through supergranules and meso-granules (which may correspond to associations of granules (Kawaguchi, 1980)) to the photospheric granulation. Would we expect smaller scales to appear as the resolution of telescopes is increased? The granular velocity field generates thermal or shear turbulence, with a Reynolds number $R \approx 10^9$ and a magnetic Reynolds number $R_m \approx 10^6$. Energy dissip-ation will occur on the viscous lengthscale ($\sim$10 cm) or the magnetic scale ($\gtrsim$ 30 m) with a field B $\gtrsim$ 30 G. Comparison with other turbulent flows suggests that small-scale motion on scales less than 100 km will be roughly homogeneous.

Finally, how do granules evolve? Most are apparently formed by fission of larger granules. The only relevant calculation demonstrates the instability of an axisymmetric cell to nonaxisymmetric modes when the Reynolds number is sufficiently high (Jones and Moore, 1979). Presumably a similar (purely hydrodynamical) process causes the break-up of polygonal granules.

3. Small-scale magnetic fields

The properties of slender isolated flux tubes have been reviewed by Spruit and Roberts (1983). The unstable stratification leads to an adiabatic instability, causing downflow, evacuation and collapse unless the field is sufficiently strong (Parker, 1979a). Within the flux tubes there may be nonlinear oscillations (Hasan, 1984), providing a source of turbulent diffusion and generating waves that heat the chromosphere (Ulmschneider, 1984). Various problems still remain. How is the filigree related to the flux tubes? Does it outline their walls, form-ing crinkles, or is it produced in points by ohmic heating? Is the adiabatic description valid? The turbulent diffusivity, $\tilde{\eta} \approx 10^{11} \mathrm{cm}^2 \mathrm{s}^{-1}$, produces a characteristic timescale $\tau = R^2/\tilde{\eta}$ in a tube of radius R; for a typical flux tube, $\tau \approx 10$ min so diffusion cannot be neglected. In order to maintain a flux tube in a steady state for periods greater than 10 min a slow flow ($\sim$100 m s^{-1}) inwards and downwards across the field is required. An isolated flux tube will have a lower temperature than its surroundings and so generate a flow around itself (Deinzer et al., 1984a,b). However, flux tubes cannot be considered in isolation from the ambient granular convection, since observations of facular points (Muller, 1983) and magnetic field measurements (Title, Tarbell and Ramsey, 1984), show that intense fields form in intergranular lanes.

Magnetoconvection has been studied in considerable detail (Proctor and Weiss, 1982). Solutions of the kinematic problem (where a steady convective velocity is prescribed) evolve to a steady state with magnetic flux confined to isolated sheets or tubes. For two-dimensional rolls there is a unique symmetrical solution, with no distinction between updrafts and downdrafts (Galloway and Weiss, 1981); in an axisymmetric cell nearly all the flux in contained in a tube around the central axis (Galloway, Proctor and Weiss, 1978). Galloway and Proctor (1983) obtained solutions for a hexagonal cell with a central updraft. At the top of the cell they found strong fields, as might be expected, at the corners but more than half the flux was distributed over a broad region around the centre of the cell.

Systematic studies of the dynamical problem have so far been confined to two dimensions. In the Boussinesq approximation motion is excluded from the stagnant flux tubes. Symmetry is lost and multiple solutions can be found, with a tendency to favour wider cells (Weiss, 1981a,b). Compressible calculations (Cattaneo, 1984; Hurlburt, 1984) show that flux tubes are evacuated and that flux concentration is limited by magnetic pressure as the Mach number approaches unity. These results are confirmed by Nordlund's (1984) three-dimensional simulation of the interaction between granules and magnetic fields.

The question, where do flux tubes form, is not yet adequately answered. Observations show that facular points (Muller, 1983) and intense magnetic fields (Title et al., 1984) occur between the granules but weaker fields (< 600 G) are just as likely to be found within granules and presumably correspond to the facular granules observed at Pic du Midi. Nordlund's numerical experiments show the formation of strong intergranular magnetic fields - but in the calculations of Galloway and Proctor most of the flux eventually ends up at the centre and not at the corners of the cell. Of course, individual granules may not survive long enough for such a state to be achieved and compressibility or dynamical effects may change the field configuration. Nevertheless, it is clear that the competition between inflow at the base of a convection cell and outflow at the top is of great importance (Parker, 1979c). In the absence of a fully compressible, three-dimensional calculation we must rely on simple models. If we assume that magnetic flux is confined to a slender flux tube then we can investigate the motion of that tube within a convective cell, including certain dynamical effects. The tube drifts relative to the ambient gas at a rate determined by magnetic buoyancy,

tension along the lines of force and turbulent drag (Parker, 1979a).
Model calculations (Schmidt, Simon and Weiss, 1985) show that small
flux tubes, with fluxes F $\lesssim 10^{18}$ mx, are swept into intergranular lanes,
where they remain. Larger flux tubes tend to be dragged inward, with
the assistance of magnetic buoyancy, and end up at the axis of the cell.
This result suggests that tubes with F $\gtrsim 10^{18}$ mx and fields that fan
outwards near the photosphere may be responsible for facular granules.
Such a tube could sit at the centre of a granule, like a sunspot
surrounded by a moat; but the fission of the granule would leave it at
a corner in the convective pattern. On the other hand, the small flux
tubes, with fields of up to 1500 G, would remain between granules,
moving along the intergranular lanes and collecting preferentially at
the corners. More precise observations are needed to confirm details
of this behaviour.

4. Sunspots

     So far we have been discussing the effect of convection on magnetic
fields when the average field is relatively weak. In pores and sunspot
umbrae strong fields extend over large areas. Strong vertical fields
inhibit overturning convection but allow oscillatory convection if the
magnetic diffusivity is less than the (radiative) thermal diffusivity.
Before considering small-scale features in sunspots we need to have
some idea of their overall structure. The discovery of 5-minute oscillations in sunspots (Thomas, Cram and Nye, 1982) implies that they are
reasonably coherent. This does not, of course, preclude a tight clus-
tered structure (Parker, 1979b) and it is inevitable that some convection must occur even within a monolithic plug (Knobloch and Weiss, 1984).
     Observations of penumbral filaments show fine structure that is
apparently associated with convection rolls. There is, however, a
conflict between observations of the Evershed flow which is horizontal,
and magnetic measurements, most of which suggest that the field is
inclined. To resolve this contradiction, it has been supposed that there
are two field components, a horizontal field in the dark interfilamentary
regions, where the Evershed flow occurs, and an inclined field in the
bright filaments. Over the years, many attempts have been made to
explain both convection in rolls and the Evershed effect with such a
two component model but none of these theories seems satisfactory. It
appears more likely that the field is nearly horizontal over most of
the penumbra, as found by Wittmann(1974). In that case, the penumbra

must be shallow, with an abrupt transition between magnetized and field-free plasma; indeed, a model calculation suggests that the magnetized layer is only 100 km deep (Schmidt, Spruit and Weiss, 1985). Moreover, shallow penumbral models offer a better opportunity of explaining bright grains and associated structures, while the Evershed flow appears as a natural consequence of pressure differences at the footpoints of a flux tube (Meyer and Schmidt, 1968). In such models the penumbra only contributes a small fraction of the total magnetic flux. Energy is principally transported by radiation, though convection occurs in rolls aligned by the almost horizontal field. The overall dynamical equilibrium of the sunspot depends on curvature forces in the transition zone at the umbral-penumbral boundary, where the Wilson depression increases from 200 to 600 km, as pointed out by Mattig (1969).

Within the umbrae, where the field is nearly vertical, only oscillatory convection can occur. It is natural to identify the ubiquitous umbral dots with time-dependent convection of this type (Parker, 1979d; Knobloch and Weiss, 1984). The bright dots have diameters of about 10 km and lifetimes of around 25 min (Adjabshirzadeh and Koutchmy, 1980, 1983; Koutchmy and Adjabshirzadeh, 1981). Their general properties can be explained in terms of highly nonlinear oscillations, with a brief burst of activity as a hot element rises, followed by a long quiescent period as it gradually cools off. The individual cells are larger than the bright dots and may have diameters of 300 km with lifetimes of, say, 8 hours. It is not clear how such cells would couple to convective motions deeper down. Oscillatory motion occurs when ζ, the ratio of the magnetic to the thermal diffusivity, is less than unity. At the photosphere (as in most of the solar interior) $\zeta < 1$, owing to the efficiency of radiative transport; at depths in the range 2000-20000 km, however, the opacity increases, owing to ionization, and $\zeta > 1$. Below 2000 km, overturning convection will occur, with natural scales of order 1000 km. Whether there are two distinct regions, or whether oscillatory motion is coupled to overturning cells below, has yet to be determined.

5. Conclusion

In this brief survey I have tried to emphasize those puzzling questions that might be answered with the aid of better observations. As solar physicists, we accept it as self-evident that we should try to understand small-scale processes at the surface of the sun. There are also more general reasons for regarding such questions as important. The solar photosphere is a unique laboratory where hydromagnetic behaviour can be observed; combining theory with detailed observations therefore extends the range of physics. Moreover, it is only by understanding the interaction between magnetic fields and convection there that we shall be able to describe the structure of the fields in stellar interiors, which are not accessible to observation.

The future holds out the prospect of higher resolution in the observations. New techniques will be employed for ground-based observations, while better measurements will be made from space. These observations provide an opportunity not only of understanding photospheric features but also of using magnetic fields as a probe to study structures deeper in the sun.

References

Adjabshirzadeh, A. and Koutchmy, S. 1980, Astron. Astrophys. 89, 88.

Adjabshirzadeh, A. and Koutchmy, S. 1983, Astron. Astrophys. 122, 1.

Bray, R.L., Loughhead, R.E. and Durrant, C.J. 1984, The Solar Granulation, Cambridge University Press.

Cattaneo, F. 1984, Ph.D. dissertation, University of Cambridge.

Deinzer, W., Hensler, G., Schüssler, M. and Weisshaar, E. 1984a, Astron. Astrophys. 139, 426.

Deinzer, W., Hensler, G., Schüssler, M. and Weisshaar, E. 1984b, Astron. Astrophys. 139, 435.

Galloway, D.J. and Proctor, M.R.E. 1983, Geophys. Astrophys. Fluid Dyn. 24, 109.

Galloway, D.J., Proctor, M.R.E. and Weiss, N.O. 1978, J. Fluid Mech. 87, 243.

Galloway, D.J. and Weiss, N.O. 1981, Astrophys. J. 243, 945.

Hasan, S.S. 1984, Astrophys. J. 285, 851.

Hurlburt, N.E. 1984, preprint.

Hurlburt, N.E., Toomre, J. and Massaguer, J.M. 1984, Astrophys. J. 282, 557.

Jones, C.A. and Moore, D.R. 1979, Geophys. Astrophys. Fluid Dyn. 11, 245.

Kawaguchi, I. 1980, Solar Phys. 65, 207.

Knobloch, E. and Weiss, N.O. 1984, Mon. Not. Roy. Astron. Soc. 207, 203.

Koutchmy, S. and Adjabshirzadeh, A. 1981, Astron. Astrophys. 99, 111.

Mattig, W. 1969, Solar Phys. 8, 291.

Meyer, F. and Schmidt, H.U. 1968, ZAMM 48, 218.

Muller, R. 1983, Solar Phys. 85, 113.

Nordlund, A. 1984, in The Hydromagnetics of the Sun, ed. T.D. Guyenne,
    p.37, ESA SP-220.

Parker, E.N. 1979a, Cosmical Magnetic Fields, Clarendon Press, Oxford.

Parker, E.N. 1979b, Astrophys. J. 230, 905.

Parker, E.N. 1979c, Astrophys. J. 230, 914.

Parker, E.N. 1979d, Astrophys. J. 234, 333.

Proctor, M.R.E. and Weiss, N.O. 1982, Rep. Prog. Phys. 45, 1317.

Schmidt, H.U., Simon, G.W. and Weiss, N.O. 1985, Astron. Astrophys.,
    submitted.

Schmidt, H.U., Spruit, H.C. and Weiss, N.O. 1985, in preparation.

Spruit, H.C. and Roberts, B. 1983, Nature 304, 401.

Thomas, J.H., Cram, L.E. and Nye, A.H. 1982, Nature 297, 485.

Title, A.M., Tarbell, T.D. and Ramsey, H.E. 1984, preprint.

Ulmschneider, P. 1984, in Small-scale Processes in Quiet Stellar Atmos-
    pheres, ed. S.L. Keil, National Solar Observatory, Tucson.

Weiss, N.O. 1981a, J. Fluid Mech. 108, 247.

Weiss, N.O. 1981b, J. Fluid Mech. 108, 273.

Wittmann A. 1974, Solar Phys. 36, 29.

Discussion

Schatzman: From the point of view of the theory, what decides between
upward and downward motion in a hexagonal cell? In Hurlburt the motions
are downwards at the axis; in the modal model (anelastic approximation)
of Massaguer et al., they cannot decide.

Weiss: In the strict Boussinesq approximation solutions are degenerate
and the preferred sense of motion is therefore determined by non-
Boussinesq effects. In the laboratory the most important effect is the
variation of viscosity, ν, with temperature: if $d\nu/dT < 0$ (as in a liquid)
motion is upward on the axis, but if $d\nu/dT > 0$ (as in a gas) the motion
is downward. In a star the most obvious effects are the density stratification and the variation of the thermal diffusivity, κ, with depth.
The density stratification favours up-hexagons (which are subcritically
unstable). I don't know what happens when κ decreases downward.

Rösch: The slide already projected on the Monday afternoon session
shows images of a 12" x 12" field obtained at approximately 90 second
intervals where the central granule increases and explodes, then one of
the pieces explodes, and so on several times. That is what Kawaguchi
has called a family of granules. His co-worker Oda has demonstrated

in a paper in Solar Physics (93, 243, 1984) that such families, and
more generally the "exploding granules" are essentially distributed
along the contours of the mesogranulation. (All pictures from Pic du
Midi.) The average horizontal velocity during expansion is of the
order of 1 or 2 km s^{-1}.

Weiss: Are the observations then consistent with the existence of a
critical size such that a granule explodes whenever it exceeds that
size?

Rösch: Quite consistent.

Muller: It must be noted that the estimated diameters and lifetimes
of facular points in the quiet sun and in umbral dots are the same,
namely 150 km and 25 mins respectively.

Weiss: That is an interesting fact, though the two phenomena seem very
different: a facular point in the network is associated with the intense
magnetic field in an isolated flux tube, while an umbral dot is associated with a field strength that is somewhat lower than the average
umbral value.

Wiehr: The line ratio measurements yield a unique range of field
strengths (e.g. kilogauss ± 30%). Hence, the variety of flux values
discussed should originate from different flux tube diameters rather
than field strengths.

Weiss: For small flux tubes in intergranular lanes, corresponding to
facular points, this should certainly be so. If there are flux tubes
at the centres of granules, however, one would expect the field to fan
out at the photosphere, where there is a radial outflow, so that the
field strength is somewhat less. For a flux of 10^{19} mx distributed
over the surface of a granule the field would still have to be about
1000 gauss.

Pneuman: I believe it was stated some time ago by Parker that flux
tubes such as the sunspots you discussed should be subject to the inter-
change instability which has a growth rate much faster than the lifetime
of the spot. What are your thoughts as to how this instability might
be suppressed?

Weiss: Curvature of the field lines does encourage interchanges but
this is counteracted by the stabilizing effect of the stratification,
with less dense fluid lying above denser plasma. F. Meyer, H.U. Schmidt
and I showed some time ago (MNRAS 179, 741, 1977) that a flux tube is
in fact stable provided the radial component of the field decreases
upwards at its outer boundary. This condition is satisfied for simple
sunspot models, which is why the spots survive.

Alissandrakis: I wish to point out that in a recent study of the

Evershed flow in Hα and the photosphere, together with observations of the magnetic field (Dialetis, Mein, Alissandrakis, Astron. Astrophys., submitted) we find that the general characteristics of the flow are consistent with siphon flow.

Weiss: That is very reassuring. I believe that this is the only satis-factory explanation of the Evershed effect.

Semel: I like the way you explain the observed non-collinearity between the velocity and magnetic vectors, in the penumbra. However, I wish to mention that you may also add a third component to explain the inward flow observed with strong spectral lines, in the chromosphere.

Weiss: The inward flow is observed at a different level from the normal Evershed effect, and presumably follows lines of force that enter the photosphere in the umbra. Since the umbral field (unlike the penumbral field) is greater than that in isolated flux tubes in the network, the siphon model of Meyer and Schmidt would predict a flow of gas into the umbra but out of the penumbra. In each case we expect the velocity to be along the field lines.

RELEVANCE OF MAGNETIC FLUX EXPULSION FROM THE LOWER SOLAR ATMOSPHERE TO THE
ACCELERATION OF THE SOLAR WIND

G.W. Pneuman
Institut für Astronomie
ETH-Zentrum, CH-8092 Zürich
Switzerland

Evidence for the continual injection of material into the corona from the lower
solar atmosphere is certainly clear from observations of spicules, macrospicules,
explosive events in the transition region, etc. It is also clear that the large
amount of magnetic flux emerging into the upper atmosphere ($\approx 5 \times 10^{22}$ Mx/day)
estimated by Golub (1980) could replenish the entire polar magnetic field of the Sun
in about 4 hours were it not rapidly dissipated or ejected outward via small-scale
magnetic reconnection. If such elements are injected into the coronal magnetic field
diamagnetically, the consequences for the acceleration of the solar wind could be
important. We use the term 'diamagnetic' here to denote material that is injected
between the lines of force of the preexisting ambient field rather than along them
as is generally assumed. As has been shown long ago by Schlüter (1957) and Parker
(1957), such material is subjected to an outward Lorentz force by the displaced external field if the field diverges outward (the melon-seed effect).

If such diamagnetic material can be coupled with the ambient medium effectively,
then a powerful acceleration mechanism for the solar wind becomes available since
the ambient coronal magnetic field (which contains most of the energy of the corona)
then plays an active role in accelerating the gas. The observed high-speed solar
wind streams (700-800 km/s) at 1 A.U. evidently cannot be produced by gas pressure
forces alone and the effectiveness of Alfvén wave pressure has not been verified on either observational (Athay and White, 1978, 1979a,b; Cheng et al, 1979) or theoretical
(cf. Leer et al, 1982) grounds. Hence, although Alfvén wave pressure cannot and should
not presently be ruled out as the accelerating mechanism for the wind, it should be
pointed out that there do exist other mechanisms which utilize the coronal magnetic
field in, perhaps , even a more direct way to accelerate coronal gas outward.

Consider a diamagnetic element of internal gas pressure p_i and magnetic
pressure $B_i^2/8\pi$ immersed in a diverging external magnetic field of strength B_a. If
the size of the body is small as compared to the scale over which B_a varies, then the
net force on the body, to a close approximation, is given by

$$\vec{F} = - \text{Vol.} \cdot \nabla \left(B_a^2/8\pi \right)$$

(1)

If the body is pressure dominated, i.e., $p_i >> B_i^2/8\pi$, then the internal gas
pressure must essentially balance the external magnetic pressure giving

$$\text{Vol.} \propto \frac{M}{\bar{\rho}} \propto \frac{MT}{\bar{p}} \propto \frac{8\pi MT}{B_a^2} \; ; \; and \; \vec{F} \propto - MT \nabla \left[\ln \left(B_a^2/8\pi \right) \right]$$

(2)

where M is the mass, $\bar{\rho}$ the average density, $\bar{p}$ the average pressure, and T the
temperature of the body. Hence, in this case, the force is proportional to the temperature and inversely proportional to the scale over which the external field diverges.
It is independent of the field strength. If, on the other hand, the body is magnetically dominated $(p_i << B_i^2/8\pi)$, the force is related to the magnitude of the field by

$$\vec{F} \propto - \nabla \left(B_a^2/8\pi \right)^{1/4}$$

(Cargill and Pneuman, 1984). In all the above expressions, of course, the external gas pressure has been neglected assuming that the plasma beta is small in the region of our interest (the corona).

To illustrate the general effectiveness of diamagnetic material in the corona towards the acceleration of the solar wind, Pneuman (1983) has calculated a simple solar wind model which consists of a mixture of ordinary ambient magnetic gas along with a certain percentage, R, of the mass flux consisting of small diamagnetic elements. Although the force, of the form given by Equation (2), acts only on the diamagnetic elements, it is assumed that it is transmitted to the wind as a whole uniformly through coupling between the two mixtures. In addition, the corona is assumed isothermal with both the magnetic and diamagnetic material having the same temperature. The resulting velocity profiles, corresponding to a coronal temperature of 1.1×10^6K, are shown in Figure 1. Noting that R = 0 corresponds to the Parker solution for an isothermal corona, we see that the speed of the solar wind is increased significantly as more and more diamagnetic material is introduced.

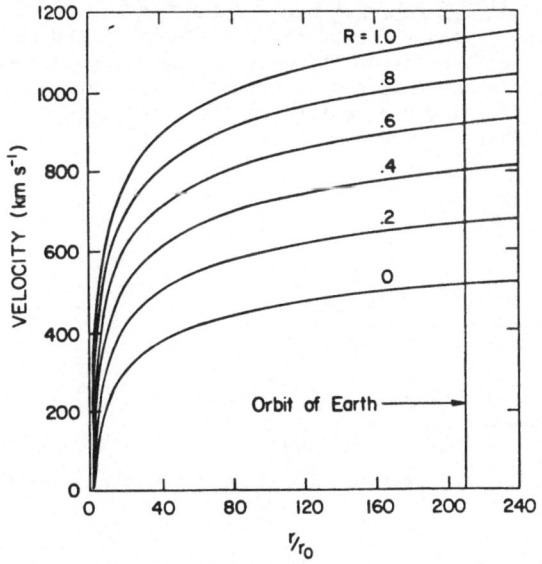

Figure 1: Velocity as a function of radial distance from the solar center for different values of R, where R is the ratio of the mass flux in diamagnetic material to the total mass flux.

The evolution of the shape of diamagnetic plasmoids as they are accelerated outward into the corona has recently been studied in some detail by Pneuman and Cargill (1984). To do this, we assume the plasmoid to be a prolate ellipsoid of width w and length l, both of which can vary with time as the body moves in the external magnetic field. In order to study the evolution of both w and l with time in addition to the motion of the center of mass located at $r_c(t)$, we utilize the tensor virial equations developed by Parker (1957) as well as the force balance equation integrated over the entire body which gives the net force. This results in three independent differential equations for w, l, and r_c, i.e.,

$$\bar{w} \frac{d^2 \bar{w}}{d \bar{\tau}^2} = \beta_0 + C_\ell \frac{\bar{\ell}}{\bar{w}^2} - \frac{\bar{\ell}^3}{k_0^2 \bar{r}^{2n}} F_1(\alpha) \tag{3}$$

$$\frac{\bar{l}}{k_0^2}\frac{d^2\bar{l}}{d\tau^2} = \beta_0 - e_l\frac{\bar{l}}{\bar{\omega}^2} + \frac{e_\omega}{\bar{l}} - \frac{\bar{l}^{-3}}{k_0^2\bar{r}^{2-2n}}F_2(\alpha) \tag{4}$$

$$\frac{d^2\bar{r}}{d\tau^2} = -\frac{\psi}{\bar{r}^2} - \frac{2n\omega_0^2\bar{l}^{-3}\bar{r}^{-2n-1}}{5r_0^2 Q_1'(\alpha)} \tag{5}$$

where $\bar{\omega} = \omega/\omega_0$, $\bar{l} = l/l_0$, $\tau = \frac{\sqrt{5}}{\omega_0}[B_0^2/8\pi\langle\rho_0\rangle]^{1/2}t$,

$\beta_0 = 8\pi\langle\rho_0\rangle a^2/B_0^2$, $e_l = 8\pi\langle\rho_0\rangle E_{l_0}/MB_0^2$, $e_\omega = 8\pi\langle\rho_0\rangle E_{\omega_0}/MB_0^2$,

$\bar{r} = r_c/r_0$, and $\psi = 8\pi G M_0\langle\rho_0\rangle\omega_0^2/5r_0^3 B_0^2$. w_0 is the initial width, l_0 the initial length, B_0 the external field strength at the reference level r_0, $\langle\rho_0\rangle$ the initial average density in the body, a the internal sound speed, E_{l_0} and E_{ω_0} the initial energies in the internal longitudinal and transverse magnetic fields respectively, $k_0 = w_0/l_0$, and n a measure of the divergence of the external field where we have assumed that

$$B_a = B_0\left(\frac{r_0}{r}\right)^n$$

Also

$$F_1(\alpha) = \frac{2 + 3[(\alpha^2-1)^2 Q_1'(\alpha)]/\alpha}{2\alpha^2(\alpha^2-1)Q_1'(\alpha)^2}$$

$$F_2(\alpha) = -\frac{[2 + 3\alpha(\alpha^2-1)Q_1'(\alpha)]}{\alpha^2(\alpha^2-1)Q_1'(\alpha)^2}$$

where $\alpha = 1/(1-k^2)^{1/2}$, $k = w/l$, and $Q_1'(\alpha)$ is an associated Legendre function of the second kind given by

$$Q_1'(\alpha) = \frac{d}{d\alpha}Q_1(\alpha) = \frac{1}{2}\ln\left(\frac{\alpha+1}{\alpha-1}\right) - \frac{\alpha}{(\alpha^2-1)}$$

The reader is referred to Pneuman and Cargill (1984) for the derivation of Equations (3), (4), and (5) and a discussion of their solutions and physical implications subject to varying initial conditions. We merely list here some of the main conclusions regarding the evolution of shape of both pressure dominated and magnetically dominated bodies as they move outward in an ambient magnetic field:

1. For totally pressure dominated bodies (no internal magnetic field, $e_w = e_l = 0$; $\beta_0 = 1$), no equilibrium with the external magnetic field exists and the body elongates along the field lines while contracting laterally. This result was previously found by Parker (1957). The velocity of elongation is of the order of the sound speed while the contraction occurs in an oscillatory manner with a period of the order of the transit time of a sound wave across the body. The time-scale for this elongation is

much shorter than that for the motion of the center of mass so that the body should quickly evolve into a true jet rooted at the bottom to the solar surface.

2. If the body is totally magnetically dominated ($\beta_0 = 0$), an equilibrium with the external field is possible. If the body is initially in equilibrium and the center of mass begins to move outward in a diverging field, it will expand self-similarly maintaining its initial shape while remaining in equilibrium with the external field. If the field strength declines outward as r^{-n}, both w and l increase as $r^{n/2}$. If the body is not initially in equilibrium, it will seek out its equilibrium shape and oscillate around it with a period of the order of the time for an Alfvén wave to cross the body. The center of mass moves outward at a speed determined by the average shape of the body.

3. For situations between 1. and 2., i.e., bodies which are neither totally pressure dominated or magnitically dominated, the deformation is more complex. An equilibrium with the external field is possible for any internal plasma β and shape but, in general, the body tends to elongate more and more as it moves outward with the rate of elongation increasing as the internal plasma pressure contribution (β_0) is increased.

The physical results discussed above can be demonstrated graphically by an examination of Figure 2 which gives a plot of k/k_0 as a function of time, τ , for different values of β_0 . Since $k = w/l$ (k_0 being its initial value), it is a good parameter in which to visualize the shape, i.e., $k = 0$ is a long thin pencil shape and $k = 1$ a sphere. Here a totally pressure dominated body corresponds to $\beta_0 = 1$ and a totally magnitically dominated body to $\beta_0 = 0$. We see that for $\beta_0 = 1$, k decreases monotonically indicating a continual elongation of the shape with time. As β_0 becomes smaller, how-

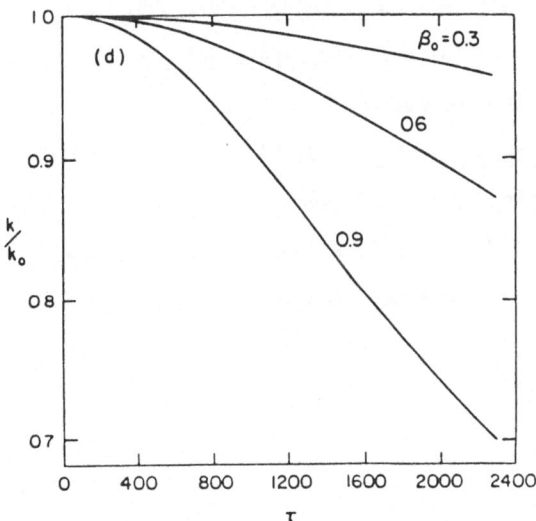

Figure 2: Variation of k/k_0 with time for different values of β_0.

ever, the rate of elongation decreases until, in the limit $\beta_0 = 0$, the solution is the horizontal line $k/k_0 = 1$ corresponding to a completely self-similar solution as discussed in 2.

In summary, we see that the ultimate fate of a diamagnetic body as it moves outward into the corona due to forces exerted by the external field is critically dependent upon how important the internal gas pressure is relative to the internal magnetic pressure. If the pressure is identically zero, the body will expand self-similarly for all time if dissipative processes do not occur. If the internal β is non-zero, however, it will ultimately deform into a jet since the magnetic energy continually declines in our model whereas the thermal energy does not. This conclusion, however, rests upon our assumption that the temperature of the body remains isothermal. If it declines as expected, the result could be quite different. Future work (Cargill and Pneuman, 1985) will remove the restriction of isothermality. Also, hydrodynamic effects could also change our picture considerably. Such effects are currently under consideration by MacGregor (1984).

REFERENCES

Athay, R.G, and White, O.R.: 1978, Astrophys. J. 226, 1135.

Athay, R.G. and White, O.R.: 1979a, Astrophys. J. Suppl. 39, 33.

Athay, R.G. and White, O.R.: 1979b, Astrophys. J. 229, 1147.

Cargill, P.J. and Pneuman, G.W.: 1984, Astrophys. J. 276, 369.

Cargill, P.J. and Pneuman, G.W.: 1985, (in preparation).

Cheng, C., Doschek, A., and Feldman, U.: 1979, Astrophys. J. 227, 1037.

Golub, L.: 1980, Phil. Trans. Roy. Soc. London, 297, 595.

Leer, E., Holzer, T.E., and Fla, T.: 1982, Space Sci. Rev. 33, 161.

MacGregor, K.B.: 1984, (in preparation).

Parker, E.N.: 1957, Astrophys. J. Suppl. 3, 51.

Pneuman, G.W.: 1983, Astrophys. J. 265, 468.

Pneuman, G.W. and Cargill, P.J.: 1984, Astrophys. J. (in press).

Schlüter, A.: 1957, in IAU Symposium 4, Radio Astronomy, ed. H.C. van de Hulst, CUP, p. 356.

MAGNETIC AND VELOCITY FIELD ANALYSIS OF A QUIET REGION NEAR THE CENTER OF THE SUN

Helen DARA-PAPAMARGARITI[1] and Serge KOUTCHMY[2]

1 - Research Center for Astronomy, Academy of Athens, GREECE, 136.

2 - Institut d'Astrophysique - CNRS - 98 bis, Bd. Arago, F 75014 Paris.

Abstract - A selected region of the quiet Sun near the center of the disc was analyzed in order to deduce parameters concerning regions of highly concentrated magnetic
field, including network elements. The present analysis covers a sequence of spectra
corresponding to a total time laps of 6 min over the same region. Spectra of both
magnetically sensitive and insensitive lines were measured in detail in order to deduce the velocity distribution around magnetic flux tubes. Large amplitude five minute oscillatory components were removed using an averaging over time procedure.
Preliminary results are discussed as well as factors which could affect their interpretation.

1 - Introduction.

High resolution observations allow the analysis of network elements which often are
resolved in tiny concentrated magnetic fields. These elements correspond to the very
typical "line gaps" on a good spectrum and to filigrees on a photospheric or minimum
temperature region picture. One kilog auss longitudinal magnetic fields are typical
for these regions. However, we do not yet know from the observations the detailed
3-dimensional distributions of the temperature, velocity and magnetic fields of these regions. This is specially difficult as there are no reasons to beleive that a
typical stationary pattern exits. The velocity field of the quiet photosphere for
example, is well known for showing an oscillatory component at the photospheric level in the vertical direction. In this paper we specifically consider the time va-

riations of different parameters at a time scale of order of the 5 min oscillations using spectra obtained with the best (typically 0.7 arcsec) spatial resolution.

2 - Observations.

The spectra of three iron lines Fe 5576 Å (Magnetically insensitive line) and Fe 6301.5 Å (g=1.5) and Fe 6302.5 Å (g=2.5) with cicular polarization analysis, were observed simultaneously with the SPO tower-telescope over a region, see figure 1, situated near the center of the disk where no activities like pores or even faint faculae were detected (June 8, 1976). Slit jaw pictures and filtergrams before and after the spectrographic observations were also performed. More details can be found in Koutchmy and Stellmacher, 1978 and Dara-Papamargariti and Koutchmy, 1983. For the present analysis, we extracted a set of 10 uniformly distributed best resolution spectra (total duration 5.5 min of time over a 120 arcsec length) obtained with the slit moved in order to follow the solar rotation (with a step of 0.1 arcsec) during the whole time sequence. Earth atmospheric induced differential refraction effects were also minimized.

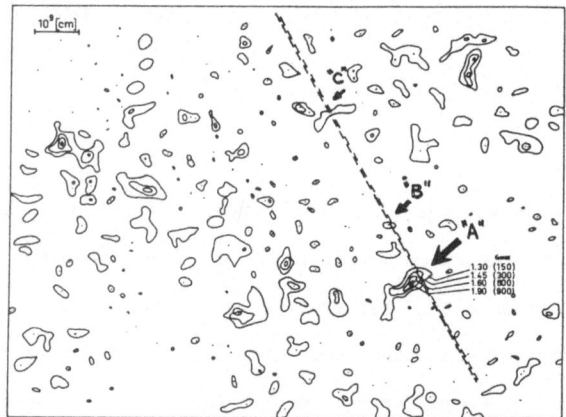

Figure 1.

3 - Analysis and Results.

Using a Video Image processing machine, each spectrum was properly enlarged and color coded with isophotes and the apparent location of the center of the line contour drawn. From the comparison of left hand and right hand circularly polarized spectra we deduced an "effective" value of the longitudinal component of the magnetic field over each region (named A,B,C, see fig.1) showing a distribution of concentrated magnetic field. Outside these regions, and inside when the magnetically insensitive 5576 Å line is considered, we measured the Doppler shifts. Assuming that the lines are formed in the same layers (an assumption supported by the fact that all three lines are coming from the same multiplet of the same atome, with a similar line strength) the deduced longitudinal (in our case it is also the vertical one) component of the velocities is presumably the same, so an average value is computed for each point along the slit: $v_{\parallel} = \langle v(5576); v_R(6301.5); v_L(6301.5); v_R(6302.5); v_L(6302.5)\rangle$ resulting in a good precision (± 0.05 kmsec^{-1}) in $v_{\parallel}$. The precision in the determination of $B_{\parallel}$ is of order of ± 50 gauss; as a profile of $B_{\parallel}$ along the structures is measured, we also considered a parameter which is presumably less sensitive to errors: $\langle B_{\parallel}\rangle = \int B_{\parallel} \, xdl$, especially when seeing effects are present, as this parameter is integrated along the smeared profile of the magnetic structure. In order to clear out the influence of the image quality we first looked for seeing related correlation.

Each spectrum was also scanned with a digital microphotometer for deducing I_K, the fluctuations of the continuum intensity near the 5576 line and $I(\lambda o)$ the fluctuations of the central intensity of the magnetically insensitive line. Figure 2 shows their average intensities over 3.5 min of time. From each scan of I_k, we computed

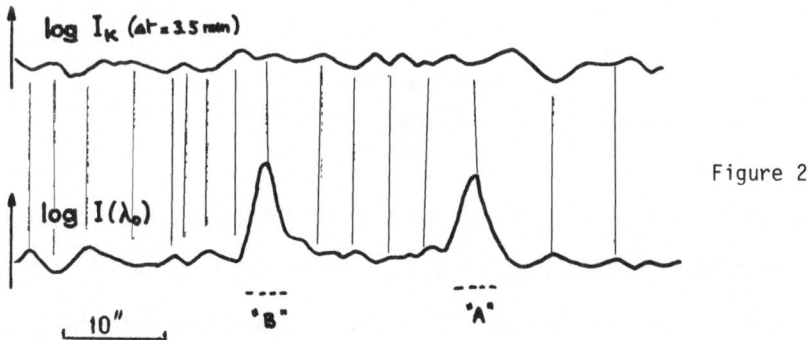

Figure 2

the RMS value corresponding to each spectrum of the time sequence. On figure 3, the measured values of B^{max}, corresponding to the best defined magnetic regions A and B, are plotted as a function of the RMS measured at the same time. No correlation exits, so the observed fluctuations of $B_{\parallel}$ seem not to be related with the seeing

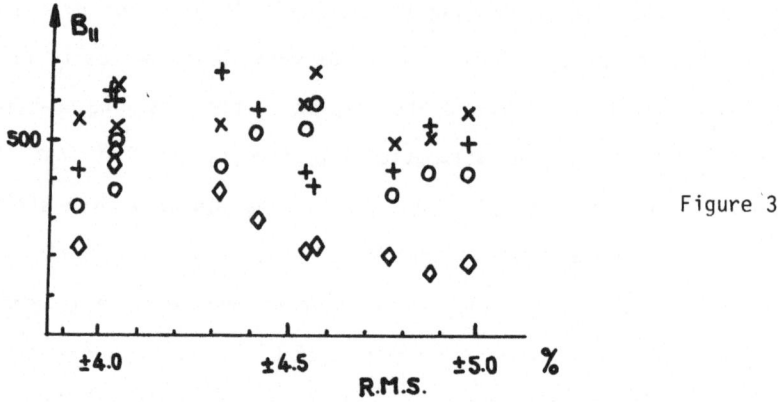

Figure 3

related parameter RMS. A further evidence is coming from the test illustrated on figure 4 where we plotted for both magnetically sensitive line, the observed values of B^{max} of one structure as a function of the value observed for the other one. There also no correlation seems to exist. We conclude from these results that real variations of $B_{\shortparallel}$ and, a fortiori, of $\langle B_{\shortparallel} \rangle$ seem to emerge from our measurements.

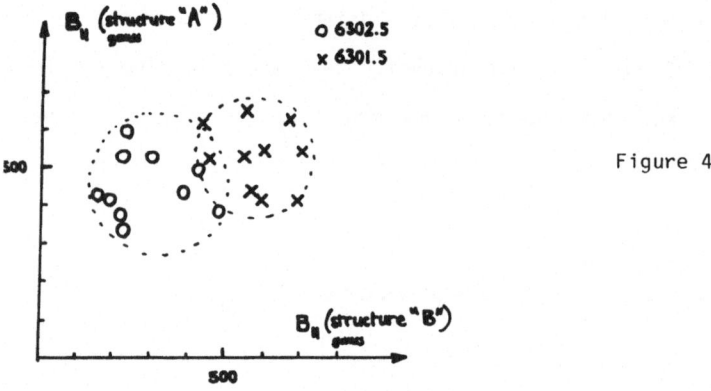

Figure 4

Finally on Figure 5 and 6 we plotted, as function of time, the measured values for both magnetic regions, log $I(\lambda o)$, $v_{\shortparallel}$, and $\langle B_{\shortparallel} \rangle$ for both 6301.5 and 6302.5 lines. $I(\lambda o)$ is the central intensity of the 5576 Å line (temperature sensitive) and $v_{\shortparallel}$ corresponds to the amplitude of the observed vertical velocities over the magnetic structure.

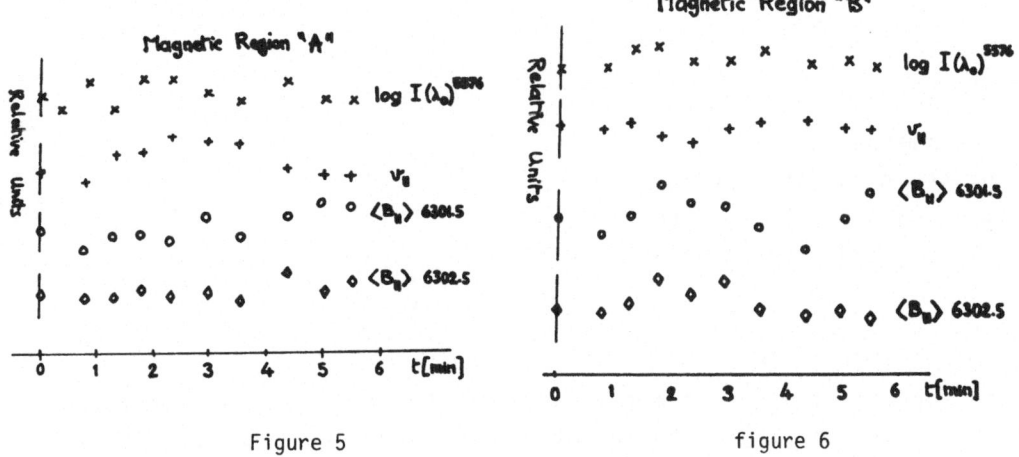

Figure 5   figure 6

4 - Discussion.

Outside of magnetic regions, large amplitude vertical velocities, up to ±1.1 km-sec^{-1}, are observed. Over the magnetic structures, time variable vertical veloci-ties, up to ±0.5 kmsec^{-1}, are also observed but the stationary component, as dedu-ced from the averaging of 5.5 min of time sequence, is less than 0.1 kmsec^{-1}. When the variable amplitudes of the magnetic field (±30%) are considere d (only the ver-tical component $B_{\parallel}$), no correlation seems to appear with the observed temperature excess (presumably related to the central intensity of the line) but a slight anti-correlation seems to emerge with the observed vertical velocity field (a "negative" velocity corresponds to an increase of the magnetic field) when figures 5 and 6 are scrutinized. Finally, we definitely noticed that no stationary large inflow nor out-flow are present over the magnetic structure with a precision better than 0.1 kmsec^{-1}. We hope soon to improve these results and we are acknowledged to Dr.G. Stellmacher who presented these preliminary results at the Meeting.

References - Dara-Papamargariti,H. & Koutchmy,S.:1983, Astron. Astrophys. 125, 280
           Koutchmy,S. & Stellmacher,G.:1978, Astron. Astrophys. 67, 93.

Discussion

G. Elste : What is the size of the magnetic regions ?
G. Stellmacher : 3 to 2 arc sec at the center of the line, but much smaller in the continuum.

A POSSIBLE MECHANISM FOR SOLAR PHOTOSPHERE BRIGHT POINT FORMATION

V.N.Dermendjiev
Department of Astronomy and NAO
Bulgarian Academy of Siences
72 Lenin Blvd., Sofia 1184, Bulgaria

This contribution concerns two problems. The first one is connected with the possibility of the creation of strong small-scale magnetic field in the subphotosphere during the formation of a short-lived, substantially unsteady magneto-hydrodynamic (MHD) vortex, and the second one is connected with the dynamical response of the photosphere to the emergence of such a magnetic field.

Some important results obtained in laboratory plasma experiments on the creation of the so-called dynamically stable current filament (DSCF) are taken as a basis for the proposed mechanism of strong small-scale magnetic field formation.

It has been found experimentally (Komelkov et al.,1962; Komelkov et al.,1960; Anderson et al., 1958) that the DSCF and the coaxial plasma are simulatneously formed in various plasma configurations. The structure of the current and magnetic fields, as they are revealed from photographic and probe measurements (Komelkov et al.,1962), is shown in Fig. 1.

In the most general case a plasma current filament and a coaxial plasma are formed in an initially toroidal eddy current. In particular, they have been observed when plasma cluster is extracted through an axial gap. More stable filaments are formed when external longitudinal magnetic fields are applied.

The filament and coaxial plasma have their own longitudinal magnetic field of the order of 10^2 to 10^4 G, Fig. 1a, which can be detected in both polarities. Fig. 1b presents the longitudinal magnetic field H_z and current density J_z distribution over two cross-sections of the chamber. The magnetic field maintains its direction over the entire cross-section; the decrease near the axis is in agreement with the distribution of the current since the source of this field is the dynamical current filament with an internal helix moving with a velocity of the order of 4 to 150 km/s.

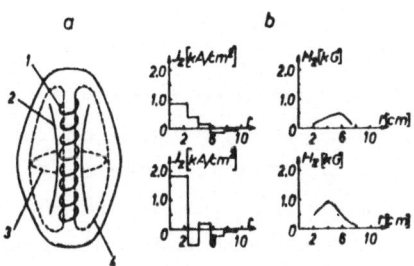

Fig.1. a) Structure of the currents and magnetic fields in a plasma jet and plasmoid; 1 - current helix, 2 - longitudinal currents, 3 - azimutal magnetic field, 4 - longitudinal magnetic field. b) Longitudinal magnetic field H_z and current density J_z distribution over two chamber cross-section (Komelkov et al., 1962).

DSCFs are stable over a wide range of time (5 to 200 s), pressure (10^2 to 10^5 dyna/cm^2) and density (10^{-10} to 10^{-3} g/cm^3), at different value of the discharge, which suggests that it is rather likely to observe similar phenomena everywhere in Nature where plasma currents exist. They will be formed in the general case when a sharp nonuniformity in the distribution of currents and magnetic pressure arises.

Three quesions may be asked: 1. Is it possible that such dynamic structures are formed in the solar subphotosphere? 2. If this is possible, do they have long enough time to reach the photosphere? 3. What will be the dynamical response of the photosphere, i.e. the main observable effects.

It is rather difficult to give answer to these questions especially to the first two ones concerning the problem of the small-scale structure of MHD turbulence.

Orszag and Tang (1979) suggest the following simple physical explanation for the production of small-scale motions by MHD flows. When the kinematic viscosity is small and equal to the magnetic diffusivity, an initial weak magnetic field is streched and convected by the velocity field and thus wrapped into tight "ropes" that closely follow the large-scale fluid flow. Neighbouring lines of forces are thus stretched into close proximity to each other and the magnetic diffusion can locally reconnect them. When the lines of force snap, their tension force reacts back on the flow field to produce small eddies on top of the larger convecting ones, giving an enhanced cascade process. Dynamical forces like those provided by magnetic fields and stable stratification have stabilizing effects on small amplitude motions but they may also have destabilizing effects. Thus a lot of small-scale motions can be reduced to smaller ones and fade away.

An analogy with the above-mentioned results from laboratory plasma experiments leads to very interesting conclusions about the evolution of such vortex structures in the subphotosphere. A small-scale vortex moving acceleratively in a magnetic rope or, in the more idealized case, in a magnetic tube, is put under the same dynamical influences as a plasma cluster, in laboratory experiments, extracted through an axial gap with longitudinal magnetic field. Thus it is rather possible that in the subphotospheric layers dynamical structures, similar to the DSCF with coaxial plasma, confined between their own azimutal and longitudinal magnetic fields can be formed.

Our analysis of experimental results leads to the following two conclusions: 1. The lifetime of DSCF is approximately equal to the time that the current needs to travel the distance $S = 2\pi (L/\delta)r$, where L is the helix length, δ is the helix spacing and r is the radius of the helix usually equal to 0.25 of the radius of the current filament. In other words, the lifetime of DSCF is equal to the time of its formation. 2. The helix velocity depends on the pressure in the following way $V = V_o/1.5(\log P - \log P_o)$.

This regularity gives us a possibility to obtain some quantitative estimates of the lifetime t and the depth of formation h of such dynamical structures. As an example, we estimate t and h for vortex of a characteristic size of 100 km, close to the dimensions of photospheric bright points (Müller, 1981).

For subphotospheric layers at a pressure of 10^6 dyna/cm^2 we obtain $V = 6.5$ km/s. It is about three times higher than the turbulent velocity $\langle u^2 \rangle^{1/2} = 2.1$ km/s that occurs there. The formation time of DSCF is 160 s. For comparison, the magnetic diffusivity time τ_η for turbulent plasma is 500 s. If we assume that the vortex moves toward the photosphere with a velocity equal to the mean value of the helix and turbulent velocity we obtain a depth of formation h = 720 km. This result suggests that the depth of ~500 km, where the turbulence is strongest, is possibly the place of formation of such vortices.

In order to investigate the dynamical response of the photosphere to the emergence of the longitudinal magnetic field of such a short-lived vortex we built a two-dimensional time dependent MHD model. The model in cartesian coordinates describes a magnetic arch system which grows linearly with time.

The behavior of the photospheric plasma above the place of increase of the magnetic flux is described by the equations of the two-dimensional MHD of a compressible media of fully ionized hydrogen plasma. (see Dermendjiev, 1983). The effects of thermal conductivity, gravity and Joule heating are taken into account, viscosity and radiation are neglected and ion and electron temperatures arc calculated separately.

For the numerical solution of the model we used the modification of the two-step Lax-Wendroff difference scheme as in Freeman and Lane (1968). The following scenario was used.

We place a two-dimensional 50x50 network with a step $\Delta x = \Delta z = 1$ km in the photosphere at a height of 100 km above the bottom of the photosphere. The initial atmosphere is assumed to be motionless with temperature T = 5330 K and density $\rho = 1.54 \times 10^{-7}$ g/cm^3 (Gingerich et al., 1971). The boundary conditions, besides the magnetic pulse at the lower boundary, are set in such a way as to saitsfy a first-order extrapolation (Chu and Sereny, 1974). In accordance with the observational and laboratory experiments data we assume an initial value of the magnetic induction at the lower boundary $B_o = 1500$ G and a growth of the magnetic field dB/dt = 10 G/min.

We confine our attention only to the result obtained in the time interval equal to the Alfvén wave travelling time after setting the magnetic pulse, since most of the characteristics of the solution appear during this time. In Fig. 2 five consecutive profiles of the temperature and density, normalized to the respective initial values are shown.They are cross-sections of the respective two-dimensional profiles taken on the central part of the studied region. The numbers associated with the profiles indicate the time.

Due to the continuing presence of the magnetic field gradients near the lower boundary, relatively hotter (T/T$_o \sim 1.6$) and denser ($\rho/\rho_0 \sim 2.6$) layer is formed. It represents the so-called T-layer (Tikhonov et al., 1967). The formation of this layer results in the efficient transformation of the magnetic energy into Joule heating of the plasma and the kinetic energy of macroscopic motions. After about 2 s an instability is developed and a shock wave with a velocity ~ 18 km/s is formed.

Conclusions: The proposed mechanism for strong small-scale magnetic field formation is rather controversial and needs a theoretical argumentation. Here it is presented as one possibility only. However, regardless of the manner of formation of such a magnetic field, the effects of its emergence in the photosphere are very interesting. The above-described numerical simulation treats only the initial phase of the emergence and simulates the process development in a rather short characteristic time; more detailed analysis will be published elsewhere.

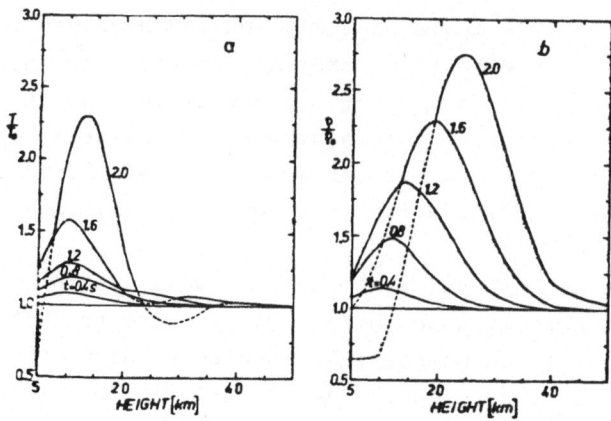

Fig. 2. Electron temperature (a) and density profiles (b) for five consecutive times t = 0.4, 0.8, 1.2, 1.6 and 2.0 s.

Nevertheless, the following conclusions can be made.

The hotter and denser region in the investigated area provides a reasonable simulation of a photospheric bright point. The shock wave, formed during the development of instability of the T-layer, may produce a small-scale transient phenomenon in the chromosphere above the bright point, as was reported by Dr. Mein at this colloquium. Due to the emerging flux, bright points will appear as a pair.

Acknowledgements: The greater part of the work was carried out during the author's visit to Mendon and Pic-du-Midi Observatories with the financial support of IAU Commission 38. The author is grateful for the kind hospitality and the financial support.

References
Anderson, O.A. et al.: 1958, in Proc.of the Second Int.Conf.on the
    Peaceful Uses of Atomic Energy, 32, 150.
Chu, C.K. et al.: 1974, Journal of Computational Physics, 15, 476.
Dermendjiev, V.N.:1983, Publ. of DebrecenHeliophys.Obs., 5, 475.
Freeman, J.R. et al.: 1968, Report LA-3990, Los Alamos Sci.Lab., New
    Mexico, paper C7, 1968.
Komelkov, V.S. et al.: 1962, in Proc.of the Fifth Int.Conf. on Ionized
    Phenomena in Gases, vol. 2, 2191, ed. H.Maecker.
Müller R.: 1981, in Proc.of the Japan-France Seminar on Solar Physics,
    142, eds. F.Moriyama, J.C.Henoux.
Orszag, S.A. et al.: 1979, J.Fluid Mech., 90, 129.
Gingerich, S.A. et al.: 1971, Solar Phys., 18, 347.
Tikhonov, A.N. et al.: 1967, Dokl.Akad.Nauk SSSR, 1173, 808.
Komelkov, V.S. et al.: 1960, in Proc. of the Fourth Int.Conf. on
    Ionized Phenomena in Gases, vol. 2, 1141, ed. N.R.Nilsson.

Discussion

<u>J.C. Vial</u> : Is your mechanism a good candidate for explaining "coronal bullets" ?

<u>Dermendjiev</u> : In the case of division of the vortex, when it penetrates the photosphere, a part of vortex ejects. However, it is more probably the ejected plasma to be observed as a surging arch than as a "coronal bullet".

THE STRUCTURE OF THE SOLAR GRANULATION

R. Muller and Th. Roudier
Observatoires du Pic du Midi et de Toulouse
65200 Bagnères-de-Bigorre, France.

Abstract - The structure of the solar granulation has been analysed
using computer-processed images of two very high resolution (0.''25)
white light pictures obtained at Pic du Midi Observatory. The narrow
range of sizes of granules is not confirmed : on the contrary, it is
found that the number of granules increases continuously toward the
smaller scales; this means that the solar granulation has no characteristic scale. Nevertheless the solar granules appear to have a <u>vertical</u>
size of 1.''37, for which drastic changes in the properties of granules
occur; in particular the fractal dimension changes at the critical
size, which is also the size of the granules providing the largest
contribution to the total granule area and radiation. The granules
smaller than the critical size could be of turbulent origin.

1 - Introduction.

Two granulation photographs of very high resolution (0.''25) obtained
with the 50 cm refractor of Pic du Midi Observatory have been computer
processed and analysed in order to study in detail the structure of the
solar granulation. The informations yielded by the image processing
include the position of the centre of gravity, the size distribution,
the area contribution of granules of given size to the total area, the
power spectrum of brightness fluctuation, δI_{rms}. In this paper we present the results obtained for the area distribution, the area contribution, as well as a fractal analysis of granules. The results are discussed in terms of a turbulent convective origin of the solar granules.

2 - Computer image processing.

The granulation photographs have been digitized, the microdensitometer
scanning aperture and step being respectively of 0.''014 and 0.''01. First,
the digitized pictures are filtered in order to reduce the high
frequency noise due to the graininess of the photograph and to bring

all the granules, including the smaller and fainter ones, located in dark intergranular spaces, to the mean photospheric intensity level (low frequency filtering). The 3dB levels of the filter, in the Fourier domain correspond, toward the high frequencies, approximately to the resolution of the instrument,while toward the low frequencies it corresponds approximately to the mean distance between granules (Figure 1). Then we define a threshold level : only the intensities above this level are kept and form closed areas which correspond to granules (Figure 2)

The level at which the original photographs are reproduced by the processed images is 1.01 x $\bar{I}$, $\bar{I}$ being the mean intensity of the photosphere. We are also forced to introduce a cut-off area A_o; features smaller than $\bar{A}_o$, which is roughly the size of the resolution of the photographs, are eliminated, because most of them are created artificially by the image processing.

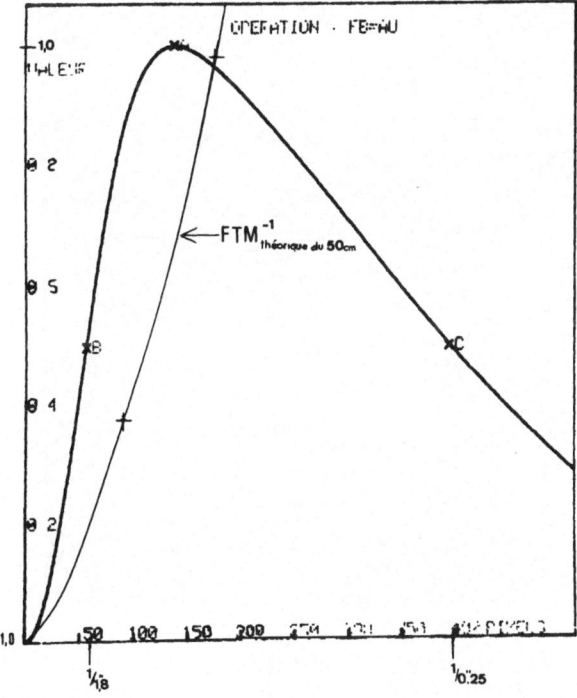

Figure 1. The filter used for the solar granulation pictures filtering.

a

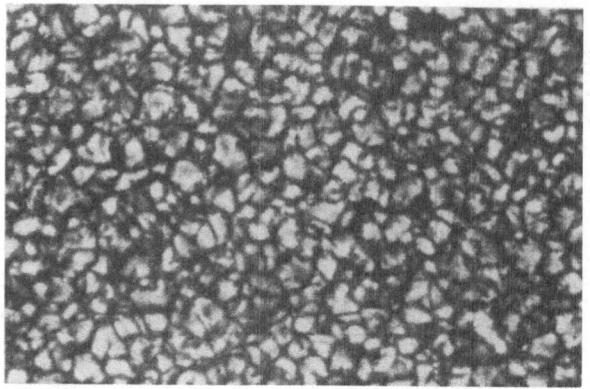

b

Figure 2.(a) original photograph
(b) processed image of the granulation, corresponding to the central square in (a), which
is not at the same scale as (b).

Finally the parameters defining the filter, the threshold level and
the cut-off area A_O, are adjusted empirically in order that the granulation pattern becomes as similar as possible on both the processed and
the original images (Figure 2). The parameters change slightly from one
picture to another, due to change of the resolution of the scale and of
the graininess of the films.

3 - Area distribution of granules.

One of the granule parameter yielded by the image processing is the
granule area. Figure 3 shows that the number of granules increases
continuously toward the smaller scales. That means that the solar
granulation has no morphological characteristic scale.

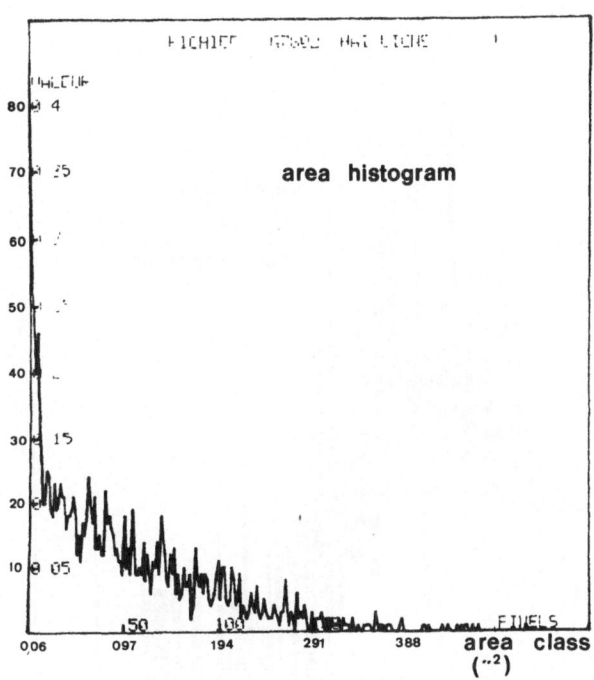

Figure 3. Area distribution of granules.

Previous determinations exhibit (Namba and Diemel, 1969), on the
contrary, a maximum and a narrow range of the size distribution, which
have been interpreted as strong evidence of a characteristic size for
the solar granulation (Bray et al.,1984). The area distribution in

Figure 3, resulting from a very high resolution picture, does not give
an evidence of such a morphological characteristic size.

4 - <u>Area contribution of granules of given size to the total area.</u>

n_i granules having an area A_i contribute $\dfrac{n_i A_i}{\Sigma n_i A_i}$ to the total area of
the granules. The main contributors to the total area, and conse-
quently to the radiation emitted by the granules are the granules of
intermediate size, around 1″4 (Figure 4). The smaller granules, although
being very numerous, contribute very little to the total area, and
still less to the radiation as they are fainter than the larger ones.

Thus the solar granulation gives the visual impression of having a
narrow range of sizes, that is to say a characteristic size, because
the small granules contribute very little to the total radiation : they
do not catch the eyes. But, in fact, the solar granulation has no morphological characteristic size.

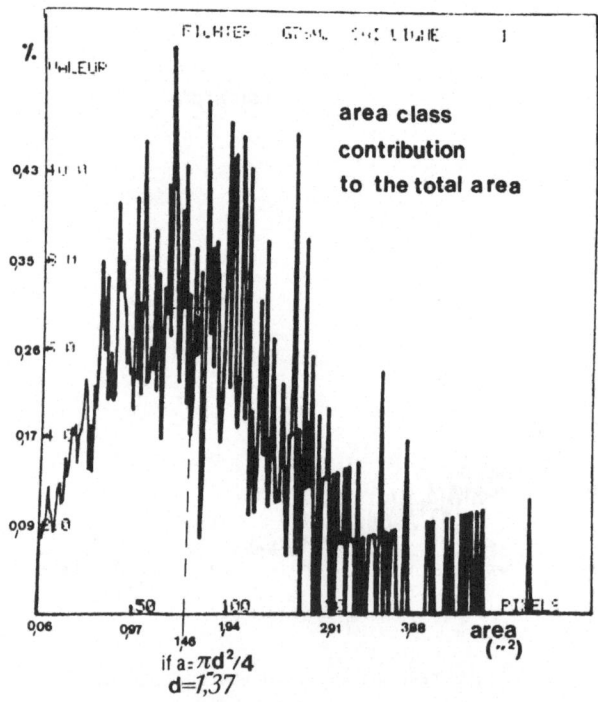

Figure 4. Area contribution of granules of given size
to the total area.

5 - Area-Perimeter relation : fractal analysis of the solar granulation.

It is evident from Figure 2 that many of the granules are irregular and even complex in shape. The Area-Perimeter relation $P \sim A^{D/2}$ (where D is the fractal dimension), is used to analyse the planar shape of the granules (Mandelbrot, 1977, Lovejoy, 1982). For smooth shapes such as circles and squares, $P \sim A^{1/2}$ and thus D = 1, the dimension of a line. As the perimeter becomes more and more contorted and tends to double back on itself, filling the plane, $P \sim A$, and D approaches the value 2. The log A vs log P relation as determined from measurements made on the processed pictures is linear (Figure 5), but a sharp change in the slope and consequently in the factal dimension, occurs for a diameter of d = 1".37 (if it is assumed that the granules are circular , $A = \frac{\pi d^2}{4}$). The fractal dimension of the granules smaller than

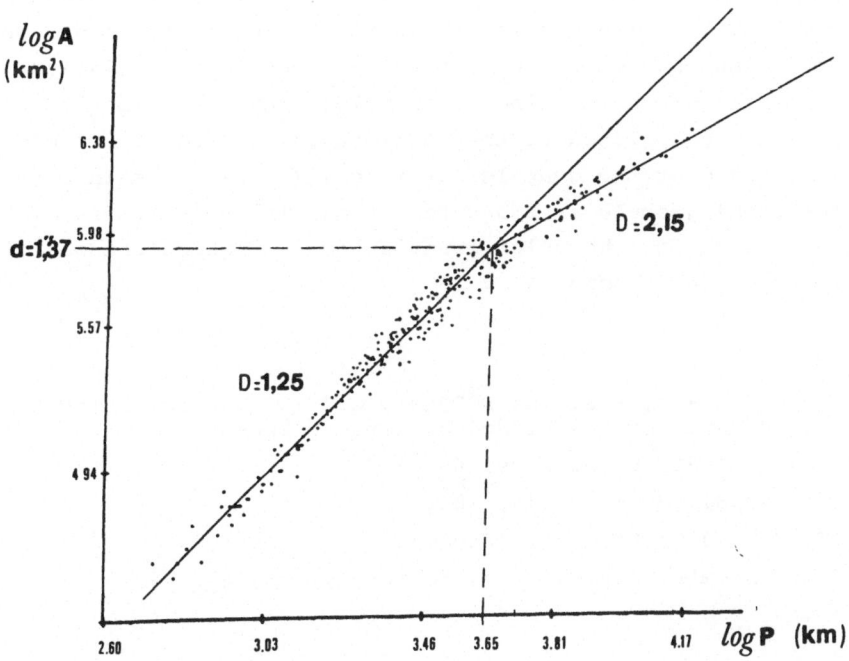

Fig.5. Area-perimeter relation of solar granules measured.
Each point represents one measured granule.

1".37 is of 1.25, while it is as high as 2.15 for the larger ones. That means that the shape of granules begins to be contorted above 1".37, probably because they are merging, splitting or exploding. As D cannot exceed the value 2, the questionnable value of 2.15 may be due to the small range of sizes of the larger granules (between 1".37 and 3").

6 - Conclusion.

The top of the convection zone being a highly turbulent atmosphere we would expect an increasing number of smaller granules, either produced by a monotonic increase of the growth rate or by a turbulent cascade. This expectation is confirmed by our results, but from the only inspection of Figure 3, it is not possible to discriminate between those two possibilities. Fractal analysis for solar granulation provides an additional important information : the value of the parameter D for the granules smaller than 1".37, is very close to the value of 1.33 predicted for isobars in a turbulent, homogeneous and isotropic atmosphere (Mandelbrot, 1977). Our results are thus in favour of a turbulent cascade of the convective energy. Kawaguchi (1980) reported that almost all the granules reaching a size of 1".5 will fragment or merge, which suggests that the change of slope of the area-perimeter relation is due to the fragmentation and merging of granules. Thus the granule size 1".37 is probably associated to the characteristic size of fragmentation of granules. Moreover the granules of size 1".37 are the main contributors to the total granule area and radiation. It thus appears that 1".37 is a critical size for the solar granulation, for which changes of the properties of granules occur.

References

BRAY, R.J., LOUGHHEAD, R.E. and DURRANT, C.J.: 1984, The Solar Granulation, 2nd edition, Cambridge University Press.
KAWAGUCHI, I. : 1980, Solar Phys. 69, 207.
LOVEJOY, S. : 1982, Science 216, 185.
MANDELBROT, B : 1977, Fractals (Freeman, San Francisco).
NAMBA, O. and DIEMEL, W.E. : 1969, Solar Physics, 87, 243.

A MODEL FOR THE RUN OF THE HORIZONTAL AND VERTICAL VELOCITIES IN THE DEEP PHOTOSPHERE

Anastasios Nesis
Kiepenheuer-Institut für Sonnenphysik
Schöneckstr. 7, 7800 Freiburg, FRG

A more detailed knowledge of the transition region between the convectively stable and unstable layers is desirable for the theoretical approach to solar convection, and for the understanding of line formation (asymmetry). It is assumed that the dynamics of this layer is determined by the overshoot. The term "overshoot" is used in Astrophysics for situations in which the convective motions disturb the convectively stable layers. This is also known as "penetrative convection" which is probably responsible for the generation of gravity waves in the photosphere, as well as for the asymmetry of absorption lines.

In the deeper photosphere we found correlation between the intensity and velocity fields but not in the middle and upper photosphere. Therefore, the atmosphere in its vertical extension is not homogeneous from the dynamical point of view. The asymmetry of the absorption lines could reflect this situation. In this investigation we tried to find out the extension up to which the velocity field is correlated with the intensity field, i.e. up to which height in the photosphere the overshoot can be observed. Therefore, we investigated the dynamics of the lower photosphere by determining the variation of the horizontal and vertical small-scale ($<3.''5$) velocities with height in the atmosphere.

We used spectrograms taken with the baloon-borne spectrograph at three different positions on the solar disc ($\mu=1.0$, 0.6, 0.2) in the wavelength range 5120-5124 Å. The small-scale horizontal and vertical Doppler-velocities were measured using absorption lines of different strength which could be ordered in the five levels, O to IV corresponding to rest intensities of 95%, 85%, 65%, 30%, and 18%. Thus, we obtained information about the variation of velocity with height in the photosphere.

To give velocity as a function of height in the atmosphere it was

necessary to specify the correspondance between the measured velocity
and the geometrical height. There are two ways of doing that.

If we assume that the motion whose velocity we have measured is located
in a specific level of the photosphere, we can try to find its geometrical height by means of contribution functions. Or, we may consider
that the motion is spread out over the entire compressible atmosphere.
Then, a method has to be used which takes into consideration the contribution of the entire atmosphere to the measured velocity.
This is the case in the method of velocity weighting functions which
is a perturbation method based on Taylor-series development of the
first order (see Mein, 1971; Canfield, 1976).

Tab. 1. Comparison between the measured velocities V_{obs} and the velocities V calculated by equation (1). $V_{mod}(z)$ model run;
V_{01}, V_{02}, H_1, H_2: 4 parameters used in the calculation.
$\mu = 1.0$: vertical velocity; $\mu = 0.2$: horizontal velocity.
Fe 5121,65 Å: absorption line for which the velocities were
calculated.

Fe 5121, 65 Å				
$\mu = 1,0$		$\mu = 0,2$		
$H_1 = 75km$ $V_{01} = 1,0 \frac{km}{sec}$ $H_2 = 130km$ $V_{02} = 0,06 \frac{km}{sec}$		$H_1 = 75km$ $V_{01} = 1,1 \frac{km}{sec}$ $H_2 = 290km$ $V_{02} = 0,1 \frac{km}{sec}$		
	V_{Obs}	V	V_{Obs}	V
95%	753	742	440	432
85%	607	670	359	363
65%	557	539	298	312 363
30%	563	556	387	
18%	553		403	
$$V_{Mod}(Z) = V_{01}\, e^{\frac{-Z}{H_1}} + V_{02}\, e^{\frac{+Z}{H_2}}$$				

The functional dependence between the observed velocities V_{obs} and the model velocities $V_{mod}(z)$ is the following:

$$V_{obs} = \int W_V(z) V_{mod}(z) dz \qquad (1)$$

The velocity weighting function $W_V(z)$ was calculated for each absorption line which was used for the measurement of velocity. The model run of velocity $V_{mod}(z)$ was determined by 4 parameters (V_{01}, V_{02} : velocity amplitudes, H_1, H_2 : scale heights) in such a way that equation (1) gave the observed velocities V_{obs}. The calculation was performed for the vertical and the horizontal velocities.

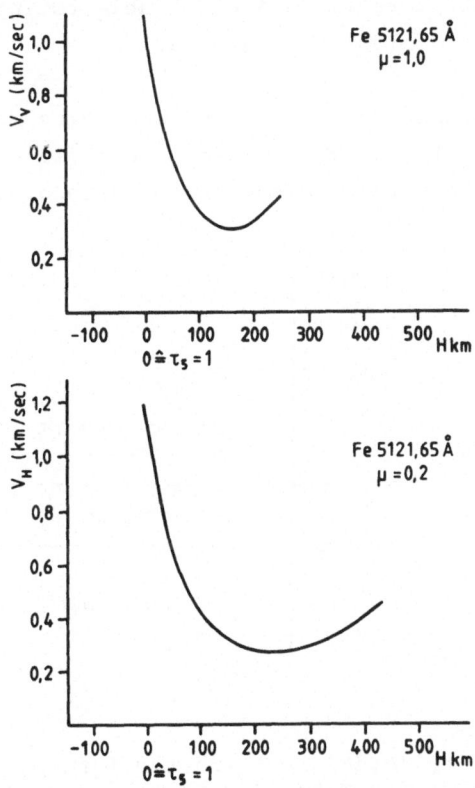

Fig. 1. Model run of vertical (above) and horizontal (below) velocity.

The measured velocities are shown together with the velocities calculated by the model in Table 1. Here, $V_{mod}(z)$ and the 4 parameters used in the calculation are also given. In Figure 1, the model run of velocity with height in the atmosphere is shown. The vertical and horizontal velocities decrease exponentially up to a height of 150 and 200 km, respectively, above τ_5 = 1. The velocities have a minimum at 150 and 200 km, and increase again but with different slopes above this height.

Based on our results of the coherence analysis (Nesis et al., 1984) we interpret this model run of velocity in the following manner: there is an overshoot up to a height of 150 km for the vertical velocity. Above this level there is a secondary motion of non-convective nature. The different locations of the minima of the vertical and the horizontal velocities and the different steepness of the model run above the minimum indicate that the overshoot of the horizontal velocity has a different behaviour. This is also inferred from our results of the coherence analysis. At present, we are not able to interpret the overshoot of the horizontal velocities satisfactorily.

The time which is necessary to bring the convective motion up to the height of 150 km beginning at τ_5 = 1 can be calculated. If we use the change of velocity with height in the photosphere given by the model,

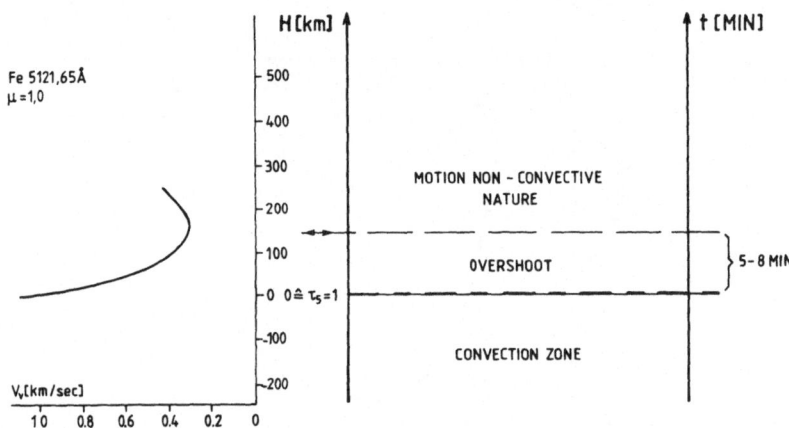

Fig. 2. Vertical velocity (model run) and the inferred dynamics of the deep photosphere. 5-8 min: life time of a granulum (see text).

we find a duration of 5-8 min. This is in good agreement with the life time of a granulum of 6 min (Mehltretter, 1978).

This agreement can be understood if we assume that the ordered velocity pattern of granulation gradually disappears within about 150 km above τ = 1. The motions which we found above this level must be due to secondary motions which are induced by convective motions either directly or indirectly. It is possible that the secondary motions are due to gravity waves. These considerations are summarized in Figure 2.

REFERENCES

Canfield, C.R.: 1976, Solar Phys. 50, 239.
Mein, P.: 1971, Solar Phys. 20, 3.
Mehltretter, P.J.: 1978, Astron. Astrophys. 62, 311.
Nesis, A., Durrant, C.J., Mattig, W.: 1984, in press.

INFLUENCE OF UMBRAL DOTS ON SUNSPOT MODELS

E.Wiehr, Universitäts-Sternwarte Göttingen, D-3400 Göttingen
G.Stellmacher, Institut d'Astrophysique, F-75014 Paris

Empirically determined umbral temperature stratifications usually are
based on measurements in darkest sunspots. However, the existence of
brighter umbrae became important since Albregtson and Maltby (1981)
found a relation between spot brightness and the phase of the solar
cycle. An explanation of this finding was given by Schüssler (1980) who
considers a cycle dependent age of the fluxtubes rising from deep layers.

We investigated umbrae of different brightness photoelectrically in three
clean continuum windows at 4365 ± 1/8; 6305 ± 1/4 and 8089 ± 1/2 $\overset{o}{A}$
strongly simultaneously in the 13th, 9th and 7th order of the f = 10 m
Echelle grating spectrograph at the Locarno observatory (Wiehr et al.
1980). Using a Ø = 1 arcsec aperture we measured the intensity minima
('dark cores') of individual umbrae in the spot groups from Sept. 29,
1982 at 12°N, 14°E and from July 31, 1983 at 22°S, 36°E.
The evacuated Gregory telescope at Locarno has a field stop in its prime
focus which largely reduces scattered light (Stellmacher and Wiehr 1970,
Fig. 2). The remaining influence of seeing was minimized by selecting
the deepest mean values from time-scans over several minutes. The high
quality of these measurements is indicated by the lowest relative intensities which amount (without any correction !) to 4.3%, 10.5% and 20.5%
in the above three windows; in agreement with the prediction of the model
M4 by Kollatschny, Stellmacher, Wiehr & Fallipou (1980). We assume these
lowest intensities to represent umbral background (i.e. inter-dot material). Simple calculations show that a given contribution of *dots with
photospheric temperature* yields brighter umbrae with a pronounced blue
excess as compared to our measurements (see Fig.1). The assumption of
cooler dot material may remove this problem. As lower limit for such reduced dot brightness one may use our highest umbral intensities (32%,
45%, 51%; see Fig.1) if one assumes that this umbra is 100% covered by
dots. Even the assumption of such *cool dots* still yields an unrealistic
blue excess for our medium bright umbrae. This excess is also not sufficiently reduced when using the flat dot model by Adjabshirizadeh and
Koutchmy (1983).

As a consequence of these considerations we state that *umbral dots can not account for different umbral brightness observed*. Instead, we propose dots to represent a less important additive for *umbrae of different temperatures*. From corresponding calculations we find that our observed umbral continuum intensities as function of wavelength (Fig.1) are well reproduced by 'down-scaled' photosphere models taking a chosen

$$\Delta\Theta = \frac{5040}{T^*_\tau} - \frac{5040}{T^\odot_\tau}.$$ The corresponding temperature stratifications (Fig.2) are considerably flatter than those which are obtained by 'up-scaling' of the steep umbral model, M4, by Kollatschny, Stellmacher, Wiehr & Fallipou (1980), since the deep layers of that model are in radiative equilibrium (r.e. $\triangleq T^4 \sim T^4_{eff} \times \tau_R$; see Stellmacher and Wiehr 1975).

These r.e. models with corresponding T_{eff} fit our highest observed umbra intensities significantly worse than do the flatter photosphere-type stratifications (see Fig. 1). For the lower umbral intensity observations, this difference becomes less significant thus allowing for a smooth transition between the more convective photosphere-type and the more radiative umbra-type stratifications. Such variable influence of convective energy transport for umbrae of different brightness might *additionally* manifest in the population of dots, although the dot brightness alone can not account for the color curves observed.

Reference

Adjabshirizadeh, A., Koutchmy, S.: 1983, Astron. Astrophys. 122, 1
Albregtson, F., Maltby, P.: 1981, Solar Physics 71, 269
Kollatschny, W., Stellmacher, G., Wiehr, E., Fallipou, M.A.: 1980, Astron. Astrophys. 86, 245
Schüssler, M.: 1980, Nature 288, 150
Stellmacher, G., Wiehr, E.: 1970, Astron. Astrophys. 7, 432
Stellmacher, G., Wiehr, E.: 1975, Astron. Astrophys. 45, 69
Stellmacher, G., Wiehr, E.: 1981, Astron. Astrophys. 95, 229
Wiehr, E., Wittmann, A., Wöhl, H.: 1980, Solar Physics 68, 207

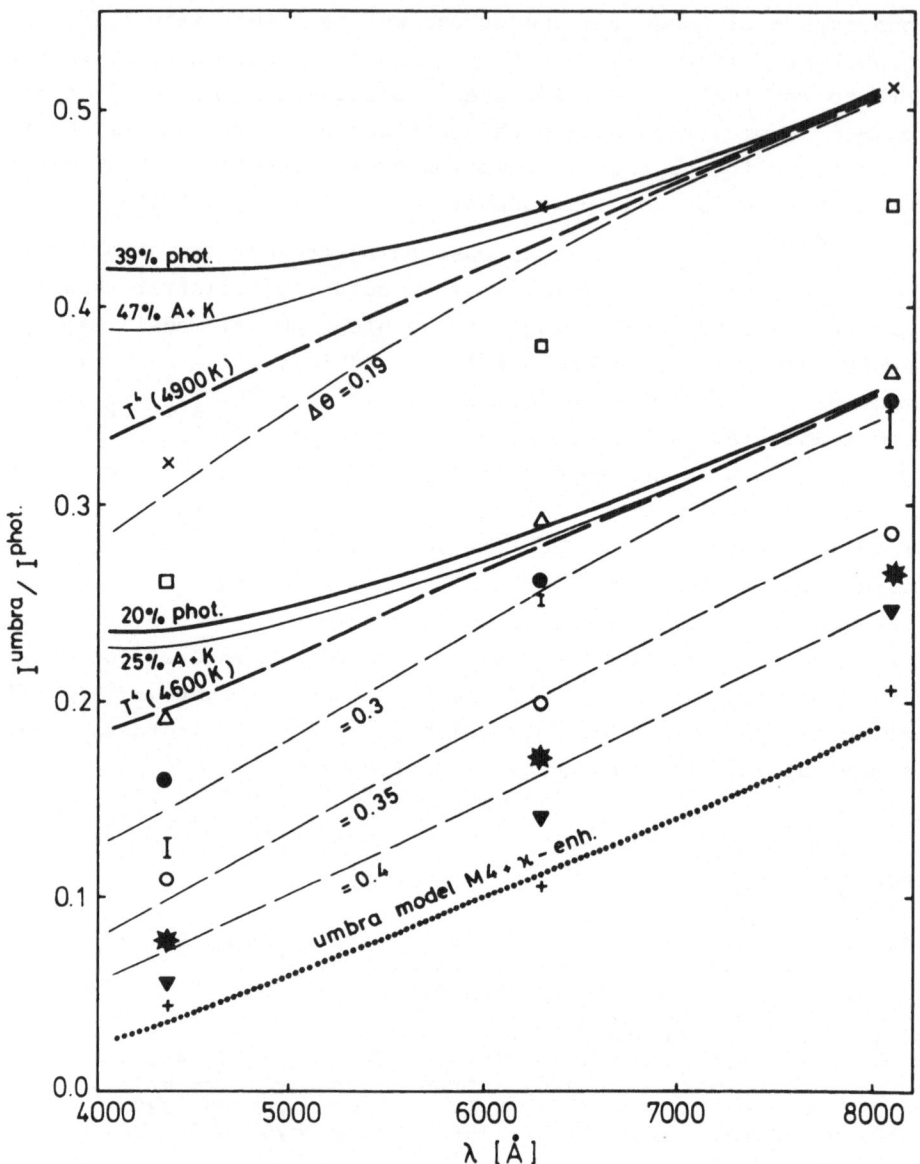

Fig.1: Relative intensities of different umbrae measured simultaneously
in three clean continuum windows, together with calculations
using scaled (a) photospheric models with given

$$\Delta\Theta = \frac{5040}{T^*} - \frac{5040}{T^{\odot}}$$, (b) radiative equilibrium models with
different T_{eff}; as well as influences of
dots with (c) photospheric and (d) temperatures given by
Adjabshirizadeh and Koutchmy.

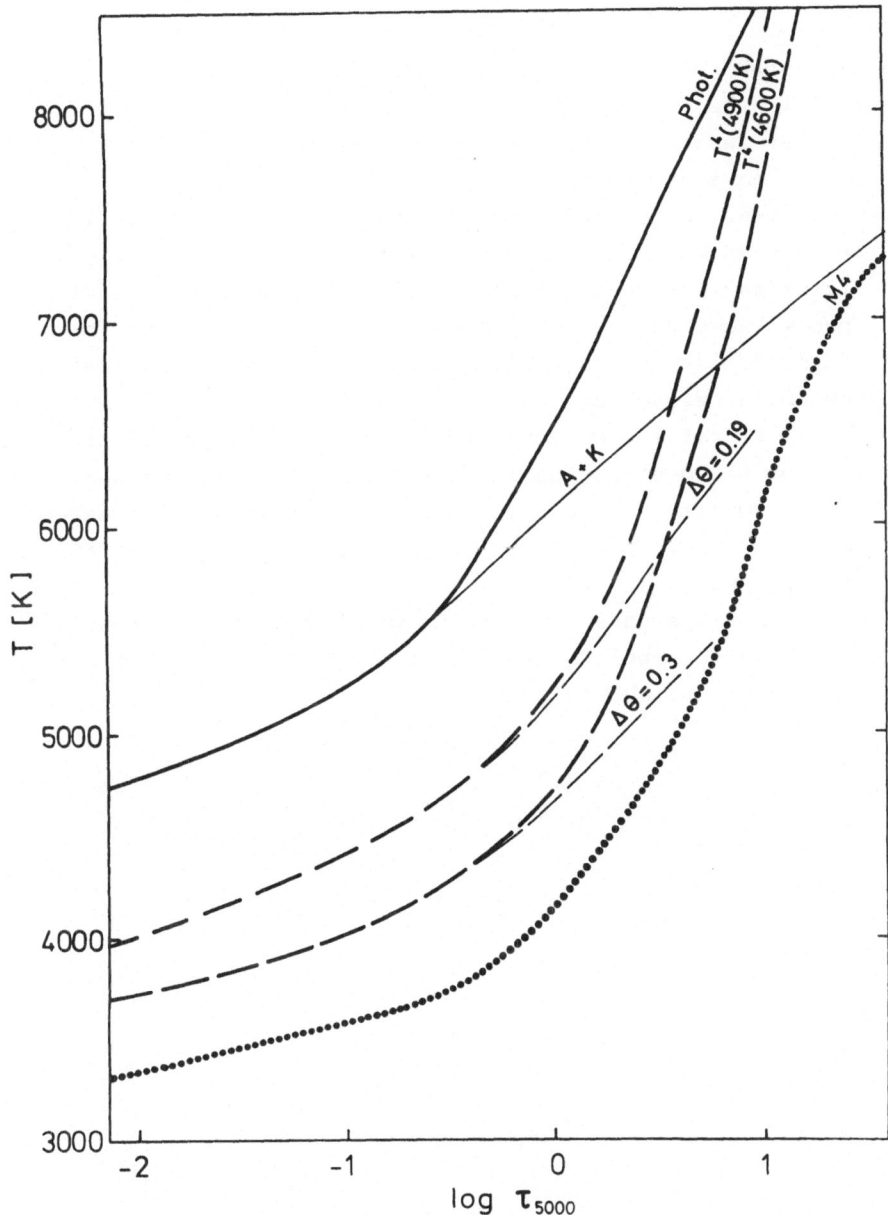

Fig. 2: Temperature vs. optical depth models used for the calculations given in Fig.1; phot. represents the Holweger (1967) model modified for deep layers by the Vernazza et al. (1973), M4 represents the umbral model by Stellmacher and Wiehr (1981; cf. other references).

Discussion

Mattig : Have you observed photoelectrically ?

What is the size of the diaphragma ?

At what zenit distance you have observed ?

Do you have an influence of atmosphere dispersion ?

E. Wiehr : We observed photoelectrically with a 1 arc sec. aperture near solar meridian passage. The maximum offset from chromatic refraction does not exceed 1 arcsec in agreement of theory. However our measuring procedure by "waiting for best moments", i.e. lowest intensities minimizes the influence of remaining refraction.

C. Zwaan : From V profiles of infrared lines a singly peaked distribution of field strengths is found, with the maximum near 1500 ± 300 gauss. Clearly we can only measure field strength above the limit set by the line width, that is above 600 gauss in our case.

E. Wiehr : I appreciate this measurement which confirms my result from the three-slit-polarimeter (A & A, 1978) for isolated field regions of variable diameter.

SOME RESULTS OF PHOTOSPHERIC FINE STRUCTURE INVESTIGATIONS AT THE
PULKOVO OBSERVATORY.

V.N. Karpinsky
Pulkovo Observatory
196140 LENINGRAD USSR.

Observations with the Soviet Stratospheric Solar Observatory and Pamirs
Solar telescope allow to make the following conclusions :
1. All the photospheric levels are strongly inhomogeneous. The
   brightness and excitation temperature differences are as large as
   $1000°K$ at small distances with horizontal gradients up to 5-10 K km^{-1}.
2. The two-dimensional spatial spectrum of granulation brightness
   field is represented by two power functions k^n with n = 1.2 and
   n = - 5 for the ascending and descending branches respectively.
   The sight velocity spatial spectrum should have in that case its
   maximum at a period of about Λ = 800 km, a steep growth ($\sim k^5$) and
   slow decay ($\sim k^{-1}$). Inertial domain of the spectrum should be
   absent and energy and dissipative domains are overlapped. The
   granulation brightness field differs significantly from the
   Gaussian field and is an ensemble of two-dimensional bright and
   dark impulses from some "starting" brightness level, which is lower
   than the mean photospheric one. "Births" and "deaths" of granules
   take place seldom. Proper motions with brightness fronts velocity
   up to 20 km s^{-1}, divisions and couplings are more typical. Two
   granules can be drawn together but may stay separated by a narrow
   steady space between them. There are some reasons to distinguish a
   special class of dot granules which are smaller than 200 km and
   unevenly distributed over the solar surface.
3. The sight velocity photospheric field is represented by homogeneous
   vertical columns. The regularity is very sharply broken in the
   very thin (20-50 km) transition stratum at a height of about
   300 km above τ_5 = 1 level. On the contrary, the "brightness" field
   has a very fine vertical structure, which is three-dimensional irregular formations with boundaries slowly inclined to the horizontal plane (< 10°).

The coexistence of the coherence minimum K = 0.5 for Λ = 1500 km and the maximum K = 0.9 for Λ = 2900 km in the coherence spectrum brightness - sight velocity is an explanation for the discrepancy between different authors and argumentation for low correlation in photospheric fine structure.

It seems that the conception of convective overshoot is not a constructive explanation for the major observing facts. Some aspects of the photospheric fine structure nature are discussed.

4. THE SURFACE FINE STRUCTURE AS A PROBE OF THE SOLAR INTERIOR

ATMOSPHERIC FINE STRUCTURE AS A PROBE FOR THE SOLAR INTERIOR

Cornelis Zwaan
Sterrewacht "Sonnenborgh"
Zonnenburg 2
3512 NL Utrecht, The Netherlands

Introduction

In order to gain more insight in the solar interior we should read the solar face intently, and interpret its features carefully. Most of our present knowledge is based on global data (diameter, mass, luminosity, neutrino flux, abundances) and on large-scale features: Doppler shifts caused by rotation and oscillations, magnetic phenoma such as sunspot groups and active regions, and the patterns in the activity cycle. The study of atmospheric fine structure promises new insight in the solar interior - below I discuss some examples. These studies are made possible by many efforts invested in site testing and in developing the instrumentation for high-resolution studies. Pioneering work has been carried out by Professor Rösch, and to him and his collaborators we owe heliograms of the best quality obtained as yet.

Magnetic elements as tracers for velocity fields

The idea is simple: small magnetic elements are passively carried by flows. However, the implicit "float-plus-sea-anchor" model (fig. 1a) is inadequate in many cases, because this model does not agree with Maxwell's law $\nabla \cdot \vec{B} = 0$: any magnetic element in the atmosphere is part of a fluxtube that has no ends (fig. 1b). Forces act on the flux tube due to (1) buoyancy (F_b may be positive or negative), and (2) curvature: $F_m = B^2/(4\pi R)$, where R is the radius of curvature. Only a submerged fluxtube may be in a static equilibrium with respect to the ambient medium (fig. 2a) but such an equilibrium is not stable (van Ballegooijen 1982a,b). Hence the fluxtubes are in a dynamical equilibrium: they move at a constant lateral velocity v_d with respect to the ambient medium such that a balance is achieved between the forces due to buoyancy, curvature of field lines, and (3) hydrodynamic drag F_d (:) $\rho\ v_d^2\ r^{-2}$ (ρ is the mass density, r is the radius of the tube), and (4) Coriolis effect $\vec{F}_c = 2\vec{v} \times \vec{\Omega}$. Let me illustrate the dynamical equilibrium with an example. Van Ballegooijen (1982a,b) has

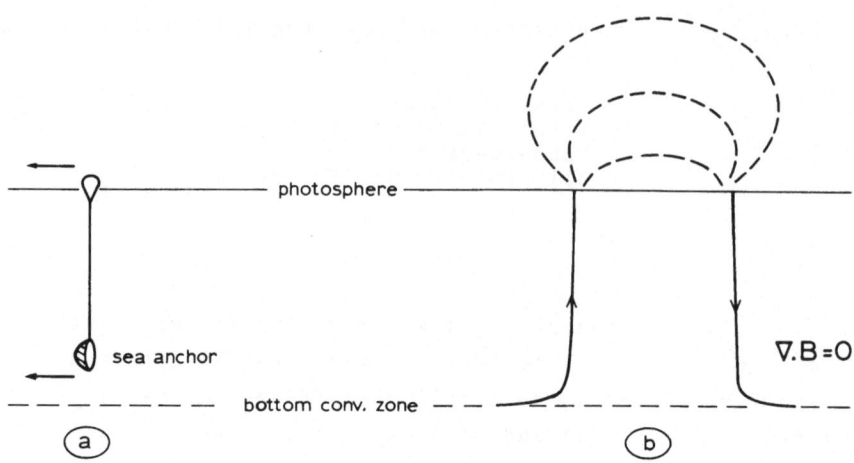

Fig. 1: Models for velocity tracers: a. float plus sea anchor, b. magnetic fluxtube.

computed models for adiabatic fluxtubes rooted in a toroidal magnetic field just below the base of the convection zone. After the top of a flux loop has emerged the fluxtube legs stand nearly vertical in the convection zone, and the dynamical equilibrium requires that the legs drift apart with a speed v_d relative to the ambient medium (fig. 2b).

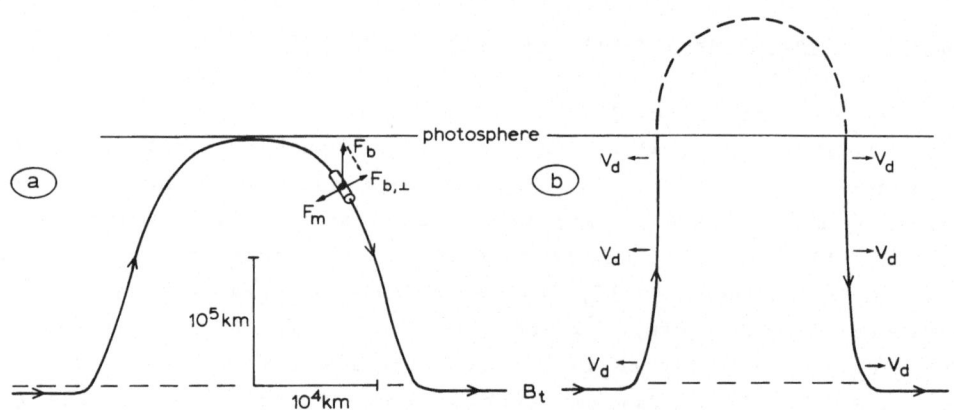

Fig. 2: a. Unstable static equilibrium of a submerged flux loop, b. dynamic equilibrium of an emerged flux loop.

This lateral velocity v_d depends on the magnetic configuration, in the example of fig. 2b: on the field strength B_t in the toroidal system. From observational data on large active regions, with $\Phi \simeq 3 \times 10^{22}$ Mx, van Ballegooijen estimates a systematic drift $v_d \lesssim 70$ m s^{-1}, and hence a field strength B_t in the toroidal system of a few times 10^4 Gauss.

We conclude that the displacements of magnetic elements in the atmosphere do not relate to motions somewhere in the convection zone in the simple float + sea anchor model. In order to interpret the motions of the tracers we need to know where the magnetic fluxtubes are anchored, the effects of the overlying velocity field, and the main characteristics of the magnetic structure in which the tracers are embedded. Probably magnetic elements may be used as tracers for large-scale velocity patterns in the convection zone but in the course of the interpretation a model for both the velocity field and the magnetic field has to be developed and improved.

The measurement of displacements of tracers is not simple either. Positions need to be measured with a precision of a fraction of an arc second relative to a frame of reference defined, for instance, by the disk center and the E-W direction. This requires either a very stiff telescope, or a clever scheme to minimize the effects of flexure and vibrations of the telescope (see Schröter and Wöhl, 1975).

Large-scale structure of the magnetic field in the solar interior

During and shortly after the peak of its development, a large active region shows the well-known characteristics:
(i) The orientation of the long axis of the regions is nearly E-W, the leading parts being slightly closer to the equator.
(ii) Most regions are bipolar, or become bipolar after some decay of the region.
(iii) During the 2 x 11 years magnetic cycle the active regions follow the polarity rules discovered by Hale.
(iv) The typical separation between the centroids of magnetic flux of opposite polarities within a large bipolar active region is about
1 x 10^5 km.

Whereas many of the rules specifying the solar activity cycle are trends, no tight relations, the above rules (i), (ii) and (iii) are well defined and they apply to the vast majority of active regions. So these rules reveal a clear pattern of the magnetic field in the solar interior. Since the heuristic dynamo models by Babcock and by Leighton we understand this pattern as the result of the differential rotation wrapping the magnetic field into toroidal flux bundles.

The well-defined orientation of the active regions indicates that the turbulent convection cannot play havoc with the magnetic field. Hence the field strength in the toroidal system B_t should exceed the equipartition field strength B_{eq} (Zwaan, 1978), which is about 1×10^4 Gauss at the bottom of the convection zone. (There are compelling reasons to situate the toroidal magnetic flux system in the interface between the bottom of the convection zone and the radiative interior, see Galloway and Weiss, 1981, and van Ballegooijen 1982a,b,c).
Note that at $B_t \gtrsim 1 \times 10^4$ Gauss the magnetic flux of a large active region $\Phi \simeq 3 \times 10^{22}$ Mx may be stored in a flux bundle with a radius $r \lesssim 1 \times 10^4$ which is about one fifth of the pressure scale height H_p near the bottom of the convection zone. In this way we may understand how large active regions emerge as coherent structure. Moreover, the measures of the whole magnetic flux loop constituting a large active region form a plausible set (fig. 3). (Probably the toroidal flux

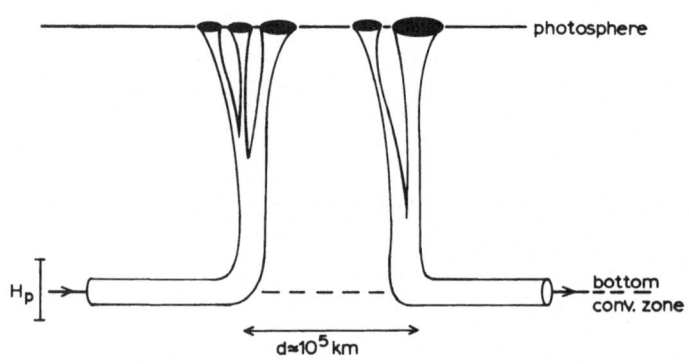

Fig. 3: Schematic picture of a large active region emerged from a toroidal strand (see text).

bundles released in the emergence of an active region are not circular – more likely these are flat ribbons scooped up by overshooting convection). Models invoking strong magnetic fields in the toroidal flux system require a mechanism to counter the buoyancy (see van Ballegooijen 1983, and Schüssler, 1984).

Emerging Flux Regions

The birth and early development of an active region is a process that is determined by the structure of the magnetic field in the convection zone prior to the emergence, and by the adjustment of the magnetic field to atmospheric conditions. To take advantage of such an event, observations are needed with an angular resolution better than an arc second.

The first sign of an emerging flux region is a tiny, compact and bright bipolar plage (Sheeley, 1969). In Hα the arch filament system is observed (Bruzek, 1967), which suggests magnetic loops connecting facular elements of opposite polarity. The facular feet move apart, at first the rate of separation exceeds 2 km s^{-1}, then it drops to values between 0.7 and 1.3 km s^{-1} (Harvey and Martin, 1973). New flux emerges somewhere in between the diverging faculae. If sufficient magnetic flux becomes available, pores and eventually sunspots are formed near the leading and following edges of the active region (see

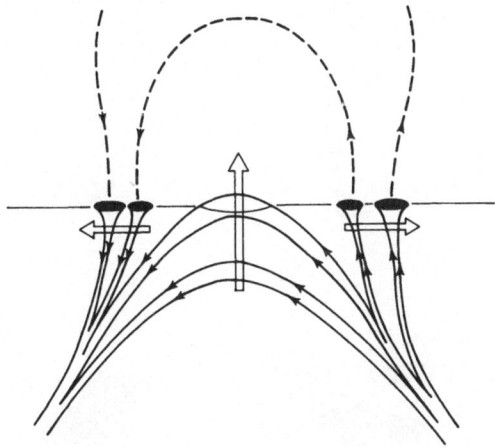

Fig. 4: Emergence of a bundle of flux loops, separation of polarities and coalescence of sunspots. Broad arrows indicate local displacements of fluxtubes. Real emerging flux regions consists of many more separate flux loops.

Zirin, 1974). Sunspots grow by coalescence of pores and faculae (see Vrabec, 1974).

The course of events in an emerging active region is well described by a rising top of a loop, consisting of many flux tubes (fig. 4). The buoyancy of the fluxtubes causes the separation of elements of opposite polarity and the coalescence of elements of the same polarity into sunspots. Apparently somewhere near to the bottom of the convection zone the bundle of fluxtubes is held tightly together for the lifetime of the sunspots.

A time series of spectrograms and slitjaw pictures (Zwaan, Brants and Cram, 1985), analysed by Brants (1985a-c), brought more insight in some aspects of emerging flux regions. The well-known downdrafts in

chromospheric and photospheric spectral lines over emerging flux
regions appear to be localized in small areas, adjacent to but not
coinciding with sunspot pores. Brants and Steenbeek (1985) describe
such a downdraft area developing into a small pore that later
coalesces with other small pores in the formation of a large sunspot
pore (fig. 5). Probably such a downdraft corresponds to the "convec-

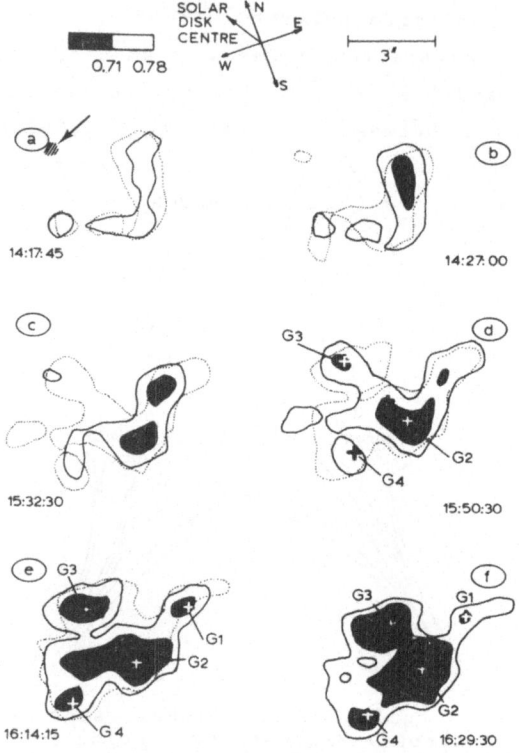

Fig. 5: The growth of a sunspot pore. Dotted isophotes represent the
isophotes in the next frame. The shaded area in frame a., indicated by
the arrow, is the site of a strong downward flow (from: Brants and
Steenbeek, 1985).

tive collapse" of fluxtubes from a field strength of several hundred
Gauss to a field strength $B \simeq 1500$ Gauss (see Spruit 1979).
Large-scale upward flows are found near the main polarity dividing
line $B_{//} = 0$ and small-scale upward flows in the vicinity of polarity
reversals elsewhere in the region. Here matter darker than the average
photosphere is observed to be rising.
We would like to know the field strength in the tops of the flux loops
at the very moment of emergence. Close to the main polarity dividing
line Brants (1985b,c) finds a very strong transverse field of at least
2000 Gauss. Probably this part of the magnetic field has emerged quite

early in the birth process of the region since it is attached to the
magnetic field in the leading sunspot mass. Moreover, the local field
is strained by a strong velocity shear in the photosphere (Brants and
Steenbeek, 1985). Hence the strong transverse field near the leading
sunspot may be is not typical for the magnetic field <u>during</u> emergence.
During the observations quite probably new flux emergence occurred in
the central part of the active region, where inclusions of opposite
polarity are found. In adjacent resolution elements with strongly
inclined field Brants (1985c) estimates field strengths of about 500
gauss. The granulation in the central part is characterized by a
system of aligned, abnormally dark intergranular lanes; such systems
have been described by Bray and Loughhead (1964, p. 67). Brants and
Steenbeek (1985) find that an individual alignment (fig. 6) lives for

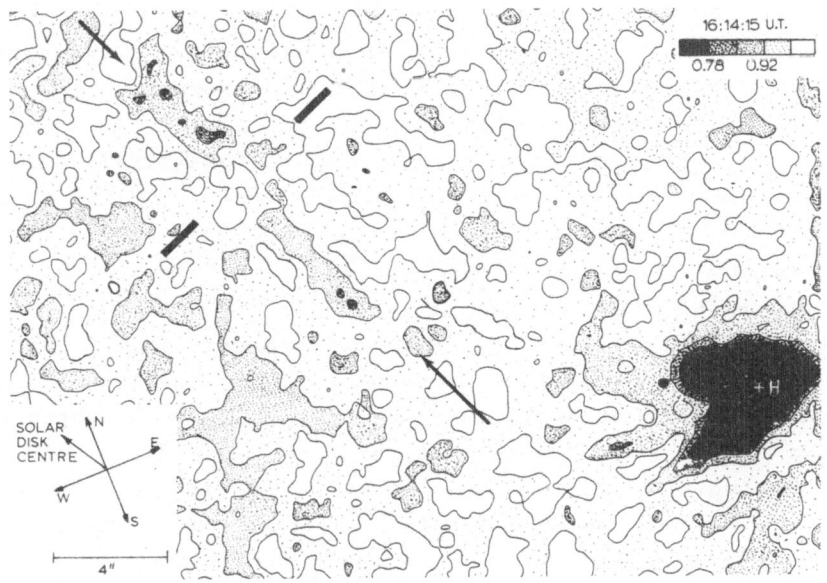

Fig. 6: Isophotes in a portion of an emerging flux region, showing a
dark alignment in the granulation, between the arrows (from: Brants
and Steenbeek, 1985).

about 10 minutes. Assuming that such an alignment is caused by the
penetration of a flux tube into the photosphere, thereby controlling
the granulation, the speed of rise is estimated from the diameter ($\simeq$
1500 km) and the lifetime to be about 3 km s^{-1}. If this speed equals
the Alfvén speed, the field strength in the tube would be about 600
Gauss and the total flux in the tube about 1×10^{19} Mx; the latter
figure agrees with estimates of the magnetic flux per loop in arch
filament systems (Born, 1974). So the present data suggest that

magnetic flux emerges as a loop of many fluxtubes; (the most conspicuous among) these tubes are characterized by a field strength of
several hundred Gauss and a flux of about 10^{19} Mx.

More precise information on the magnetic field in the top of the convection zone, just prior to emergence, requires time series of white-
light photographs, and Ca II and Hα filtergrams at an angular
resolution better than an arc second, supplemented by at least a few
magnetograms. The scope of such data were greatly enhanced by a
collection of high-resolution spectrograms in the two directions of
circular polarization.

Structure and evolution of magnetic elements

Sunspots are formed exclusively in recently emerged flux, during the
finale of the emergence process. Once a sunspot is formed the
desintegration process sets in immediately. After a period of some
days to some weeks, or, rarely, some months, exclusively facular and
network elements remain. Apparently sunspots are fabricated and for
some time maintained by processes in the deep convective zone (Zwaan,
1978). From the observational data known to me I infer that the same
applies to sunspot pores, and perhaps even to magnetic knots (i.e.
magnetic elements of about 10^{19} Mx, or somewhat larger, that do not
show up as dark in the photosphere). In any case, the magnetic
network, consisting of small elements of high field strength, is the
only long-lived magnetic feature on the face of the Sun.

The intranetwork field, discovered by Livingston and Harvey (see
Harvey, 1977) has recently been confirmed by videomagnetograms
obtained at Big Bear Solar Observatory (see Martin, 1984). As yet
little is known about this magnetic field, except that its pattern
seems irregular, and that it is variable on a time scale of about half
an hour.

For umbrae of sunspots the most plausible model resembles a composite
flower (fig. 7): in the convection zone the magnetic field is a bundle
of separate fluxtubes, with more or less field-free plasma between the
tubes, wheras in the atmosphere the fanning fluxtubes meet. Such a
composite model can accomodate the umbral dots and other bright structure. Moreover, the composite model may solve the puzzle how the umbra
may radiate as much as 20% of the photospheric radiation (see van
Ballegooijen 1982b).

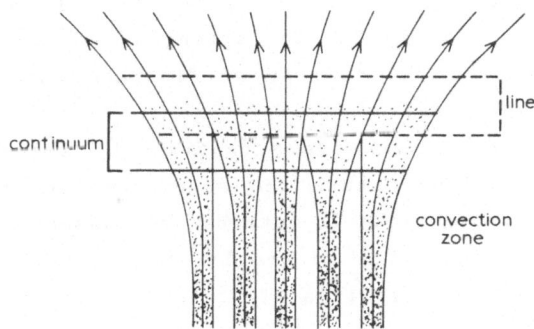

Fig. 7: Composite model for a sunspot umbra. In the convection zone the constituting fluxtubes (gray) are separated by field-free plasma.

Note that the bright structure in different (parts of) umbrae may differ strongly in density and in brightness. Rapid decay of a sunspot starts with the appearance of many very bright dots (Zwaan, 1968), which suggests that the confinement which held the bundle of fluxtubes together is suddenly released.

Spectrograms obtained at sufficient angular resolution to show some intensity structure in umbrae present a problem: there is no correlation between the small-scale brightness variation along the slit and the magnetic field strength. The field stength varies quite smoothly across the umbra despite acute intensity variations (see fig. 3 in Zwaan, Brants and Cram, 1985). A possible explanation is indicated in fig. 7: the individual fluxtubes would fuse in the photosphere such that the very tops of the field-free hot columns are still visible in the continuous spectrum. In the slightly higher layer where the Zeeman lines are formed the magnetic field would be fairly homogeneous. There are no conspicuous Doppler shifts, nor variations in Doppler broadening associated with intensity variations either.

The discovery of the umbral brightness depending on the activity cycle (Albregtsen and Maltby, 1978, see also Maltby et al., 1984) raises the question: is it the packing of the constituting fluxtubes that varies, or does the structure of the individual fluxtubes vary?

The "Wilson depression" of the umbra with respect to the surrounding photosphere is an important integral measure of the thermodynamic structure of the magnetic structure in the top of the convection zone. Its value is only crudely known to be about 600 ± 200 km. We ask for a more precise value, which probably depends on the average brightness and so on the phase of the cycle.

For the study of the individual elements in faculae and network an extremely high angular resolution is required. The discovery that most of the magnetic flux through the photosphere is present in small elements characterized by field strengths between one and two kilogauss has prompted the concept of discrete fluxtubes. There are two extreme possibilities:

(i) The facular and network field consists of magnetic fibers of a long lifetime, much longer than the supergranular lifetime. In this case the changes in the magnetic network are caused by the reshuffling of the flux fibers. Note that at smaller time scales variation in the thermodynamic conditions within a fluxtube may be caused by convection acting on the tube.

(ii) The fluxtubes diffuse and are concentrated on time scales smaller than the granular or supergranular time scales.

The first alternative suggests a fundamental difference between the network field, consisting of fibers firmly rooted in the convection zone, and the intranetwork field, which would be a skin effect. In the second case the intranetwork field would be the field recently escaped from the fluxtubes, that may be reconcentrated again in the network field.

Clearly, a thorough high-resolution study is needed to establish the nature of the network field and the intranetwork field, and to deter-mine the interaction between the convection and these magnetic fields. The lifetime of the elements in the magnetic network is also of practical interest in the use of these elements as tracers of velocity fields. Up to now clusters of magnetic elements have been used as tracers. Clusters in plages and in enhanced network live for about four days, and in the quiet network for at least one day.

Removal of magnetic flux

Magnetic flux must disappear from the photosphere at time scales ranging from a few hours to several years. In an ephemeral region the polarity opposite to that of the background field disappears within a day. Inclusions of opposite polarity in young active regions disappear, tranforming a magnetically complex region into a bipolar decaying region. Particularly large rates of flux disappearance are required in activity complexes, in order to compensate for the large rate of flux emergence (Gaizauskas et al. 1983). From a magnetograph study Howard and LaBonte (1983) conclude that magnetic flux is replaced in about 10 days. The polarity reserval at the polar caps requires flux disappearance at the cycle period.

Despite the large amounts of magnetic flux involved, flux disappearance is much less conspicuous than flux emergence. Apparently the actual disappearance occurs at very small scales and without fireworks. Let us consider how magnetic flux may disappear from the photosphere.

Diffusion does not work to remove magnetic flux from the photosphere - it merely spreads the flux over a larger area. Flux removal requires that loops of flux are pulled out of the photosphere - upwards or downwards. There are three possible modes (Zwaan, 1978) - see fig. 8:

(i)   Retraction of an existing flux loop, without reconnection;

(ii)   Loops pulled upwards from the photosphere after reconnection below the photosphere;

(iii)   Loops pulled downwards after reconnection above the photosphere.

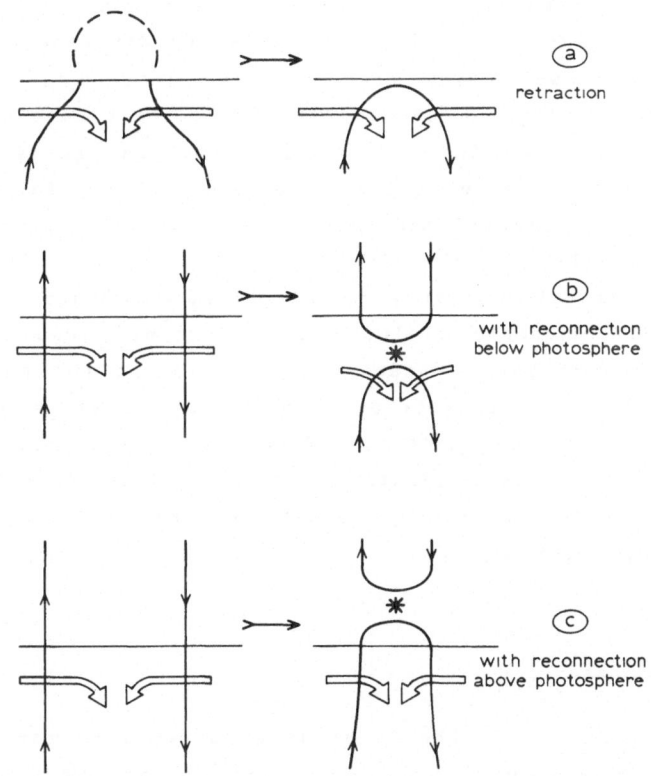

Fig. 8: Three modes of removal of magnetic flux from the photosphere (horizontal line). The broad white arrows symbolize the required converging flows, the asterisks indicate sites of field reconnection.

Hence disappearance of magnetic flux from the photosphere requires flux loops; in two of the three modes reconnection is needed as the first step. In order to make magnetic flux disappear over large areas, or at large rates, large-scale flows converging on polarity dividing lines $B_{//} = 0$ must occur. Note that submergence of flux loops in a descending flow is helped by the convective instability. If flux disappearence happens via numerous tiny flux loops too small to be resolved in magnetograms, then clever combinations of the mechanisms (ii) and (iii) may result in flux disappearence "at a distance" (Parker, 1984; Van Ballegooijen, Title and Spruit, 1984). Such a scheme is required to explain the disappearance of magnetic flux from decaying sunspots (Wallenhorst and Howard, 1982)

Recently direct indications of flux removal have been found on video magnetograms obtained at Big Bear Solar Observatory. In a quiet region magnetic elements approach each other and eventually disappear by cancellation (Martin, 1984).

The procedure to find the mechanisms of the removal of magnetic flux is clear: scrutinize the vicinity of polarity dividing lines $B_{//} = 0$ for converging flows, using the magnetic elements as tracers, and search for approaching and eventually cancelling elements of opposite polarity. In order to find whether reconnection occurs and, if so, whether that happens above or below the photosphere, follow the chromospheric features associated with the magnetic elements before, during and after the cancellation.

Magnetic flux removal involving reconnection, and hence changes in the topology of the magnetic field, hold important clues on the solar dynamo. Whereas published dynamo theories locate the magnetic field reconnection in the convection zone, the outer atmosphere may contribute to the reconnection. Promising zones are around the polarity dividing lines separating the polar caps from the adjacent high-latitude belts of opposite polarity, that is where the polar-crown prominences are located.

Concluding comments

Many of the processes of the solar interior calling for observations at high angular resolution, require time series over many hours to several days, thus over time spans much larger than the period of excellent seeing at one of the existing observatory sites. There are

three ways to overcome this limitation: (i) by an observatory in
space, for instance SOT, (ii) a network of dedicated instruments at
good sites around the world, or (iii) one observatory near one of the
poles. Either effort demands a well - organized collaboration involving many colleagues. The progress we can make in this challenging
field of research depends on the means, skills and ingenuity we can
bring together, and on our dedication to the common goal.

References

Albregtsen, F. and Maltby, P.: 1978, Nature 274, 41
Born, E.: 1974, Solar Phys. 38, 127
Brants, J.J.: 1985a, Solar Phys. 95, 15
Brants, J.J.: 1985b, Thesis, Utrecht
Brants, J.J.: 1985c, Solar Phys., in press
Brants, J.J. and Steenbeek, J.C.M.: 1985, Solar Phys., in press
Bray, R.J. and Loughhead, R.E.: 1964, Sunspots (Chapman and Hall,
    London)
Bruzek, A.: 1967, Solar Phys. 2, 451
Gaizauskas, V., Harvey, K.L., Harvey, J.W. and Zwaan, C.: 1983,
    Astrophys.J. 265, 1056
Galloway, D.J. and Weiss, N.O.: 1981, Astrophys. J. 243, 945
Harvey, J.W.: 1977, in E.A. Müller (ed.), Highlights of Astronomy
    4, part II, 223
Harvey, K.L. and Martin, S.F.: 1973, Solar Phys. 32, 389
Howard, R. and LaBonte, B.J.: 1981, Solar Phys. 74, 131
Maltby, P., Barth, S.B., Lilje, P.B. and Vikanes, F.W.: 1984, in
    The Hydromagnetics of the Sun, ESA SP-220, 233
Martin, S.F.: 1984, in S.L. Keil (ed.), Small- Scale Dynamical
    Processes in Quiet Stellar Atmospheres, p. 30
Parker, E.N.: 1984, Astrophys. J. 281, 839
Schüssler, M.: 1984, in: The Hydromagnetics of the Sun, ESA SP-220, 67
Sheeley, N.R.: 1969, Solar Phys. 9, 347
Schröter, E.H. and Wöhl, H.: 1975, Solar Phys. 42, 3
Spruit, H.C.: 1979, Solar Phys. 61, 363
Van Ballegooijen, A.A.: 1982 a, Astron. Astrophys. 106, 43
Van Ballegooijen, A.A.: 1982 b, Thesis, Utrecht
Van Ballegooijen, A.A.: 1982 c, Astron. Astrophys. 113, 99
Van Ballegooijen, A.A.: 1983, Astron. Astrophys. 118, 275
Van Ballegooijen, A.A., Title, A.M. and Spruit, H.C.: 1984, in
    preparation
Vrabec, D.: 1974, in R.G. Athay (ed.), Chromospheric Fine Structure,
    IAU Symp. 56, 201
Wallenhorst, S.G. and Howard, R.: 1982, Solar Phys. 76, 203
Zirin, H.: 1974, in R.G. Athay (ed.), Chromospheric Fine Structure,
    IAU Symp. 56, 161
Zwaan, C.: 1968, Ann. Rev. Astron. Astrophys. 6, 135
Zwaan, C.: 1978, Solar Phys. 60, 213
Zwaan, C., Brants, J.J. and Cram, L.E.: 1985, Solar Phys. 95, 3

DISCUSSION

Rösch: How do you imagine the "transition" between bright penumbral
grains (in shape of little fishes with bright head inwards and thin

tail outwards), which still appear at the rim of the penumbra but as through an absorbing cloud, and the umbral dots? Could it be that the bright structures are sinking head-on into the umbra, inclined along the "Wilson effect", which would be in agreement with your sketch of the flux-tubes bundle opening outwards at increasing altitude?

Zwaan: I imagine that the penumbral fluxtubes in the sunspot bundle are strongly inclined. In this model the bright fish-shape features may be the (variable) spaces in between the darker fluxtubes.

Schröter: Your finding of the absence of magnetic field strength variations across umbral dots seems to me not convincing by two reasons. Firstly the picture you showed indicated that the "umbral dots" you investigated belong to a light-bridge which are not representative for umbral dots. Secondly, we certainly agree that the spatial resolution of your spectra are approx. 1", so with a diameter of 0.3 arcsec you have a "dilution" factor of ~ 10 and hence a real ΔB = 500 gs is reduced down to 50 gs or less.

Zwaan: Clearly, we did not resolve "umbral dots". The figure shown illustrates our general finding that, wherever there was a marked intensity peak because of (unresolved) bright structure, there was no corresponding variation in the magnetic field strength, Doppler shift, or Doppler broadening.

Pneuman: I would like to emphasize the importance of expelling or dissipating this emerging flux you speak of on a rapid time-scale. Golub, from his X-ray bright point measurements, estimates a flux emergence rate of 5 x 10^{22} Mx/day in small elements alone. For a polar field strength of 1-2 gauss, this rate could replenish the polar magnetic field in less than 4 hrs. Hence we must get rid of this flux rapidly to maintain a steady-state average field strength in the solar atmosphere.

Zwaan: I agree.

Wiehr: I would like to comment that your explanation of cycle variation of spot intensities (Oslo group) by variable flux tube package is at variance form the explanation by Schüssler who discusses the influence of heating for flux tubes of different age through the cycle.

Zwaan: That may be - I will look into Schüssler's argument again.

VARIABILITY OF THE QUIET PHOTOSPHERIC NETWORK

R. Muller & Th. Roudier
Observatoire du Pic-du-Midi
65200 Bagnères-de-Bigorre, France.

Abstract - High resolution photographs of the photospheric network taken in the CaIIK 3933 Å line and λ 4308 Å are analysed in order to study the variation, in latitude and over the solar cycle, of its density (the density is defined as the number of network elements - also called facular points - per surface unity). It appears that the density of the photospheric network is not distributed uniformly at the surface of the Sun : on september 83 it was weakened at both the low (equatorial) and high (polar) active latitudes, while it was tremendously enhanced toward the pole. The density at the equator is varying in antiphase to the sunspot number : it increases by a factor 3 or more from maximum to minimum of activity. Implications for the latitude and cyclic variation of the magnetic flux in the quiet Sun are discussed.

1 - Introduction.

High resolution filtergrams, taken in the CaIIK 3933 Å line (bandpass

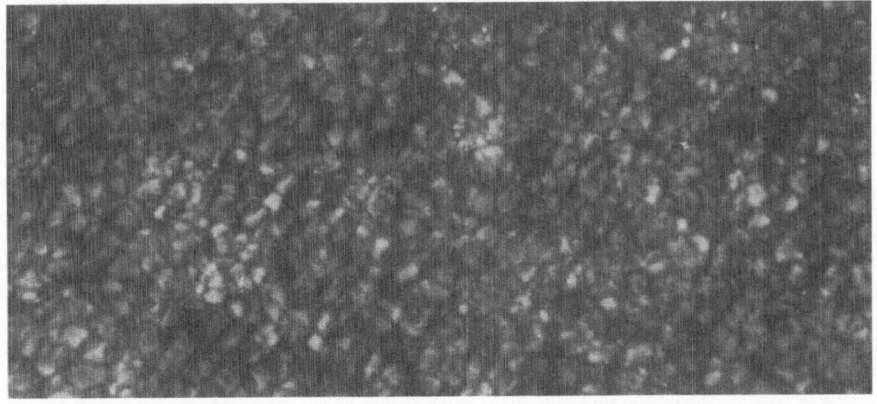

Fig. 1. 4308 Å filtergram showing facular points in intergranular spaces.

15 Å) from 1975 through 1979, and in the CH band at 4308 Å (bandpass 10 Å) from 1980 through 1983, have been used to study the variation in latitude and over the solar cycle of the "density" of the photospheric network. The density is defined as the number of network elements - also called facular points - per surface unity. Facular points appear on the filtergrams as tiny, bright and well shaped points located in intergranular spaces (Figure 1).

2 - <u>Variation   over the solar cycle.</u>

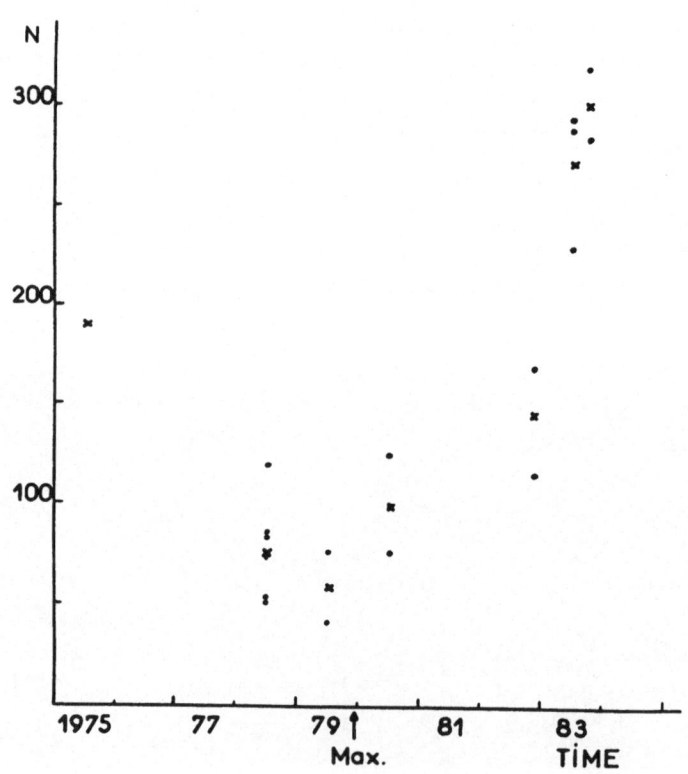

Fig. 2. Variation of the number density N of facular points over the solar cycle.

Figure 2 clearly shows that, at the disk center, the number density N
of facular points in the quiet photosphere (that is, in fact, the
density of the photospheric network) varies over the solar cycle, out
of phase with the sunspot number : the two quantities appear even to
be anticorrelated. The number varies by a factor of 3, at least.

3 - Latitude variation.

Figure 3 shows the variation of the number density of facular points
along the northern meridian, on september 17, 1983. From this figure
it is evident that the density of the quiet photospheric network is
not uniform over the solar surface. The photospheric network exhibits
two minima, at 22° and 45°, which seems to be associated respectively
with the sunspot latitude and the high latitude active band. Moreover
a tremendous increase of the number of facular points toward the limb
is revealed. The N values indicated on figure 3 are corrected for the
loss of visibility of the photospheric network toward the limb;

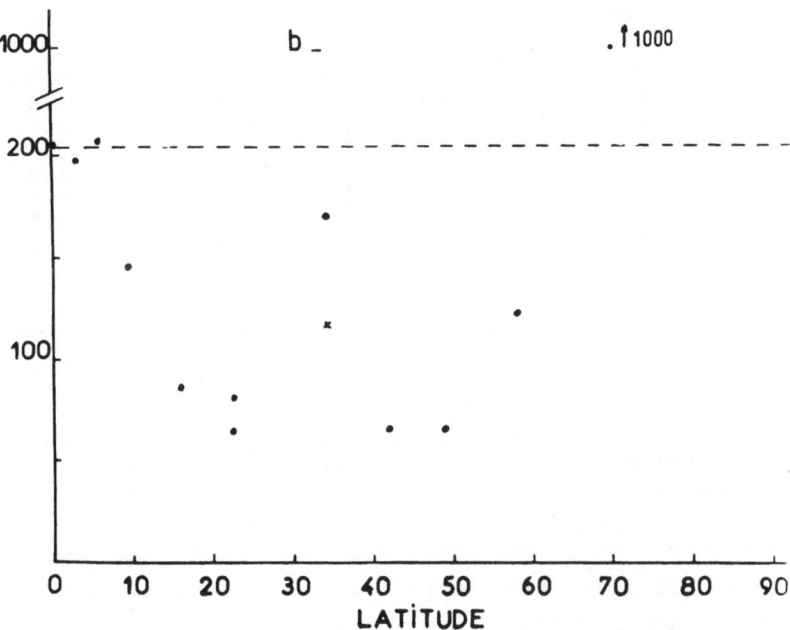

Fig. 3. Latitude variation of the number density N of
facular points.

that means that, at any latitude, the number density is compared to the number found on the equator at the same distance from the center of the disk.

4. Magnetic flux and field strength of the quiet photospheric network.

Facular points are associated with magnetic flux tubes, each of them carrying one quantum of magnetic flux of 2,5 x 10^{17} Mx. Thus a search of variability of the photospheric network density comes to a search of variability of the total magnetic flux of the quiet photospheric network.

Figure 1 implies that the cyclic variation of the magnetic flux in active regions is partly balanced by the anticorrelated variation of the flux in the quiet Sun.

The table gives the latitude variation of magnetic flux of the photospheric network, in the northern hemisphere, on september 27, 1983.

Table.

L	N	F 10^{21} Mx	%	B
5°	307	3.83	18.8	1.4 G
15	142	1.75	8.6	0.7
25	112	1.32	6.5	0.5
35	255	2.78	13.6	1.2
45	97	0.94	4.8	0.5
55	150	1.22	6.0	0.7
65	300	1.92	9.4	1.4
75	> 1500	> 6.65	32.5	> 7.1
85	?	?	?	?

total > 2.0 x 10^{22} Mx

L = mean latitude of the latitude bands : N = number of facular points /100"x 100"; F. = total magnetic flux; % : percentage of flux; B = mean magnetic field strength.

It shows that the flux of the quiet photospheric network is weakened at both the low latitude (equatorial) and high latitude (polar) active belts. There is a strong increase of flux toward the pole, which implies a high polar field strength (> 7 G).

Conclusion.

The observations described in this paper will be continued at Pic du Midi Observatory, in order to get the latitude and time variations of

the density of the photospheric network and the total magnetic flux in
the quiet Sun, over a complete solar cycle. Of particular interest are
the variation of the polar field strength, the variation of the depth
and position of the minima associated with the low latitude (equatorial)
and high latitude (polar) active belts, the variation of the total flux
of the quiet network.

Our observations should provide important informations for the understanding of the generation of magnetic field and of the solar activity cycle.

Discussion

J.C. Pecker.- The first slide (N (facular points) as a function of
cos θ)) strongly suggests that foreshortening and seeing
do not play a part in what is observed, but that the
facular points are not physically the same near equator
and at higher latitude; near equator, they might be
"deeper" (whatever it means) than the granular structure
of photosphere, as observed in the same wavelength where
you do observe the facular points.

R. Muller .- Foreshortening and seeing play an important part in the
visibility of facular points toward the limb : along the
equator the density of the observed points is reduced by
a factor 18 at the limb as compared to the centre of the
disk. A latitudinal change in the physical properties
would have negligible effects compared to the foreshortening and seeing effects.

N. Weiss .- How do your results relate to Golub's observation that
the emergence rate of X-ray bright points also varies
in antiphase with the solar cycle ?

R. Muller .- The number of facular points varies in phase with XBPs,
which seems to indicate that both features are related.
But the latitude variations appear to be different ?
Coordinated observations are required to understand the
connection between network facular points and XBPs.

SEARCH FOR GIANT CONVECTIVE CELLS FROM THE ANALYSIS OF MEUDON SPECTROHELIOGRAMS

E. RIBES and P. MEIN
DASOP - Observatoire de Meudon
92195 - Meudon Principal Cedex

One of the key problems for the understanding of the solar cycle deals with the large-scale dynamics of the convective zone. The existence of meridional flows and giant convective cells has not yet been firmly established. The differential rotation itself, for which reliable data have been obtained seems to change with time (Livingston and Duvall, 1979 ; Gilman and Howard, 1984). Moreover, the equatorial rotation rates obtained by using different methods and tracers agree within the 5% level. The magnitude of the velocity fields associated with various features of the global dynamics of the convective zone is small ($\lesssim 30$ ms^{-1}). So, one needs to measure velocity fields within this accuracy fiability.

With this aim of view, we have started the digitization of the spectroheliograms K IV (violet wing of the Ca II line) of the Meudon Collection. These spectroheliograms have the advantage over the white-light pictures of showing simultaneously two magnetic tracers (sunspots and facula). On the other hand, spectroheliograms are not instantaneous pictures and are subject to various geometric and photometric distortions due to the scanning period. (A detailed procedure will be given in a forthcoming paper, Mein and Ribes, 1984.)

Sunspots are detected by algorithms using isocontour brightness levels. Facula are detected by cross-correlation of brightness. Daily positions (latitude and longitude) of sunspots and facula have been measured whenever observations have been possible (almost daily during the summer time).

Figure 1 represents the daily motions of a stable sunspot (a) and a more complex one (b) during their transit across the solar disc. The East-West fluctuations (> 30 ms^{-1}) are usually much larger than the North-South motions (< 10 ms^{-1}). An error of the timing will affect the East-West drift more than the North-South drift. So, one has to be very cautious before claiming that the fluctuations of the rotational rates are of solar origin (e.g. supergranular motion ?).

1. ACTIVE LONGITUDES

The appearance of magnetic regions does not seem to occur randomly. Trellis (1971), among others, has suspected the existence of active longitudes (in both hemispheres) where new magnetic flux arise preferentially. We have recorded the number of sunspots visible on our K IV spectroheliograms at different periods during the cycle n° 20 and 21. We found that the sunspot number distribution versus longitude exhibits two well-defined peaks 180° apart at the maximum of the cycles. In the ascending phase, the situation is not clear, and in the descending phase the two longitudes become blurred.

Our interpretation is that it takes about half a cycle to build up a toroidal component of the magnetic field (see also, Schröter, 1984).

2. ROTATIONAL RATES

a. Sunspots

Our statistical sample is small and does not enable us to derive laws of the differential rotation. So we have compared our values to the mean rotational law based on two cycles (Balthasar and Wöhl, 1981). There is no evidence of higher rotational rates associated with 2 active longitudes   . We can only notice that the dispersion in the 1969 data is higher than in the 1972 data. If convective velocity fields (giant cells) bring up new magnetic flux, they have not been detected with our observations.

B. Facula

As stated above, the rotation of facula has been obtained by cross-correlation of brightness. For a young group, leading or following facula surround sunspots. So the brightness contribution comes mainly from the sunspots. We can eliminate their effect by monitoring the brightness level, but then we introduce a bias in the center of gravity of the facula. To get around this difficulty, we have selected at first only spotless facula, i.e. old facula. We found   that spotless facula rotate faster than sunspots by 2%.

This result can be interpreted in the following way : when a new magnetic field emerges from the deep convection zone, the surrounding plasma exerts some viscous drag which slows down the initial rotation. In the case of a sunspot which consists of a compact bundle of tubes, the viscous drag is more efficient than for facula. This effect has also been proposed by Gilman and Howard (1984) to explain the difference of rotational rates between large and small sunspots.

An alternative interpretation could be a depth difference in the anchorage of facula and sunspots.

3. MERIDIONAL FLOWS

Mean North-South motions have been calculated for sunspots during their transit across the solar disc (4 rotations in 1969 and 4 rotations in 1972). No net meridional circulation has been found when observing old and new sunspots (in agreement with Balthasar and Wöhl, 1981) and spotless facula. However, newly born sunspots which are supposed to be still anchored deep in the convection zone show a four zonal meridional circulation (Fig. 2). The pattern changes through the solar cycle. We have evidence of angular momentum transport probably operating in the deep convection zone only. With older spots (or facula) -he magnetic field has become detached from the deep anchorage and floats about independently (for more details, see Ribes and Mein, 1984).

CONCLUSION

The digitization is a powerful method to analyze the rotation of magnetic tracers since it allows the detection of velocity fields with high accuracy ($\lesssim 10$ ms^{-1}). Positions of faculae can be studied as well.

Among the main results :

- Sunspots rotate faster than old facula by 2%.

- Sunspots can be tracers of angular momentum transport at the very beginning of their emergence. Afterwards, they become detached from the depth of anchorage and float with the photospheric plasma. A complex meridional circulation, with same analogy with the zonal belts observed on the major planets has been detected, and changes through the solar cycle.

- The existence of two active longitudes together with the meridional circulation pattern suggests that the solar dynamo works as a non linear oscillator in $e^{im\phi} P_\ell(\cos \theta)$ with $m = 2$ and $\ell = 2$ for the azimuth and the degree respectively.

Acknowledgments

The spectroheliographic plates have been digitized with the P.D.S. microdensitometer (Institut d'optique, Orsay) and analyzed with a vax computer (Meudon Observatory). This work has been supported by an INAG grant.

REFERENCES

Balthasar, H., and Wöhl, H. : 1981, Astron. and Astrophys. 92, 111
Gilman, P. and Howard,   : 1984, Astrophys. J. in the press
Livingston, W., and Duvall, T.L. : 1979, Solar Phys. 61, 219
Mein, P., and Ribes, E. : 1984, in preparation
Ribes, E., and Mein, P. : 1984, Nature, submitted
Schröter, E.H. : 1984, Astron. and Astrophys. in the press
Trellis, M. : 1971, Comptes rendus de l'Académie des sciences, Paris, 272, 1026

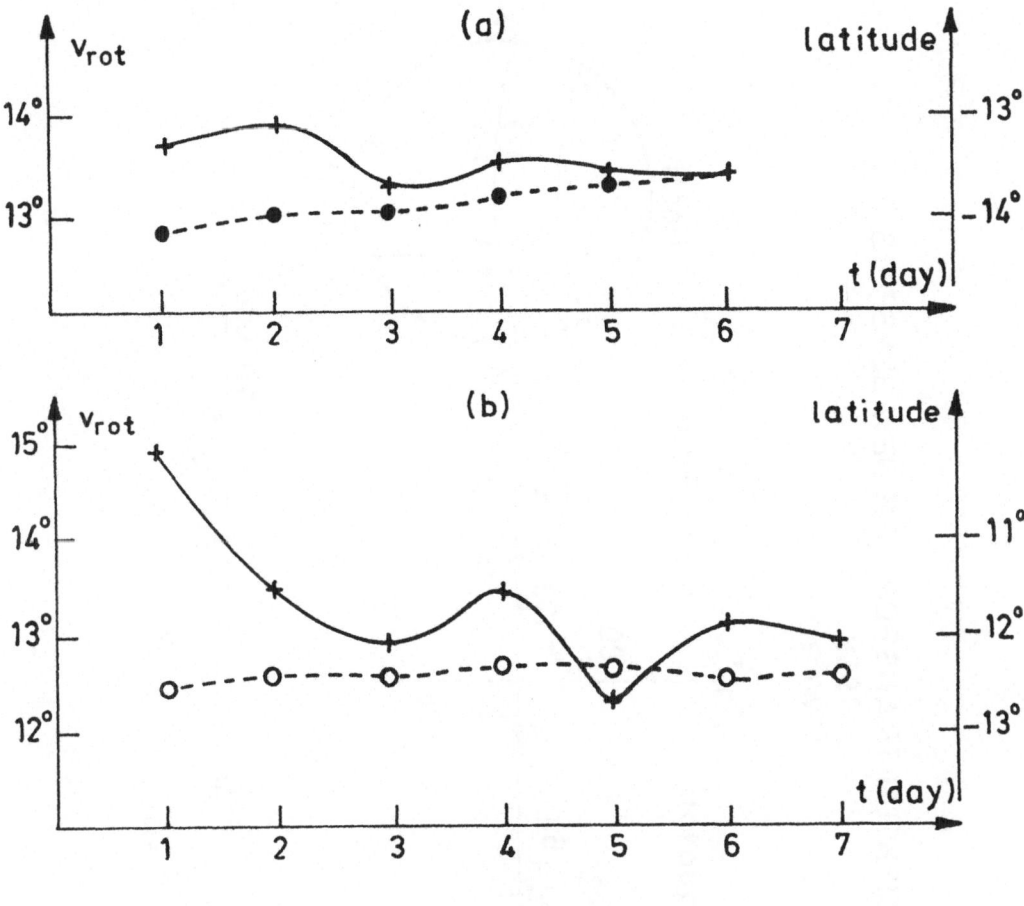

FIG.1

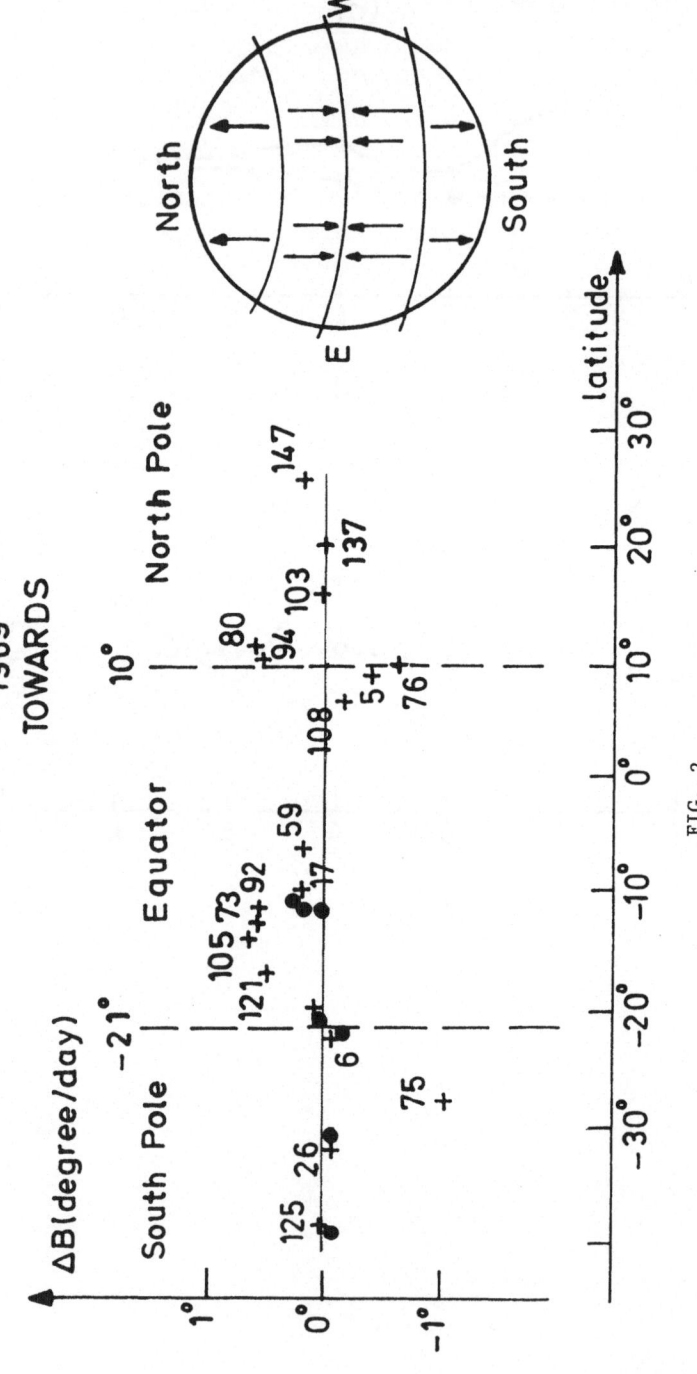

FIG. 2

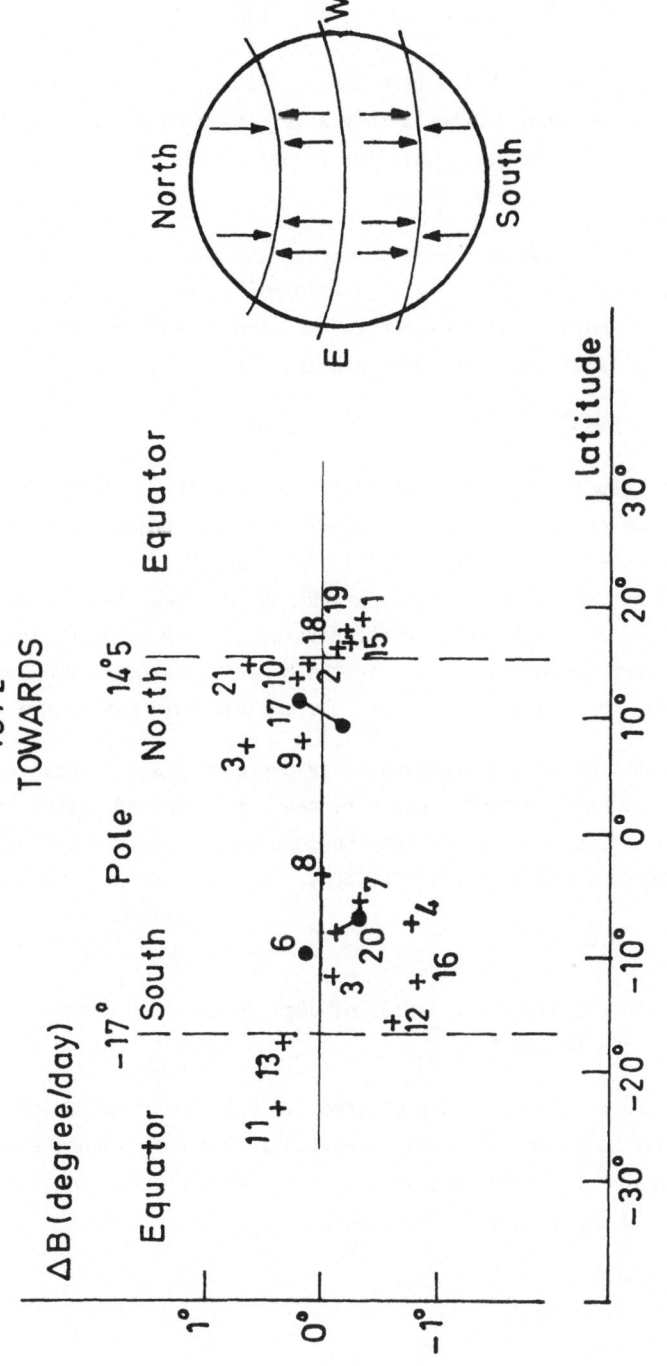

FIG. 2

QUESTION FROM : SCHROTER   TO : RIBES

QUESTION : You can imagine that I have many questions regarding the meridional flow
from sunspots, but we should discuss this after the session. Here I would only ask :
what average velocities do you obtain for the meridional flow and what are the error-
bars ?

ANSWER FROM RIBES : For newly-formed sunspots, the average meridional drift would
be of the order of 0°3/day (that is ∼30 - 40 ms^{-1}). The velocity field which can
be detected by our method is a few meters per second But the quality of the spectroheliograms is such that the error-bars are about 10 ms^{-1}.

QUESTION FROM : PECKER   TO : RIBES

QUESTION : Your results about (1) the longitude location of activity during 5 cycles
(2) the "independance" of old active regions from deep phenomena please me enormous-
ly ! Not only do they confirm the Trellis' data (and Maunder's which concerned mostly young spots) but they give a bigger weight to the idea that the bottom of the convective zone turns at about the Carrington value solidly (indeed, they appear also
to confirm the Carrington's value as did the M regions studied by Roberts and myself,
the Maunder data, the Trellis' data, and the coronal hole behaviour) !

ANSWER FROM RIBES : We have no conclusive evidence of a solid rotation from the analysis of young magnetic sunspots. However, there is some indication that the differential rotation is less pronounced for the sunspots at their early stage. A larger
sample is needed to confirm this indication.

QUESTION FROM L.M.B.C. CAMPOS   TO : RIBES

QUESTION : Can you explain the reversal of angular momentum transfer after 3 years ?
Can you estimate the torque ?

ANSWER FROM RIBES : No, we simply observe it. But 3 years after the maximum of
activity, coronal holes appear at high latitudes. The coïncidence should be studied.
As for the estimate of the torque, we should assume the depth and thickness of the
layer involved. So far it has not been done.

THE VARIABILITY OF PHOTOSPHERIC GRANULATION AND TOTAL RADIO FLUX OF QUIET SUN IN THE CENTIMETRIC WAVEBAND DURING A SOLAR CYCLE.

D. Dialetis, C. Macris, Th. Prokakis.
Astronomical Institute, National Observatory of Athens.
Athens, 11810, P. O. BOX 20048 GREECE.

Abstract.

We studied the variability of the total solar radio flux in eight frequencies, in the centimetric range during more than a solar cycle (1964-1978), and the possible relationship it has with the variability of the photospheric granulation.

1. Introduction.

Many years ago, one of us (Macris, 1955) suggested the existence of a close relationship between the solar cycle and the morphological and temporal evolution of photospheric granulation. From this time a number of publications have supported this conception (Macris et all. 1983a, 1983b). In this paper, we examine the existence of a possible relation-ship between the variability of the mean distance of granules and the variability of the slowly varying component in the centimetric waveband during a solar cycle. Statistical evidence is provided to support the proposed relationship. Such a relationship may be interpreted as a high dependence of two different kinds of solar variability or as the close relation of the two different phenomena with the solar cycle mechanism.

1. The total radio flux of the quiet sun during a solar cycle.

In a first stage, we have studied the variability of the total solar radio flux in eight frequencies in the range 1000-9400 MHz during more than a solar cycle (1964-1978). Our intention was to determine the back-ground component by eliminating the contribution of active regions on the total radio flux. It is well known that this radiation is caused by thermal emission of the solar atmosphere, predominantly on centimetric wavelengths where it originates mostly from the chromosphere (Kundu, 1965). The available published information on this subject is contradictory and incomplete. Generally, though, the results indicate that at centimetric wavelengths the values of quiet sun flux density at

maximum activity are systematicaly higher than the corresponding values
at minimum activity (Zheleznyakov 1970). In the region of centimetric wavelenthgs, the calculation of the "quiet" sun flux density is a
complex problem, because the slowly varying component is intense, particularly in years of high solar activity. In this waveband, the indensity
of the enhanced radio emission correlates well with the area of the spots
that can be seen on the sun's disk. Therefore, the flux density emission
(s) of the quiet sun is generally found from a graph of $S_\lambda(A)$, where A
is the corresponding uncorrected area of the spots by extrapolating the
area of the spots to zero. We have used two different methods for the
elimination of the contribution of active centers.

1. A linear regression method between monthly mean values of the flux
and sunspot areas.

2. A multiple linear regression between daily values of sunspot areas
for three successive rotations and daily values of the flux density.
From the computed mean monthly values of the total solar flux density
of the "quiet" sun we noticed a remarkable stability of the quiet sun
emission at two successive minima for every frequency. From the values
of flux density we compute the reduced effective temperature $T_{eff_\odot}$
of the sun. It is easy to find the connection between the "true" effective temperature and the "reduced" temperature of the source by the relation

$$\int T_{eff} d\Omega = T_{eff_\odot} \Omega_\odot \qquad (1)$$

where $\Omega_\odot$ is the solid angle subtended by the optical disk. From the
values of $T_{eff_\odot}$ we compute the frequency spectrum of the "quiet" Sun's
radio emission in the centimetric waveband (3-30cm) for the different
periods of solar activity (Fig. 1a).

From the study of the frequency spectrum as a function of time we state
the follows :

a. There are important changes of the $T_{eff_\odot}$ at the higher wavelenghts
   during a solar cycle. These changes are about:
   80% for λ=30cm, 74% for λ=21.2cm, 66% for λ=15cm, 57% for λ=11,1cm
   56% for λ=10.7cm, 50% for λ=8cm, 47% for λ=6cm and 6% for λ=3,19cm.

b. We can clearly distinguish two different families of spectra. The
   first one corresponds to the high activity period and the second to
   the low activity period (Fig. 1a). By using methods of discriminant
   analysis we can find a typical spectrum for a low activity period
   and a typical spectrum for a high activity period (Fig. 1b).

c. There is remarkable stability of the values of $T_{eff_\odot}$ for the years
   of low activity and low differences between the years of high activity.

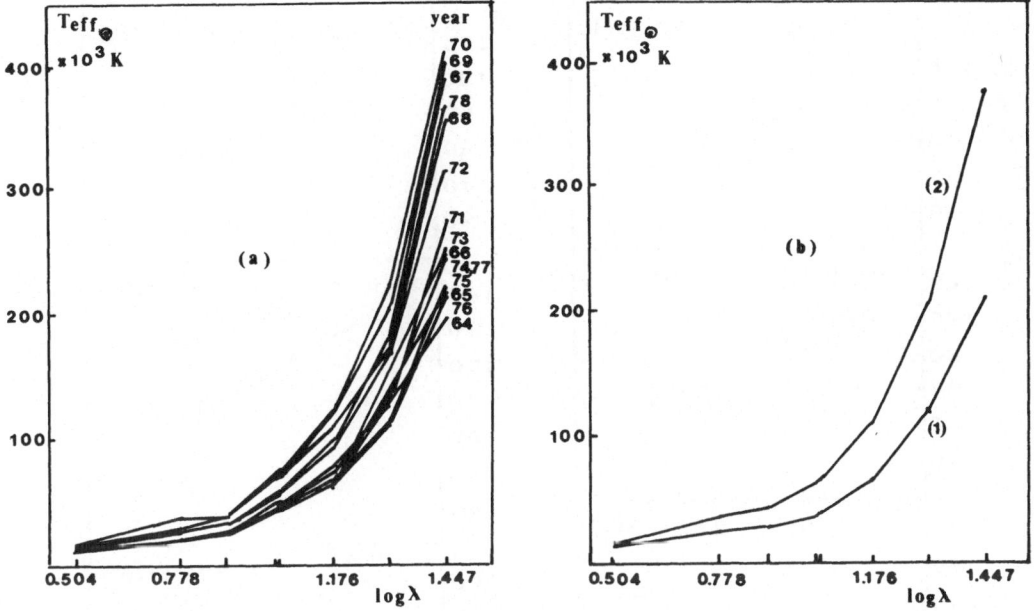

Figure 1. (a) Frequency spectrum of the "quiet" sun as a function of time. (b) Typical spectrum for a low activity period (1) and for a high activity period (2).

2. *The variations of mean distance between granules during a solar cycle.*

In a second stage, we have studied the relationship between the variations of the solar flux for zero level of activity (1964-1978) and the varia-tions of the mean distance between granules for the period 1966-1978. We give, in Table 1, the mean distance of granules in sec of arc for this period. We consider the mean distance between individual granules, as a good index of the variability of the solar granulation. We suppose that this index is also, in a general way, related with the convective heat transport during the cycle, because what we see as granulation is the instantaneously ordered large-scale pattern. The organization of this pattern is imposed by the need to carry heat, hence the granulation is a direct manifestation of the convective heat transport. We have computed the regression functions and the correlation coefficients of these two types of activity. We give, in Fig. 2, the regression lines for the mentioned relations as well as the correlation coefficients. The statistical de-pendence of the two variables is presented as a high anti-correlation.

Table 1

Year	Mean distance
1966	2".12
1967	1.92
1968	1.84
1969	1.89
1971	2.08
1973	2.19
1975	2.20
1976	2.28
1977	2.19
1978	1.84

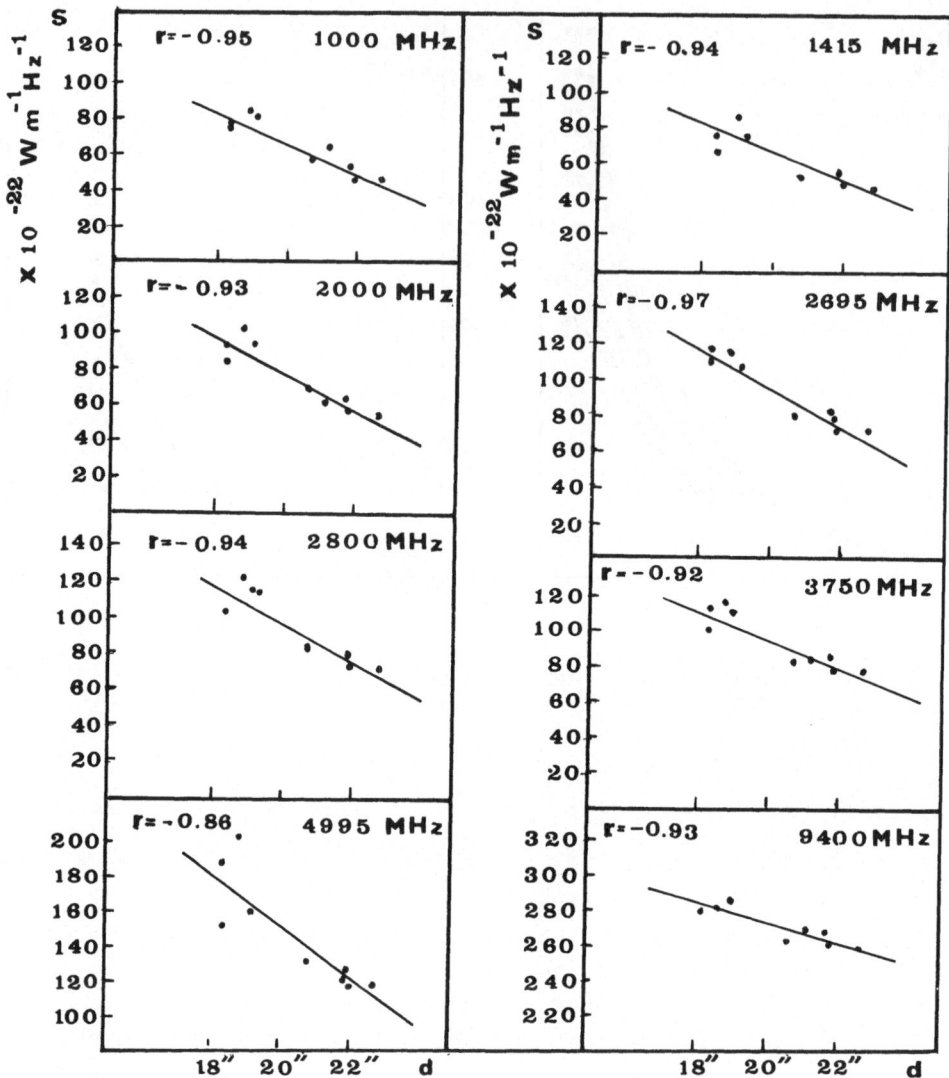

Figure 2. Regression lines and correlation coefficients (r) between the mean distance of granules (d) and the solar flux density (s) for different frequencies.

3. *Discussion.*

We can explain the changes of the total radio flux of the quiet sun during a solar cycle by two different ways.

1. Variations of the effective solar radius.
2. Variations of the brightness temperature produced either by variations of the electron temperature or the electron density.

But differences of $T_{eff\odot}$ between a maximum and a minimum of solar activity are too large to be explained by variations of the effective solar
radius. So that leads us to attribute these variations to the second case.
From the other side granules are ascending stream cells, resulting from
the turbulent convection carrying hot gases upward. The upward moving
pressure disturbances turn into shock waves heating thus the outer layers. By
the close relationship of the granulation pattern with the solar cycle
we can assume that the variation of the mean number of granules per area
unit has as a consequence, substantial quantitative changes of these
processes. Our intention is to examine if there is a close relation
between these two different kinds of activity, or if each activity is
related with the solar cycle.

References

Kundu M. R. :1956, Solar Radio Astronomy, Interscience. New York.
Macris C. J. :1955, Observatory, 75, 122.
Macris C. J.,Rösch J. :1983a, C. R. Acad. des France, 296. II, 265.
Macris C. J., Müller R., Rösch J., Roudier Th. : 1983b, Commun. by
     J. Rösch in the Meeting at Sacramento Peak, July 25, 1983.

Discussion.

J. C. Pecker : How would your data evolve with increased resolution?
Which will stay valid, which won't? In what sense will they evolve?

D. Dialetis : The increased resolution don't change our results because
we measure the center to center distance of granules.

DYNAMIC PHENOMENA IN THE CHROMOSPHERIC UMBRA AND PENUMBRA OF A SUNSPOT

C. E. Alissandrakis[1], D. Dialetis[2] and C. J. Macris[2]

[1]Section of Astrophysics, Astronomy and Mechanics, Department of Physics, University of Athens, 15771 Athens, Greece.

[2]Astronomical Institute, National Observatory of Athens, 11810 Athens, Greece.

ABSTRACT

We observed a prominent, short lived (70 sec) umbral flash in Hα with the 50 cm "Tourelle" refractor at Pic du Midi. The main component of the flash was small (1" by 0.5") with a maximum intensity 25% above the background. It occurred in a region which showed irregular intensity variations. Umbral oscillations were observed in an adjacent region and running penumbral waves around part of the superpenumbra.

1. INTRODUCTION

During the last fifteen years a number of dynamic phenomena has been detected and studied in the chromosphere of sunspots (see reviews by Moore, 1981a, 1981b). These include umbral flashes, umbral oscillations and penumbral waves.

Umbral flashes are best visible in the H and K lines (Beckers and Tallant, 1969) while they are very rarely observed in Hα; they are small, short lived brightenings of repetetive nature, associated with predominantly upward material motions. Umbral oscillations (Bhatnagar and Tanaka, 1972; Giovanelli, 1972) involve intensity and velocity fluctuations with a period near 3 min; they are commonly observed in chromospheric lines (Kneer et al, 1981) and appear to be related to photospheric umbral oscillations. Finally the running penumbral waves are often observed around part of the chromospheric superpenumbra (Zirin and Stein, 1972; Giovanelli, 1972) with a period of 3.5 to 5 min and an outward velocity of 10 to 20 km/sec.

Although these phenomena appear to be associated with one another, the details of this association are not clear. In this paper we present preliminary results of high spatial resolution Hα observations of these phenomena in the same spot and we discuss their properties and interrelations.

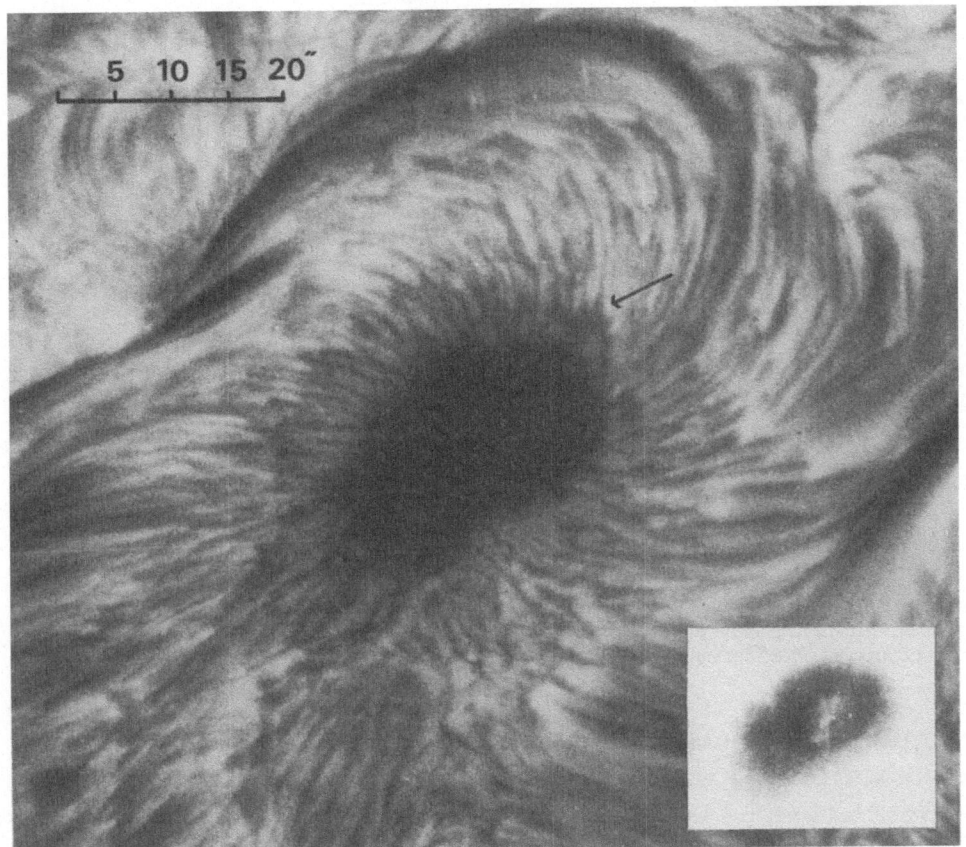

Figure 1: A sunspot observed on August 21, 1979 at 085650 UT at the center of Hα at
Pic du Midi. The insert shows the interior of the umbra, printed from the same
negative. The arrow shows the penumbral wave.

2. OBSERVATIONS AND RESULTS

The observations were obtained with a Halle filter (0.5 A passband) mounted on
the 50 cm "Tourelle" refractor of the Pic du Midi Observatory. On August 21, 1979 we
observed a large, isolated sunspot located at 18W 25S under seeing conditions ranging
from excellent to very good for approximately one hour; photographs were obtained
mainly at the center of Hα and occasionally at ± 0.5 A. In the course of these observations we detected a prominent bright structure in the umbra (Figure 1). The brightening was elongated with a length of about 4" arc and consisted of three interconne-
cted components separated by about 2" arc (Figure 2). The main component was 1" long
by 0.5" wide, with a maximum intensity about 25% above the local background.

The evolution of the brightening was very fast. Figure 3 shows mocrophotometric
tracings across the umbra from three consecutive photographs. There is almost no trace
of the structure on the photograph taken 50 sec before that of figure 1, while very

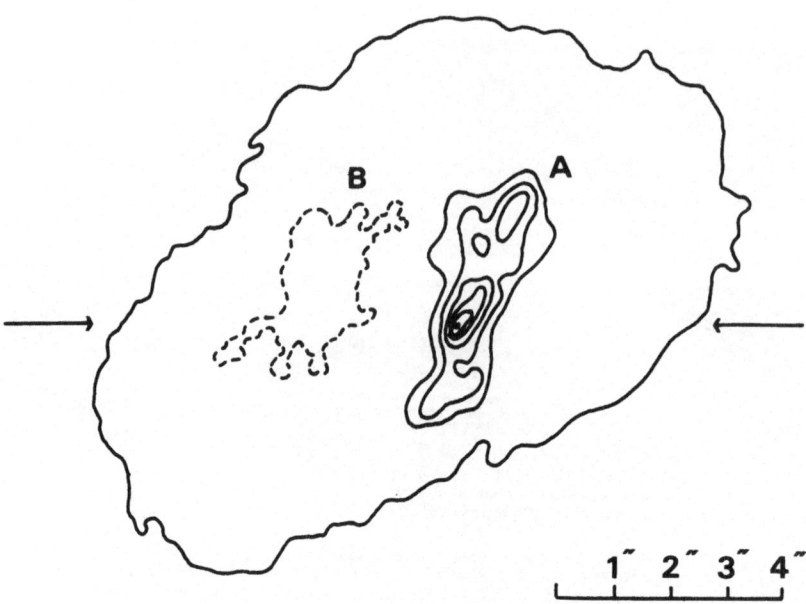

Figure 2: Contours of equal density for the umbral flash of Figure 1. The dashed contour shows the darkest region of the umbra. The arrows show the position of the tracings of Figures 3 and 4.

little is left in the photograph taken 20 sec later. This sets an upper limit of 70 sec for its lifetime. On the basis of its observable characteristics the brightening can be identified with the umbral flashes usually observed in the Ca II H and K lines.

This umbral flash was unique during our observing period as far as its intensity and structure are concerned. However, it was not the only time varying phenomenon in the umbra. In order to study the time variations of the umbra we used microphotometric tracings such as those of Figure 3 to compute a contour map of the intensity as a function of time and position along the tracing (Figure 4). The umbral flash shows up as a prominent peak, 10.6 min after the beginning of the observation at the location A. In the course of the observation the intensity at that location varied in an irregular fashion; the peak to peak intensity variation was about 40% (this includes the flash), while the rms variation was about 8%. More often than not the maxima appeared

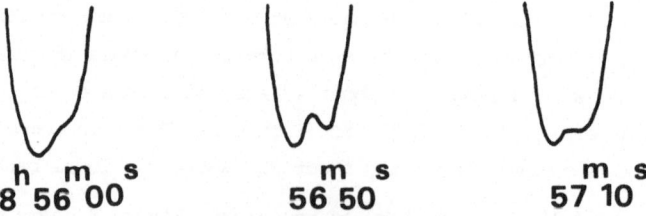

Figure 3: Microphotometric tracings through the umbra for three consecutive photographs around the time of the umbral flash.

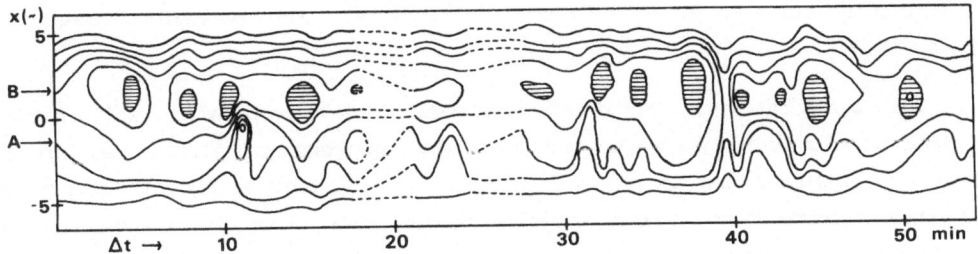

Figure 4: Countours of equal density as a function of time and position along a tracing through the umbra. Dashed contours show gaps in our observations. Hatched
regions correspond to intensity minima.

as extensions of the penumbral emission into the umbra rather than distinct structures.
It is difficult to assign a period to these time variations, however the more prominent
intensity peaks suggest an average repetition rate of about 500 sec which is about
three times slower than that of the K line flashes. If all peaks are considered as
real the average repetition rate is 220 sec.

The darkest region of the penumbra (location B in Figure 4) showed a more regular time behaviour. The peak to peak intensity variation was about 20% with an rms of
about 5%. The time separation between consecutive minima ranged from 132 to 324 sec
with an average value of 210+64 sec; this is close to the period of umbral oscilla-
tios (Bhatnagar and Tanaka, 1972; Giovanelli, 1972). The characteristic spatial scale
of the oscillations was of the order of 3 to 4" arc.

The well known penumbral running waves are quite prominent in part of the superpenumbra (arrow in figure 1). It is interesting to note that we found no trace of such
waves in the lower part of the spot where the superpenumbral fibrils had an irregular
structure. Our estimate of their period is 210 ± 40 sec, which is the same as the period of the umbral oscillations, while their velocity was estimated to be 7 ± 1 km/sec.
Unfortunately we could not study the association of the waves with the umbral oscillations due to the sharp intensity gradient between the umbra and the penumbra.

3. DISCUSSION

The bright structure that we observed at the center of the Hα line in a sunspot
umbra had the characteristics of umbral flashes commonly observed in the H and K lines.
It occurred in a region (A) with irregular intensity variations and was adjacent to
a region (B) where normal umbral oscillations were observed. The flash had no detectable effect on the oscillations. In addition the intensity maxima in region A did
not occur at a particular phase of the oscillations; in Figure 4 there are cases
where such maxima occurred just before, during, right after or in between the inten-

nsity minima of region B. Thus the intensity variations in the two regions appear to
be independent.

Running penumbral waves were detected in part of the superpenumbra, while no waves
were observed in a region where the pattern of the superpenumbral fibrils was irregular; this may indicate that the waves do not propagate in regions where the magnetic
field lines of force do not have a regular structure. The period of the waves was the
same as that of the umbral oscillations; however, the sharp intensity gradient between
the umbra and the penumbra did not allow us to study their assotiation.

AKNOWLEDGEMENTS; One of the authors (C.E.A.) is grateful to Prof. J. Rosch, Dr R.
Muller and the staff of the Pic du Midi Observatory for their invitation, their warm
hospitality and their assistance with the observations.

REFERENCES

Beckers, J. M. and Tallant, R. E. : 1969, Solar Phys. 7, 351
Bhatnagar, A. and Tanaka, K. : 1972, Solar Phys. 24,87
Giovanelli, R. G. : 1972, Solar Phys. 27, 71
Kneer, F., Mattig, W. and v. Uexkull, M. : 1981, Astron. Astrophys. 102, 147
Moore, R. L. : 1981a, Space Sc. Rev., 28, 384
Moore, R. L. : 1981b, in "The Physics of Sunspots" (L.E.Cram and J.H.Thomas,eds),
    p. 259, Sacramento Peak Observatory
Zirin, H. and Stein, A. : 1972, Astrophys. J. (letters), 178, L85

DISCUSSION

Mattig : I believe that you have shown the typical oscillation in the umbral
chromosphere. The very strong brightening is a special phase in the oscillation (as
the umbral flashes in K).

Alissandrakis : The brightening had unique characteristics and occurred in a
region with qualitatively different time behaviour from the region of normal umbral
oscillations. In our opinion it was not related to the umbral oscillations.

Zwaan : In SPO observations I have come across some tiny but intense brightening
in Hα that lasted a very brief time and did not correpond to anything in K.

PHOTOMETRY OF LIGHT-BRIDGES IN SUNSPOTS

M. Collados; E. Marco; M. Vázquez.
Instituto de Astrofísica de Canarias,
Tenerife,
Islas Canarias, Spain.

ABSTRACT.

The decay of a sunspot region was studied using high resolution white-light pictures. The photometric and geometric parameters of a light-bridge showing granular structure were determined. No difference is found comparing it with the quiet photosphere.

1. INTRODUCTION.

Light-bridges are often observed in the decay phase of sunspots. Some of them show a granular structure similar to that of the photosphere. Bray and Loughhead (1964) and Vazquez (1973) have studied the morphological aspects, having measured Beckers and Schröter (1969) a magnetic field of a few hundred gauss.

The main aim of our work was to compare the light-bridges granulation with that of the quiet photosphere in order to confirm or not its possible deformation near magnetic structures (e.g. Macris (1979); Parfinenko (1981)).

2. OBSERVATIONS.

The pictures were obtained during the interval 4-12 of June 1980 at the Observatorio del Teide (Izaña) using the Kiepenheuer Institutes' 40cm vacuum reflector telescope. In fig. 1 we present the studied pictures (one per day) showing the evolution of the region during its passage across the disk.

3. DATA REDUCTION.

The selected pictures were analyzed in the IAC's PDS-1010A microdensito-meter. The slit aperture was 10u x 10u corresponding to 0.05 x 0.05 on the sun. The calibration was made using a procedure based on the half-filter method (Collados and Bonet, 1984)

We applied the optimum filter method to correct: a) microdensitometer

HALE   16884

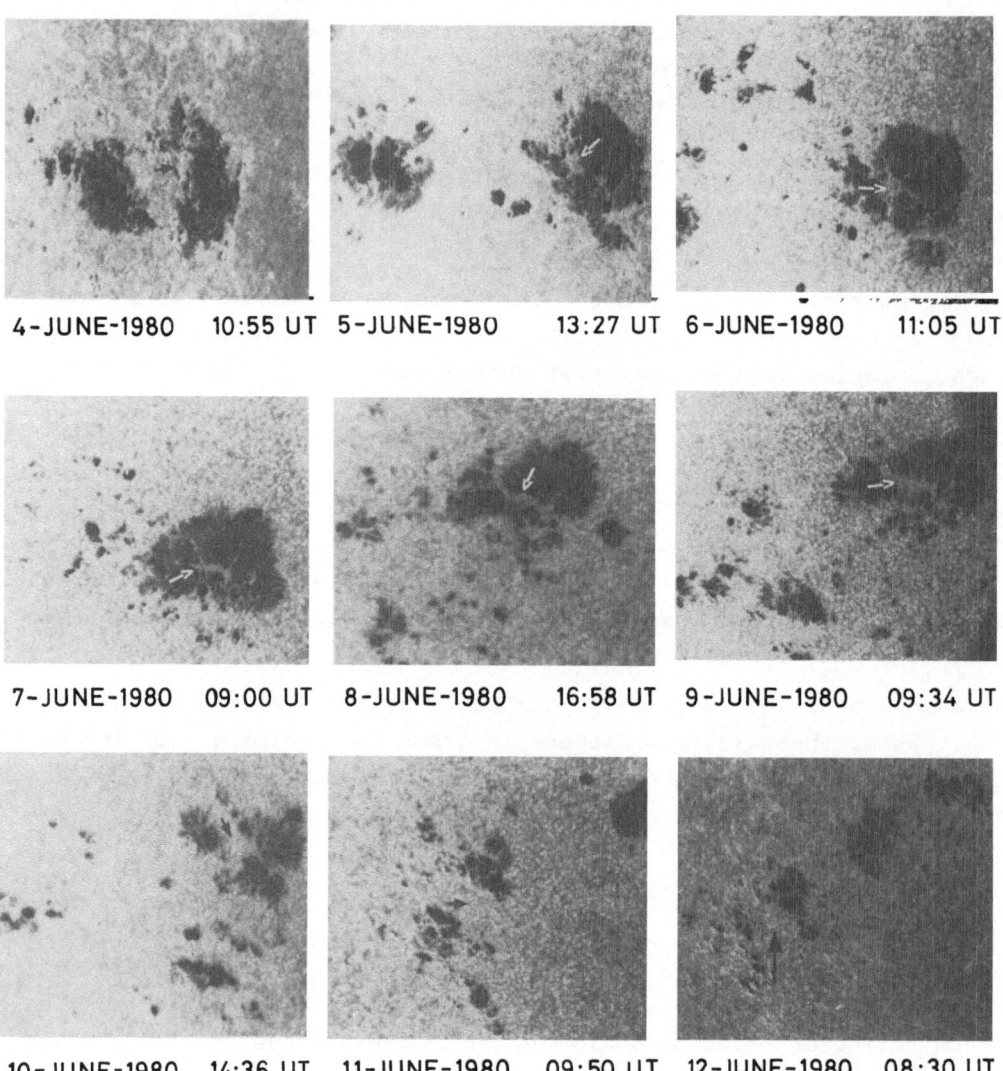

4-JUNE-1980   10:55 UT   5-JUNE-1980   13:27 UT   6-JUNE-1980   11:05 UT

7-JUNE-1980   09:00 UT   8-JUNE-1980   16:58 UT   9-JUNE-1980   09:34 UT

10-JUNE-1980   14:36 UT   11-JUNE-1980   09:50 UT   12-JUNE-1980   08:30 UT

Fig.1:
Pictures used to study the evolution of the group. The arrows point at
the studied light-bridge zone. The spot seen at the right in the two
last photographs belongs to another region.

distortion; b) atmospheric distortion, following Schmidt et al (1981),
and c) telescope aberration.

4. DATA ANALISIS AND RESULTS.

4.1 Photometric properties.

4.1.a Mean intensity.

The intensity ratio $\bar{I}_{LB}/\bar{I}_{PH}$ (where LB denotes light-bridges and PH photosphere) does not show remarkable variation during the passage as seen in
fig. 2, and being always very close to one: $0.92 \leqslant I_{LB}/I_{PH} \leqslant 1.04$, in
agreement with Korobova results (1966) for photospheric light bridges.

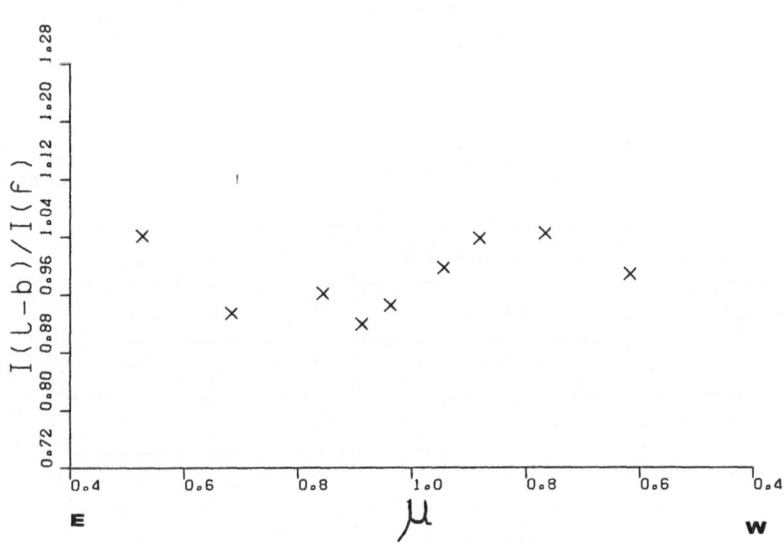

Fig.2:
Intensity ratio $\bar{I}_{Ib}/\bar{I}_{PH}$ vs. cosv. In the plot I(f) denotes the mean
photospheric intensity and I(l-b) the mean intensity in the light-bridge.
Although the ratio is a little lower when the bridge is closest, the difference is not large enough so as to attribute it to an evolution effect.

4.1.b ΔI_{rms}

Table one shows both uncorrected and corrected ΔI_{rms} values for the photosphere and light-bridge samples. No significant variation during the
evolution of the spot can be observed.

4.2. Morphological properties.

Table one shows the values of the parameters for each one of the light-bridge samples, which are similar to the mean values for the photosphere, also shown in the same table

The histograms of the granule-to-granule distances for one light-bridge region, the one resulting from the sum of all the light-bridge regions and the one resulting from the sum of all the photospheric zones can also be seen in fig. 3.

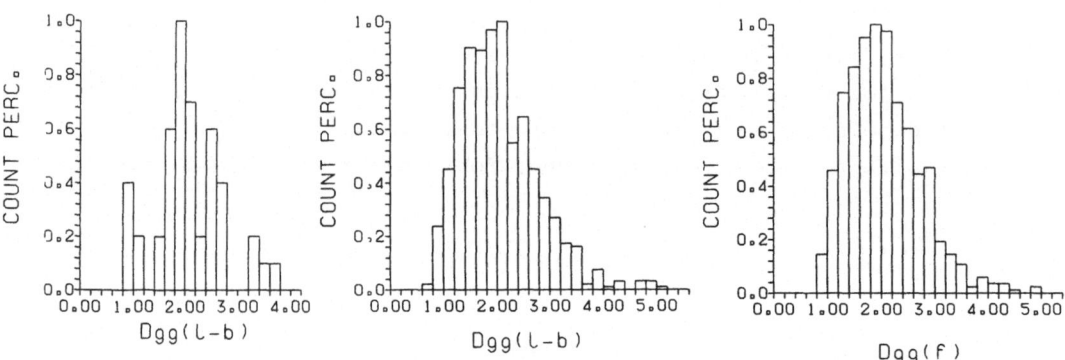

Fig.3:
Histograms of the granule-to-granule distance for a) a light-bridge, b) all the light-bridge regions together, and c) all the photospheric regions together.

DAY	ΔI_{rms}(ph)		ΔI_{rms}(l-b)		$\bar{r}$	$\bar{d}_{gg}$
4	0.07	(0.03)	0.12	(0.10)	0.70	2.04
5	0.10	(0.04)	0.13	(0.11)	0.72	1.99
6	0.11	(0.05)	0.10	(0.08)	0.67	2.07
7	0.06	(0.05)	0.10	(0.08)	0.71	2.19
8	0.10	(0.05)	0.06	(0.06)	0.78	2.10
9	0.10	(0.05)	0.10	(0.10)	0.75	2.07
10	0.09	(0.05)	0.06	(0.05)	0.77	2.16
11	0.10	(0.04)	0.05	(0.04)	0.68	1.87
12	0.03	(0.01)	0.03	(0.02)	0.71	2.04
PHOTOS-PHERE					0.72	2.07

Table I:
This table shows the values of the parameters for each one of the light-bridge regions and the mean photospheric values. The values in brackets are the uncorrected ΔI_{rms}, standing $\bar{r}$ and $\bar{d}_{gg}$ for the mean granular radius and the mean granule-to-granule distance.

CONCLUSIONS.

According to our work the parameters characterizing the granulation in the studied light-bridge and the normal photosphere do not differ. We agree with Muller (1979) and Parfinenko (1981) that there are no specific structures in light-bridges. This photospheric light-bridge seems to be the reestablishment of the normal photospheric conditions. More analysis is now in progress from the Izaña material to investigate other types of light-bridges (penumbral and streamers) and the granulation around the spots.

ACKNOWLEDGEMENTS.

We thank Dr. Fernando Moreno for critical comments and Miguel Briganti and Momica Murphy for their help in the edition of this contribution.

REFERENCES.

Beckers, J.M.; Schröter, E.H.; 1969, Solar Physics, 10, 384.
Bray, R.J.; Loughhead, R.E.: 1964, "Sunspots", London: Chapman and Hall, Ltd.
Collados, M.; Bonet, J.A.: 1984, Appl. Opt., 23, 2827.
Macris, C.J.: 1979, Astron. Astrophys,, 78, 186.
Muller, R.: 1979, Solar Phys. 61, 297.
Parfinenko, L.D.: 1981, Soln. Dan., 12, 79.
Schmidt, W.; Knölker, M.; Schröter, E.H.: 1981, Solar Phys., 73, 217.
Vázquez, M.: 1973, Solar Phys., 31, 377.

Discussion

Schroeter : Did you find a significant difference in the properties of the light-bridge granules between its early stage (when the light-bridge was shallow) and in its later stage, when the light-bridge was almost a broad photosphere-like structure ?

M. Collados : You can see a slight difference in the plot light-bridge granules radius vs. cos θ for the first stages and the last ones. But the differences are not significant enough so as to be sure of them, therefore the answer is no.

OBSERVATIONS OF ELLERMAN BOMBS IN Hα

Th. Zachariadis[1], C.E. Alissandrakis[2] and G. Banos[3]

1. Research Center for Astronomy and Applied Mathematics, Academy of
Athens, 14 Anagnostopoulou Str., 10673 Athens, Greece.
2. Section of Astrophysics, Astronomy and Mechanics, University of
Athens 15771 Athens, Greece.
3. Division of Astro-Geophysics, Department of Physics, University of
Ioannina, Ioannina Greece.

Abstract : A developing active region near the center of the solar
disk was observed for 80min at the center and the wings of Hα. Ellerman
bombs lying below an arch filament system (AFS) and near sunspots were
studied at $H-1\overset{o}{A}$ and $H-0.75\overset{o}{A}$. We found that 50% of the bombs appear
and desappear in pairs, the axis of each pair forming a small angle with
the magnetic field lines of force as evidenced by the AFS, in addition the
members of each pair show proper motions with an average relative velocity of 600m/sec. This behaviour suggests their association with bipolar emerging magnetic flux tubes. The average contrast (1.29), lifetime (19 minutes), size (0".96 x 0".58) and flux of some 70 bombs was
also determined. We find evidence of an increase of the size with
height and a fluctuation of the flux with time.

1. Introduction

The Ellerman bombs (mustaches) are small (~1" arc) bright structures
observed in the wings of Hα and other chromospheric lines which appear
in active regions both around sunspots and under arch filament systems
(AFS) (Bruzek 1968, 1972). In a recent study near the limb, Kurokawa et
al (1982) found that their shape is elongated with an average lenght
(height) of 1.1" arc and an average width of less than 0.6" arc. Estimates of their lifetime range from a few minutes to about one hour (Bruzek, 1972˙ Roy and Lepraskas, 1973). They show a fast rise to maximum
intensity, significant intensity fluctuations which may even lead to
their disappearance for a few minutes and a rapid decay (Kurokawa et al
1982).

Several authors (Rust, 1968, 1972˙ Roy and Leparskas, 1973˙Kitai and
Muller 1984), have found that the Ellerman bombs are apparantly related
to other chromospheric structures such as surges, flares, AFS and facular granules. There is some uncertainty about their association with

magnetic fields due to the low resolution of the magnetograms (Howard
and Harvey 1964, Koval 1965, Rust, 1968, 1972).

In this paper we study the morphological characteristics of Ellerman
bombs, observed both under AFS and near sunspots, near the center of
the disk and we investigate their relations with the magnetic field.

2. Observations

The observations were obtained by one of us (C.E.A.) at the center
and the wings of Hα with the 50cm " Tourelle" refractor of the Pic du
Midi Observatory and a Halle filter (0.5 Å band pass) provided by Prof.
C.J. Macris. The photometric calibration was done by means of a step
wedge. We selected an 80 minute long series of photographs of a young,
developing active region near the center of the disk (Mc Math number
16315, N 18°,E 12°) on September 26, 1979 (figure 1). The spatial resolutions of the photographs was about 0.5"arc, while the average time
difference between successive photographs at the same wavelenght was
2 min. We studied Ellerman bombs in photographs obtained at -0.75 Å
and - 1.00 Å off the center of Hα in two regions, one below the prominent AFS and another under superpenumbral fibrils near one of the
large spots. The chromospheric structure was compared with a Kitt Peak
magnetogram (courtesy of Dr. J. Harvey) obtained four hours after our
observations.

3. Results

The size of the bombs was measured both on isodensity maps obtained
with a Joyce-Loebl isodensitometer and directly on the original films ;
the results obtained by the two methods did not show any systematic
difference. The shape of the bombs is elongated with a tendency of
orientation along the direction of the overlying AFS; taking into account
the results of Kurokawa et al (1982) this is probably the result of
projection of elongated structures, inclined with respect to the vertical. At Hα-1.00Å the average value of the length is 0.96" arc with
a standard deviation of 0.41" arc, while the average value of the width
of the bombs is 0.58(±0.21) " arc. These values are the same both for
Ellerman bombs under AFS and for those near sunspots. We found slightly larger sizes at H -0.75Å than at -1.00 Å, which indicates an increase of size with height, as expected if the bombs were part of
diverging magnetic flux tubes (cf Kitai and Muller, 1984). No systematic difference was found between bombs located under AFS and under
superpenumbral fibrils.

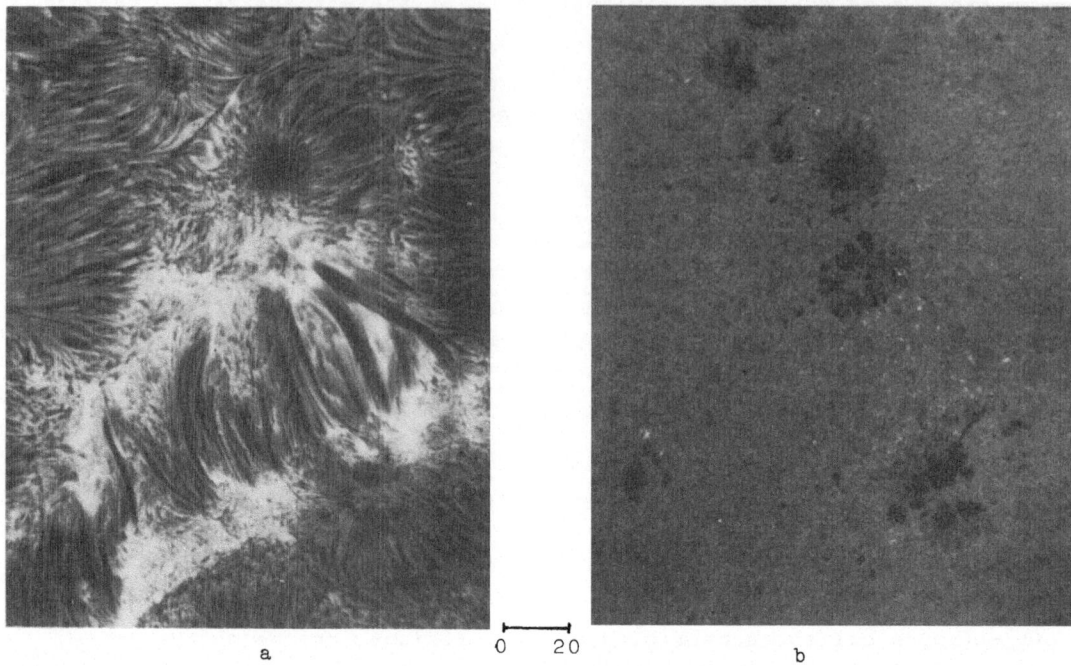

a   0   20   b

Figure 1:Active region Mc Match 16315 observed on September 26, 1979
          at the center of Hα (α, 12 12 37 U.T) and at H −1.00Å (b,12
          36 42 UT)

    The intensity of the Ellerman bombs relative to the background was
measured at Hα−1.00 Å at the time of their maximum. The values for 67
bombs located under AFS range between 1.11 and 1.48, with an average
value of 1.29 and a standard deviation of 0.10 . We found no systematic
difference between measurements at −1.00 Å and −0.75 Å or between bombs
under AFS and under superpenumbral fibrils. There is a slight increase
of intensity with size, but this could be due to the effects of finite
spatial resolution.

    In addition to the size and the maximum intensity we studied the
time evolution of the bombs , both by visual inspection of the photographs and photometrically. The average lifetime of 69 bombs located
under AFS was 19 minutes. For the photometric study we measured the
flux above the background rather than the relative intensity, in order
to avoid fluctuations due to small changes in the seeing. The flux
was computed from the measurement of the area and the relative intensity level of the isodensity contours.

    We give in figure 2 two typical time curves which show fluctuations
of the flux with time; since our measurements are very little affected
by seeing these fluctuations are real. On several occasions we observed

a repetetion of the time curve at the same location (figure 2b). We
believe that such repetetions are due to the appearance of a new bomb
very close to the old one rather than due to a second flaring of the same
bomb. This was clearly observed in some cases within groups of Ellerman
bombs, were new bombs appeated very close to (sometimes in contact with)
old bombs. The Typical rise time of the curves was 4 minutes, while the
characteristic decay time was 9.5 minutes.

The study of the time curves revealed that some bombs had almost
identical time evolution. Using this as a criterion we found that more
than 50% of the bombs within a region that was studied under the AFS
appeared in pairs showing a common time evolution. This percentage is
actually a lower limit since some bombs may form pairs with bombs outside the studied region and , moreover, in some cases their appearence
or disappearence is outside our time series. The members of a pair
appear and disappear in most cases simultaneously and in a few cases
within 2 minutes. The members of a pair have an average distance of
approximately 3" arc and they show proper motions both approaching and
receding, with relative velocity of the order of 600m/sec. It is interesting to note that the orientation of their axes is almost perpendicular to the neutral line of the magnetic field and almost parallel
to the magnetic field lines of force as delineated by the AFS.

4. Discussion and Conclusions

Our study showed that there is no difference between Ellerman bombs
which occur under AFS and those near sunspots, under superpenumbral
fibrils, at least as far as their size, relative intensity and lifetime are concerned. In addition, our photographs show that the filigree
sometimes extend below the AFS, in regions where we observe Ellerman
bombs, which may indicate a relationship between these two structures
(cf Kitai and Muller, 1984).

The relative proper motion of the bombs which are members of pairs
is similar to the proper motion of magnetic knots in active regions
(Frazier, 1972). Moreover, the orientation of the axes of pairs with
respect to the magnetic field suggests that the bombs might correspond
to the feet of emerging magnetic arches. Depanding on the exact shape
of the arch, its footpoints may recede or approach each other as the
arch rises above the photosphere.

Further observations with better time resolution are required for
a more detailed study of the time evolution of the bombs, in particular
for the study of the evolution of pairs. Moreover, high spatial reso-

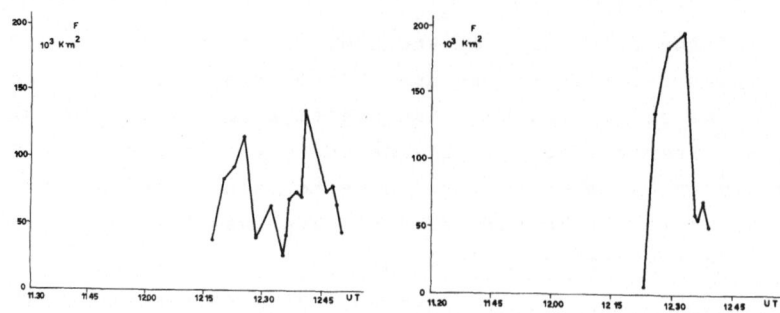

Figure 2 : The flux of two Ellerman bombs as a function of time ; a :
         simple bomb, b : bomb with double maximum

lution magnetograms will be extremely useful for a positive identification of the bombs with the feet of emerging magnetic flux tubes.

5. Aknowledgements

   One of the authors (C.E.A) wishes to express his thanks to Prof. J.
Rosch, Dr. R. Muller and the staff of the Pic du Midi Observatory for their
invitation and warm hospitality. The authors wish to thank Prof.C.J.Macris
for providing the Hα filter and Dr. J. Harvey for the Kitt Peak Magnetogram

References

Bruzek, A. : 1968, in Kiepenheuer (ed) "structure and Development of Solar
         Active Regions " IAU, Symposium 35, 293.
Bruzek, A. : 1972, Solar Phys. 26, 94.
Frazier, E.N.: 1972, Solar Phys. 26, 130
Howard, R. and Harvey , J. W. : 1964, Astrophys. J. 139, 1328
Kitai, R. and Muller, R. : 1984, Solar Phys. 90, 303
Koval, A.N.: 1965, Irv. Krym .Astrofiz.Obs. 34, 278.
Kurokawa, H. Kawaquchi, I. Funakoshi, Y and Nakai, Y.: 1982, Sollar Phys.
         79, 77.
Roy, J.-R. and Leparskas, H.: 1973, Solar Phys. 30, 449.
Rust, D.M. : 1968 in Kiepenheuer (ed) "structure and Development of
         Solar Active Regions ", IAU, Symposium 35, 77.
Rust, D. M. : 1972, Solar Phys. 25, 141.

Discussion

MATTIG : Kitai has published observations about the Ellerman bombs observed with the new Hida-instruments. As I remember the sizes elongations and life-times are the same, but the lightcurves differ remarcably. What are the differences between these both observations ?

ALISSANDRAKIS : I should point out that we have a ⁻lower time resolution, consequently we are not sensitive to very fast changes ; Moreover by using the flux rather than the relative intensity we have reduced the effects of variable seeing.

FALCIANI : Even though " Ellerman bombs" are usually called bright stuctures around sunspot penumbra, what is the criteria to couple the 50% of the events you observed (e.g., to define a pair ") and what is the behaviour of the remaining 50% of the selected events ?

ALISSANDRAKIS : Two bombs were considered to be members of a pair when they appeared and disappeared on the same frame or in two consecutive frames and had similar time curves. This criterion could not be applied to bombs that started or ended outside our observing period. It is also possible that some of the bombs that are studied formed pairs with bombs outside the region that we studied.

PEEKER : Have you computed in the studied area the α priori probability for the distribution of Ellerman bombs (and pairs of them·)to occur by pure chance.

ALISSANDRAKIS : I agree that this is an important point, however we have not made such a computation yet.

L.M.B.C. CAMPOS : Considering the number of Ellerman Bombs, and their average energy content, the overall energy can affect the global balance of the solar atmosphere, or is it a local effect ?

ALISSANDRAKIS : We have not measured the energy content of the Ellerman bombs ; however , it is interesting to note that there appears to be a continuum of amall bright structures such as filingree, Ellerman bombs, brightenings in AFS , penumbral brightenings , brightenings in fibrils etc which involve energy release and may contribute to the heating of the chromosphere.

SUMMARY

SUMMARY

H.U. Schmidt
Max-Planck-Institut für Astrophysik
8046 Garching b. München, FRG

The "colloquium on high resolution in solar physics" was dedicated to
Professor Rösch and there was a very good reason to do so. During 5
decades he strived hard with diligence and patience for observations
of higher and higher resolution on the Pic du Midi, because he was
dead sure that only with such progress in resolution we would be able
to build up real progress in our understanding of thermal convection
and of the dynamo processes which produce the observable fine
structure on the surface of the sun.

Present _instrumental developments_ begin to deal with techniques of
image stabilisation and image restoration which are very promising for
solar observation as there are plenty of photons available which allow
even instantaneous corrections with active optical systems. We learned
about test runs of a hexagonal multimirror array of 19 little
elements, each one being tilted in two angles and simultaneously
retracted into as smooth a common surface as possible. This array is
being developed by Smithson, Tarbell, Title and others at Lockheed.
Some movies and posters demonstrated sucessful restoration techniques
developed at Sacramento Peak, at the Kiepenheuer Institute in Freiburg, and at Oslo for use on a computer. Local image motion and image
stretching can so be taken off the data after the observation and
correlation tracking can thus be simulated.

The poster of Damé, Verrieres-le-Buisson, illustrated beautifully a
new interferometric method to produce a high precision corrector for
an aberrated primary mirror by use of an UV laser in a system which
includes the aberrated primary. The laser light engraves the corrector
which consists of deep UV photoresist material. In the laser illumina-
tion a nearly exact replica of the wave front deformation by the primary is burned into the corrector so that the wavefront can be flat-
tened to less than 10 nanometer. The test applied the method to a 10
cm mirror successfully. The method could e.g. allow the correction of

the Space Telescope to an accuracy limited solely by diffraction even at Ly_α, where at present such a resolution is missed by more than one order of magnitude.

Quite a number of papers dealt with the <u>solar granulation</u>. Several times it was stated that the birth and death of a granule are seldom events and proper motions seem to determine most of the evolution with time. Muller and Roudier (Pic du Midi) presented the profile of average diameter versus circumference of individual granules measured in an objective way from isophotes. The slope of the profile changes at 1000 km average diameter. The smaller granules may represent the inertial domain in a Kolmogoroff cascade down to a few hundred km. Above 1000 km the roughness of the isophote increases much more rapidly and larger granules are also observed to be unstable and to explode, often repeatedly into rings of smaller fragments. Karpinski (Pulkovo) seems to come to a different conclusion. From his two-dimensional spatial power spectrum of the granulation he argues for the absence of an inertial domain.

Detailed profiles of the velocity from spectra with center to limb variation led Nesis from Freiburg to determine an upper limit of 100 to 150 km for a convective overshoot whereas Karpinski argues that overshoot is not a constructive explanation of the observations in view of strong inhomogeneities on the smallest scales. Elste presented data on such inhomogeneities in the temperature gradients.

Several recent investigations deal with variations correlated with the solar cycle. A long-term collaboration of the solar astronomers in Athens with those at Pic du Midi has also delivered results in this area. A significant variation of the average photometric contrast in the granules within recent years was found also a variation of the number density of granules with the cycle. These new results are hints that the magnetic dynamo may have side effects on the photosphere which are even harder to interpret from present theory than the by now well established brightness changes in sunspot umbrae with latitude and phase in the cycle. Muller and Roudier presented such variations also in the density of facular points or photospheric network.

Magnetic finestructure

There seems to be now a general agreement that the distribution of
magnetic fields in and for some distance below the photosphere is
intermittent so that weak fields are absent and that there is a rather
narrow peak from say 1000 Gauss to 1600 Gauss for small fluxes and up
to over 3000 Gauss for large sunspots. This agreement on intermittency
is now manifested in the widespread use of the word fluxtube.
Theoreticians argue since several years that buoyancy of magnetic flux
ensures the coherence and even near verticality of such fluxtubes
probably even down to the bottom of the convection zone. But above the
photosphere the magnetic stress cannot be balanced by any material
stresses and so it is argued that here the intermittency gets lost and
the magnetic field is everywhere in the chromosphere and corona.
Though weak fields are absent the analysis of Zeeman profiles shows
that it is very hard to prove exact figures for the limiting fieldstrength mainly because of the disturbing influences of possible
Doppler motions. Certainly we look forward to higher resolution but
the difficulties of the analysis will not automatically go away with
higher resolution.

Impressive new spectroscopic instrumentation is being developed, e.g.
the MSDP by Mein and his group in Paris for use at Pic du Midi and at
the Canary Islands. Rayrole gave an overview of what will be available
at this new Eldorado for European Astronomy which is promising not
only during the day but also at night.

High 2D-spatial resolution is obtained with MSDP simultaneously for
full H_α profiles. One can e.g. see filigree subarcsec size bright
points in the H_α wing in active regions and associate them with
elongated absorbing features near line center.

The subject of the bright points and the filigree as detected by Dunn
at 2A from line center many years ago with the vacuum tower at Sac
Peak was brought up time and again at the colloquium.

The filigree is still the smallest phenomenon detected on the sun. Are
these crinkles, as Dunn calls them, made up out of almost circular
segments in different orientations or do they instead consist of indi-

vidual bright points put stochastically into the intergranular lanes? In the first case the circular segment may be the intersection of a fixed height above the photosphere with the funnel which engulfes a spreading circular fluxtube under different orientation to the line of sight. In the second case the sequential individual points may outline such intersections for a more or less stochastic sequence of individual fieldlines in an intergranular lane selected from the rest by thermal processes.

The Theory of <u>magnetoconvection</u> to a large degree is developed over recent years by Weiss and his collaborators in many detailed studies of physically well-defined model calculations, answering one question at a time. Now with larger computers it became feasible to do more and more complete simulations of the complex interaction of the many effects of nonstationary convection in the magnetized compressible solar atmosphere up to the photosphere with realistic treatment of the radiative transfer. Although such simulations became impressively successful it seems that in many cases the full analysis of simpler and more symmetric models which allow the discussion of the consequences of separate changes of the basic parameters will still be needed in order to understand the underlying physics. One such model calculation, on the magnetoconvection in a cell with the shape of a hexagon was recently done by Galloway and Proctor and showed some surprising effect. The evolution of the flux distribution in the cell was calculated. At the base of the cell the initially homogeneous flux rapidly coagulates in the middle and at the top it does so at the corners of the cell. But also at the photosphere develops a wide central hump of additional flux with less fieldstrength. This hump competes in total flux with the strong fields in the corner. This surprising persistence of flux in the cell center hints that in a granule there may also be flux left in the middle of the cell (although in such a calculation the magnetic Reynolds number cannot yet be pushed up to a realistic value for solar conditions).

Finally the <u>high resolution study of strong fields</u> in active regions and sunspots.

Here a photospheric (white light) observation was shown by Zwaan which is a clear manifestation of the tremendous influence of emerging flux at the birth of an active region on the photospheric convection in granules. Certainly this can stimulate theoretical studies. There are many open questions in this area. One discussed by Zwaan is the question whether the sunspot flux is made up of a cluster of individual strands which separate into individual tubes surrounded by unmagnetized normal plasma almost immediately below the photosphere or whether the strands stay together at depth. There are contradicting hints in both directions, and thorough theoretical modelling is needed to reach a decision.

The fine structure in the penumbra was thoroughly investigated by Muller. As Professor Rösch commented it shows up like little fish with bright heads and elongated tails which all swim towards the umbra. Within a complex penumbra between adjacent spots these fish can change orientation. The penumbra seems to be a rather thin layer between magnetized and unmagnetized plasma with an uniquely oriented structure. The most interesting and the smallest fine structure on the sun seems to show up always in such a kind of interface: the penumbral filaments, the spicules, the facular bright points, the filigree. All these phenomena are not static but in rapid motion. Probably there is a common cause. The physical constraints which put such magnetic interfaces into a certain place are just flux conservation and mechanical equilibrium of the total stress of magnetic field and plasma across the interface. But the energy transport across the layer by radiation and convection is an additional constraint and if the layer is not exactly horizontal or exactly vertical there is no reason to assume that both equilibria can be achieved so that hydrostatic equilibrium along the layer is also guaranteed, may be even stationary flow cannot be achieved. This situation gives an interesting hint. The interface itself wherever it has an intermediate inclination to the vertical presents already a cause for motions along the interface which in turn may make up the finestructure in it and the observable contrast.

LIST OF PARTICIPANTS

AIME Claude, Université de Nice - France.

ALISSANDRAKIS Constantin, University of Athens - Grèce.

ARNAUD Jean, Observatoires du Pic du Midi et de Toulouse - France.

ARTZNER Guy, LPSP - Verrières le Buisson - France.

BOCCHIA Roméo, Observatoire de Bordeaux - France.

CAMPOS L.M.B.C., Instituto Superior Tecnico - Lisboa - Portugal.

COLLADOS Manuel, Instituto de Astrofisica de Canarias - Espagne.

DAME Luc, LPSP - Verrières le Buisson - France.

DELBOUILLE Luc, Université de Liège - Belgique.

DERMENDJIEV V, Bulgarian Academy of Sciences - Sofia - Bulgarie.

DIALETIS Dimitris, National Observatory of Athens - Grèce.

DOLLFUS Audouin, Observatoire de Paris-Meudon - France.

DUMONT Simone, Institut d'Astrophysique de Paris - France.

DUNN Richard, National Solar Observatory - USA.

ELSTE Gunther, University of Michigan - USA.

ENGVOLD Oddbjorn, University of Oslo - Norvège.

FALCIANI Roberto, Osservatorio di Arcetri - Italie.

GREVESSE Nicolas, Université de Liège - Belgique.

HARVEY Chris, Observatoire de Paris-Meudon - France.

KUSOFFSKY Ulf, Observatorio del Roque de los Muchachos - Canarias.

LEROY Jean-Louis, Observatoires du Pic du Midi et de Toulouse - France.

MACRIS Constantin, National Observatory, Athens - Grèce.

MALHERBE Jean-Marie, Observatoire de Paris-Meudon - France.

MARCO Enric, Instituto de Astrofisica de Canarias - Espagne.

MATTIG Wolfgang, Kiepenheuer Institut, Freiburg - R.F.A.

MEIN Pierre, Observatoire de Paris-Meudon - France.

MOURADIAN Zadig, Observatoire de Paris-Meudon - France.

MULLER Edith, Observatoire de Genève - Suisse.

MULLER Richard, Observatoires du Pic du Midi et de Toulouse - France.

NESIS Anastasios, Kiepenheuer-Institut, Freiburg - R.F.A.

NOËNS Jacques, Observatoires du Pic du Midi et de Toulouse - France.

PECKER Jean-Claude, Collège de France, Paris - France.

PNEUMAN Gerald, Institut für Astronomie, Zurich - Suisse.

RAYROLE Jean, Observatoire de Paris-Meudon - France.

RIBES Elisabeth, Observatoire de Paris-Meudon - France.

ROBILLOT Jean-Maurice, Observatoire de Bordeaux - France.

RÖSCH Jean, Observatoires du Pic du Midi et de Toulouse - France.

ROUDIER Thierry, Observatoires du Pic du Midi et de Toulouse - France.

SAMAIN Denis, LPSP - Verrières le Buisson - France.

SCHARMER Göran, Stockholm Observatory - Suède.

SCHRÖTER E.H., Kiepenheuer Institut, Freiburg - RFA.

SCHMIDT Herman, Max-Planck Institut, Munich - R.F.A.

SCHMIEDER Brigite, Observatoire de Paris-Meudon - France.

SEMEL M., Observatoire de Paris-Meudon - France.

SHE Zhen-su, Observatoire de Paris-Meudon - France.

STELLMACHER Götz, Institut d'Astrophysique de Paris - France.

VIAL Jean-Claude, L.P.S.P. Verrières le Buisson - France.

VON DER LÜHE O., Kiepenheuer Institut, Freiburg - France.

WEISS N.O., Cambridge - Angleterre

WIEHR E.,Universitäts Sternwarte, Göttingen - RFA.

ZACHARIADIS Theodosis, Research Center for Astronomy, Athens - Grèce.

ZAHN Jean-Paul, Observatoires du Pic du Midi et de Toulouse - France.

ZWAAN Kees, Sterrewacht Utrecht - Pays-Bas.

Lecture Notes in Physics

Selected Issues from

Lecture Notes in Mathematics